Time

    1 year $= 3.1536 \times 10^7$ se

Temperature

    degrees Kelvin (K)   = degrees Celsius

Heat

    1 calorie (cal)   = 4.184 joules

Heat Flow

    1 heat-flow unit (HFU) $= 1 \times 10^{-6}$ cal cm^{-2} sec^{-1}
                                   $= 4.19 \times 10^{-2}$ watts m^{-2}

Viscosity

    1 poise   = 1 g cm^{-1} sec^{-1} = 0.1 Pa sec

Kinematic Viscosity

    1 stoke   = 1 cm^2 sec^{-1} = 1 poise/density

Igneous Petrology

Third Edition

Alexander R. McBirney
University of Oregon, Eugene

JONES AND BARTLETT
PUBLISHERS

BOSTON   TORONTO   LONDON   SINGAPORE

World Headquarters
Jones and Bartlett Publishers
40 Tall Pine Drive
Sudbury, MA 01776
978-443-5000
info@jbpub.com
www.jbpub.com

Jones and Bartlett Publishers
Canada
6339 Ormindale Way
Mississauga, Ontario L5V 1J2
CANADA

Jones and Bartlett Publishers
International
Barb House, Barb Mews
London W6 7PA
UK

PRODUCTION CREDITS
Chief Executive Officer: Clayton Jones
Chief Operating Officer: Don W. Jones, Jr.
President, Higher Education and Professional Publishing: Robert W. Holland, Jr.
V.P., Design and Production: Anne Spencer
V.P., Sales and Marketing: William Kane
V.P., Manufacturing and Inventory Control: Therese Connell
Acquisitions Editor, Science: Cathleen Sether
Managing Editor, Science: Dean W. DeChambeau
Editorial Assitant: Molly Steinbach
Senior Production Editor: Louis C. Bruno, Jr.
Marketing Manager: Andrea DeFronzo
Text and Cover Design: Anne Spencer
Illustrations: Elizabeth Morales
Composition: SPi Publisher Services
Printing and Binding: Malloy
Cover Printing: John Pow Company
Cover Photo: © Jeff Corwin/Stone/Getty Images
About the cover: The geology of the Frenchman Hills, Quincy, Washington, is a product of erosion of lava flows by glacial flood waters. Layers of basalt are exposed in towering 800-foot cliffs, mesas, stair-stepped benches, box canyons, and potholes.

Library of Congress Cataloging-in-Publication Data

McBirney, Alexander R.
    Igneous petrology / Alexander R. McBirney. — 3rd ed.
        p. cm.
Includes bibliographical references.
ISBN-10: 0-7637-3448-9
ISBN-13: 978-0-7637-3448-0 (alk. paper)
    1. Rocks, Igneous. I. Title.
QE461.M46 2007
552′.1—dc22   2006013042

Printed in the United States of America
10 09 08 07 06   10 9 8 7 6 5 4 3 2 1

Contents

Preface

The need for yet another edition of this book testifies to the remarkable pace at which igneous petrology continues to advance. I can recall a time not so long ago when many geologists shared the view that we had already learned essentially everything there was to know about igneous rocks and that plate tectonics held the answer for everything. What more was there to say?

Now, twenty years later, many of the theories that seemed so secure are the subject of lively debate. Do the lavas of Hawaii come from a plume rising from the deep mantle? Are the rhyolites of Iceland and other oceanic islands differentiates of basalt or products of partial melting? Do we really understand the mechanisms of crystal fractionation? After seventy years of intense scrutiny of the Skaergaard Intrusion, the most thoroughly studied body of rock on Earth, we can't even agree on its trend of differentiation! Many of these topics were scarcely mentioned when the first edition came out in 1984, and others that we thought had long since been properly explained have taken on entirely new interpretations.

Much of the recent progress comes from new research techniques that have opened fresh insights into igneous processes. Studies of oceanic basalts, for example, now draw on seismic tomography to examine their mantle origins, and space probes exploring other planets have given us a much broader perspective of flood basalts. Andesites, which only recently were explained in terms of simplistic ad hoc hypotheses, can now be interpreted with the aid of sophisticated geochemical tools that enable us to quantify the large-scale cycling of material by subduction and its related magmatism.

Igneous Petrology, Third Edition, is intended for a one-term course for students who already have an elementary background in petrography and petrology. I assume a basic knowledge of mineralogy, structural geology, and college-level chemistry, physics, and mathematics. In dealing with the principles of phase diagrams, for example, I have assumed that the student is already familiar with the elementary systems. Likewise, if the student has had a modern course in inorganic chemistry, I assume that he or she understands the fundamental aspects of solution, chemical potential, and reaction coefficients.

I cannot claim to have mastered all the arcane ramifications of these complex topics of igneous petrology—the field is simply too vast. This, of course, is why it has become so difficult to maintain the balanced perspective needed in a textbook of this kind. And yet, it seems to me that the trend toward increasing specialization makes it all the more important to do so. To that end, I have relied on the guidance of many friends and colleagues who have patiently guided me through the maze of current research. To name only a few, Kent Condie has been especially helpful in bringing me up to date on the new knowledge on the early history of the Earth. Bruce Marsh has kept me abreast of recent work on the physical aspects of magmatic processes. Adolphe Nicolas and his colleagues at Université Montpellier introduced me to the ophiolites of Oman and gave me a new appreciation of magmatism at ocean ridges. Doug Toomey, Jim Natland, and Gene Humphrey have helped me catch up with the wealth of information that geochemistry and geophysics provide on mantle processes, and Paul Wallace has helped me understand new developments in subduction-related magmatism. And, finally, I must express my gratitude to the gang of layered-intrusion enthusiasts who have exchanged their thoughts with me at a series of informal conferences organized by Derek Bostok. To these and all my other hot-rock comrades, I extend my sincerest thanks for a stimulating, intellectual exchange throughout my forty-year struggle to understand igneous rocks.

Alexander McBirney
Eugene, Oregon
Fall, 2006

Pl. X.   Pag. 298.

CRATERE DE LA MONTAGNE DE LA COUPE, AU COLET D'AISA,
Avec un Courant de Lave qui donne naissance à un pavé de basalte prismatique.

In 1763, Nicolas Desmarest traced a flow of columnar basalt to its origin in a volcanic crater near Clemont Ferrand in central France. In so doing, he demonstrated that, contrary to a widely held belief, basalts are not precipitates from sea water but products of volcanic eruptions. This simple piece of field work, together with studies elsewhere in the Auvergne, marked the demise of the Neptunist doctrine and the beginning of modern igneous petrology. The highly idealized view shown here is from an illustration published by Faujas de Saint-Fond in 1778. (Courtesy of the Bibliothèque Nationale de Paris.)

1 The Earth and Its Magmatism

Much of what we know about the composition, evolution, and internal workings of our planet has been learned through studies of igneous rocks. By drawing on many realms of science, igneous petrology integrates the results of field observations, geophysical measurements, and laboratory experiments in order to gain a better understanding of complex natural processes, few of which can ever be directly observed.

The Earth is basically an enormous igneous rock, composite in form and covered with a thin, weathered surface. Its interior is a high-temperature assemblage of crystals and magmatic fluids and its crust a layer of volcanic and intrusive rocks and the products of their weathering, erosion, and metamorphism. Each year about 50 square kilometers of its surface are covered by new lava, while 20 to 30 cubic kilometers of mantle-derived magma are added to the crust. These igneous rocks may be weathered and reconstituted into new sedimentary and metamorphic forms, only to be buried and eventually returned to the crust as new magmas. Even the waters of the seas and the gases of the air we breathe are largely of igneous origins. They have accumulated through degassing of the Earth's interior during prolonged periods of magmatic activity.

■ The Early Evolution of the Earth

The lavas erupted by Kilauea volcano in Hawaii or by Mount St. Helens in the Cascade Range represent only the latest products of a long, continuing process by which our planet has attained its present structure and form. Indeed, the igneous activity we witness today can be understood only in terms of much earlier processes through which the Earth first took form, gained mass and heat, and proceeded to differentiate into a core, mantle, and crust. Like the other dense planetary bodies in orbits close to the Sun—Mercury, Venus, Mars, and the asteroids—the Earth accreted from the hot inner part of a great disk-like cloud of gas and dust rotating slowly about the Sun. As separate bodies aggregated from their dispersed state into growing clots, they gained thermal energy, first from their inward falling mass, then from self-compression and decay of radioactive elements that were then far more abundant than they are today.

The temperatures of the planets close to the Sun were high enough during their early stages of accretion to vaporize many of the more abundant elements. Only the most refractory metals, aluminum, titanium, calcium, and their oxides, could condense and be retained, while most of the more volatile elements were dispersed into space. Farther from the Sun, iron, nickel, and silicates were able to accumulate, and only at much greater distances could large amounts of the more volatile components, such as water, ammonia, and methane, condense as parts of the giant outer planets, Jupiter, Saturn, Uranus, and Neptune.

The present rate of magmatism on Earth is far greater that that of any of the other inner planets. Mercury seems to have had little if any volcanic activity since very early in its history. Mars shows evidence of abundant volcanism in the recent past but less than that of Venus, which is second only to the Earth in the scale of its volcanism. Much of the surface of Venus has been flooded by immense flows of basaltic lava. Its average rate of production of magma is estimated at about 19 cubic kilometers per year compared with 20 to 30 for the Earth. After an initial pulse soon after they were formed, Mars and the Moon are thought to have had rates of less than 1 cubic kilometer per year.

The fact that Io, a moon of Jupiter, shows such vigorous activity today and that Mars, Mercury, and our own moon do not may tell us something about magmatic processes on Earth. The scarcity or total absence of volcanism on the innermost terrestrial planets today can logically be attributed to their compositions and sizes. When first formed in orbits close to the Sun, they retained only small amounts of volatile, heat-producing radioactive elements, and their small sizes allowed them to dissipate heat more quickly. This cannot be the only factor, however. Being farther from the Sun, Io probably retains more heat-producing elements, but in addition, the magnetic and gravitational forces due to its proximity to the enormous mass of Jupiter must continue to generate large amounts of thermal energy through tidal friction.

Meteorites and the Composition of the Earth

Much of our knowledge of the compositions and early history of the planets comes to us in the form of exotic fragments falling from outer space. Many tons of meteorites, mostly dust-size particles, reach the Earth each day. Although highly varied, this material can be divided into three broad types, known as *stones*, *stony irons*, and *irons*. Among specimens found on the ground long after their fall, stony meteorites are much less common than irons because, having the appearance of common terrestrial rocks, they are less conspicuous and unlikely to attract the attention of the farmer who overturns one with his plow. They are also more fragile. They break up in the atmosphere or on impact and deteriorate quickly by weathering. Among meteorites seen to fall, however, they are much more important; they account for by far the largest fraction, roughly 94 percent. This great preponderance among observed falls must be a more accurate reflection of their original proportions in the bodies from which they come.

The iron meteorites, or *siderites*, consist of an alloy of metallic iron with about 11 percent nickel and minor amounts of other components, most notably the sulfide mineral troilite, a form of pyrrhotite found only in meteorites. The metal is a coarsely crystalline intergrowth of nickel-rich lamellae in a nickel-poor host that developed during a long period of gradual cooling. When polished and etched with acid, it reveals a characteristic structure known as Widmanstatten figures (**Fig. 1-1a**). Intermediate between irons and stones are the stony irons, or *siderolites*, consisting of coarse crystals of olivine and, less commonly, pyroxene in a matrix of nickel-iron (**Fig. 1-1b**).

Stony meteorites are composed almost entirely of the same silicate minerals, plagioclase, pyroxene, and olivine, that are the most abundant constituents of the Earth's crust and upper mantle. They are of two types, *chondrites* and *achondrites*, which are distinguished by the presence or absence of distinctive spheroidal aggregates known as *chondrules* (**Fig. 1-2**). Ranging up to a centimeter or so in size, these curious features are intergrowths of differing proportions of pyroxene and olivine that must have nucleated and grown at elevated temperatures in globules that formed from the accreting asteroidal bodies from which most meteorites are derived (**Fig. 1-3a**). They are set in a matrix of small crystals of olivine, hypersthene, diopside, plagioclase, and rounded particles of nickel-iron and troilite.

Many chondritic meteorites have loose porous textures and contain small amounts of carbon and even water that seem to have condensed in the outermost layers of the accreting mass. These *carbonaceous chondrites* must represent the shallow parts of an early planetary body that, for some unknown reason, escaped the strong surficial heating that affected the parental bodies of other meteorites. They are of great interest because they are believed to be the closest examples we have of the original, undifferentiated material from which the Earth was formed.

About 10 percent of the stony meteorites belong to the group known as achondrites, which are distinguished, as their name implies, by an absence of chondrules. Achondrites also lack volatiles and metallic nickel-iron, and some have large crystals that appear to have been formed from a high-temperature silicate melt under conditions of slow cooling (**Fig. 1-3b**). Where well preserved, their textures resemble those of gabbroic rocks formed at moderate

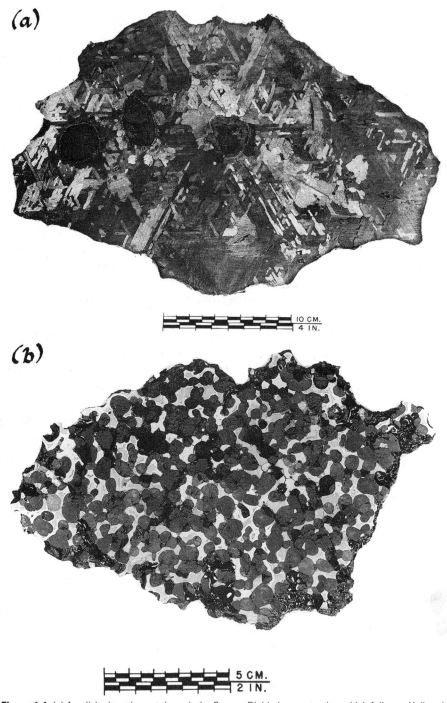

Figure 1-1 (a) A polished section cut through the Canyon Diablo iron meteorite, which fell near Holbrook, Arizona. The pattern of crystals, known as Widmanstatten structure, results from a slowly cooled intergrowth of two forms of nickel-iron. The rounded clots are troilite (FeS). (b) The Thiel Mountain siderolite of Antarctica consists of large crystals of olivine in a matrix of metallic nickel-iron. (Both photographs are through the courtesy of Dr. Brian Mason, Smithsonian Institution of Washington, DC.)

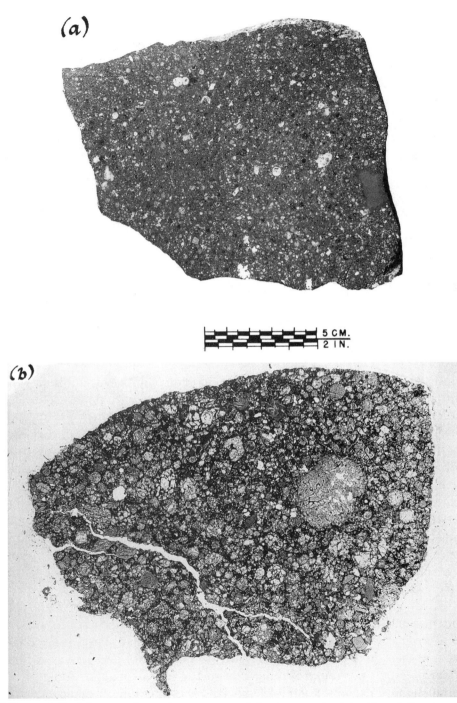

Figure 1-2 The Selma meteorite contains numerous round clots, known as chondrules. Shown in the enlarged view below, the chondrules have an average diameter of about 1 mm and are made up mainly of olivine (white) and pyroxene (light grey). (Courtesy of Dr. Brian Mason, Smithsonian Institution of Washington, DC.)

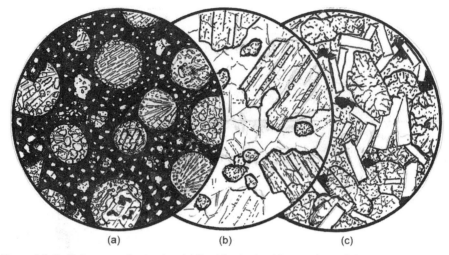

Figure 1-3 Typical textures of meteorites. (a) The Allende chondrite contains well-formed chondrules of olivine and pyroxene in an opaque iron-rich matrix. (b) The Moore County achondrite consists of pigeonite and plagioclase in a coarse gabbroic texture. (c) Shergotty, a Martian achondrite, contains crushed pigeonite and maskelynite, an isotropic form of plagioclase produced by intense shock. Diameters of fields are about 3 mm.

depths in the Earth's crust, but most are so severely crushed that their original textures have been obliterated. This pervasive crushing attests to the violent deformation that must have occurred at the time of their parent body was disrupted. The effects are too strong to have been caused by impact with the Earth. Some of the plagioclase crystals, for example, have an isotropic vitreous form known as *maskelynite* that has also been found in terrestrial rocks subjected to an intense shock, such as that of a nuclear explosion.

Ages measured from decay of the radioactive elements in chondrites fall between 4.56 and 4.55 billion years and are thought to be nearly as great as the solar system as a whole. Although the exact relations of the different types of meteorites are still uncertain, a convenient way of thinking of them is as parts of multiple bodies, smaller but compositionally similar to the Earth, that were suddenly disrupted soon after differentiating.

Most chondritic meteorites come from the asteroidal belt between Mars and Jupiter. Being farther from the Sun, their parent bodies must have differed from the inner planets. They could not have been as large as the Earth because their mineral assemblages show no evidence of ever having formed under pressures even approaching that of the Earth's mantle. Some of the common minerals, most notably plagioclase, are stable only at shallow depths, and no high-pressure forms, such as garnet or spinel, have been found to indicate that the mineral assemblages crystallized in the deep interior of a large body. Although chondrites vary somewhat in bulk compositions, proportions of chondrules, and oxidation states, their differences are not random. They can be related to differing degrees of heating and metamorphism at a range of depths within the outer part of their parental bodies. Carbon, sulfur, and water contents, for example, vary inversely with the inferred original temperatures and relative depths of burial deduced from the textures and mineral assemblages of individual chondrites.

Apart from these differences in texture and volatile content, the average composition of chondrites (Table 1-1, no. 1) is not far from that of the Earth-like

Table 1-1

The composition in column 1 is that of an average bronzite chondrite and can be thought of as roughly equivalent to the bulk composition of a primitive earth. The composition in column 2 was obtained by subtracting from column 1 the composition of a core having a mass equal to 32 percent of the Earth and a composition the same as that of iron meteorites. It can be compared with two types of rocks that are thought to be typical of the Earth's mantle. Column 3 is the average composition of 9 garnet peridotites from African diamond pipes, and column 4 is an Archean ultramafic lava (komatiite) thought to have been produced by near-total melting of the mantle.

	1.	2.	3.	4.
SiO_2	36.57	48.0	46.5	45.23
TiO_2	0.10	0.13	0.3	0.20
Al_2O_3	2.3	3.0	1.8	3.66
FeO	9.67	13.0	6.7	10.98
MnO	0.33	0.4	0.1	0.22
MgO	23.69	31.0	42.0	32.16
CaO	1.77	2.3	1.5	5.28
Na_2O	0.86	1.1	0.02	0.44
K_2O	0.10	0.13	0.2	0.17
P_2O_5	0.26	0.34	0.02	0.02
Cr_2O_3	0.42	0.55	0.4	0.47
Fe	16.90			
Ni	1.64			
FeS	5.30			

Source: 1 and 2 from B. Mason, 1962, *Meteorites,* New York: John Wiley and Sons., 3 from Bostok, personal communication, and 4 from R. W. Nesbitt and S. S. Sun, 1976, Geochemistry of Archaean spinifex-textured peridotites and magnesium and low magnesium tholeiites. *Earth & Planet. Sci Ltrs.,* 31:433–453.

planets before they formed a core and differentiated into their present stratified compositions. Many of the elements are present in proportions similar to those of the Earth. The abundances of heat-producing radioactive elements, for example, are almost exactly those that would produce the observed surface heat flow of the Earth if they were distributed in the same proportions as they have in the Earth's core, mantle, and crust. It is these remarkable consistencies that lead us to conclude that the initial bulk composition of the Earth resembled that of chondritic meteorites and that the present abundances of elements in different parts of the Earth resulted from their subsequent redistribution between the core, mantle, and crust. By removing from chondrites their small clots of nickel-iron and sulfides and recrystallizing the remaining constituents at moderate depths, it would be possible to produce rocks with most of the distinctive features of achondrites. Thus, the main differences between the two classes of stony meteorites can be related to segregation of a core of nickel-iron, a mantle of achondritic composition, and a thick chondritic crust.

Not all meteorites come from the asteroidal belt. Of the 25,000 or more specimens that have been cataloged and described, at least 20 can be closely correlated with compositionally identical rocks on the Moon, and another 27 must come from Mars. These lunar and Martian meteorites were probably ejected from their parent bodies by the impacts of other large bodies. The Martian meteorites are of special interest because they come from the planet that most closely resembles Earth. Unlike chondrites from the asteroid belt, they contain no metallic nickel-iron. Instead, they contain oxidized ferric and ferrous iron in about the same proportions as terrestrial igneous rocks. In addition, radiometric age measurements show that they formed over an extended period of time. Some have ages of about 4.5 billion years corresponding to the earliest stages of the solar system, but others are as young as 160 million years. The gases trapped in the younger Martian specimens are nearly identical to those that space probes have measured in the thin atmosphere the planet has today. The broad spread of ages suggests that the Martian rocks formed by prolonged processes similar to those of our own planet. This is in contrast to stony meteorites from asteroidal bodies that have not changed since they first cooled and crystallized within a relatively brief period of only 10 million years or so.

The Martian specimens span a wide range of compositions, including basalts, gabbros, peridotites, and dunites. The Shergotty meteorite shown in Figure 1-3c could easily be mistaken for a gabbroic terrestrial rock, but it has a composition close to that of the lavas found on the surface of Mars. Plagioclase is somewhat less abundant than in most terrestrial basalts, and the olivines and pyroxenes are more iron rich. These differences are thought to reflect a basic difference between the Martian mantle and that of the Earth. The iron-rich character of the mafic rocks of Mars may mean that less nickel-iron was separated into the core than was the case for the Earth. The core of Mars accounts for about 20 percent of the mass of the planet, whereas that of the Earth is closer to 32 percent.

Ultramafic meteorites from the Martian mantle contain more alkalies than their terrestrial equivalents, possibly because these relatively volatile elements were not depleted by a large-scale heating event, such as the one associated with the huge impact that ejected the Moon from the Earth. Nevertheless, the Martian rocks show abundant evidence of shock. It is estimated that fragments have been ejected from Mars by at least eight strong impacts in the last 20 million years—a frequency similar to that of large impacts on Earth.

Formation of the Earth's Core and Mantle

The Earth's long and complex evolution is the result of an extraordinary combination of circumstances: its size, composition, and position in the solar system. Were it smaller or closer to the Sun, it would now be as barren as our Moon or Mercury. Were it farther from the Sun or close to a much larger body, it might still be as volcanically active as Io, the moon of Jupiter that has such intense volcanism it can be observed from Earth-based telescopes. Thanks to its unique setting, the Earth has evolved a core, mantle, and diversified continental crust, and it has just the right conditions to develop oceans, continents, and, most important, an atmosphere with moderated temperatures capable of supporting life.

The ages of the oldest known rocks on Earth (4.03 billion years) are not much younger than those of the meteorites we think represent the time of accretion of the planets, but even at that early stage, the accretion of the Earth was already complete and the core had been formed. Accretion of the Earth must have taken place over a relatively short period of about 30 million years, and the Moon was detached only 5 million years later, probably when the Earth collided with another large body. Much of what we can infer about this period of intense impact comes from studies of the Moon where the record is better preserved. After it was detached from the Earth, the Moon passed through the same early phases of development that affected the Earth, but it has long since dissipated much of its internal energy and is now cold and "dead" in the sense that the record of the last 3 billion years has not been obscured by the tectonic activity and volcanism that are constantly reshaping the face of the Earth. Even though its surface was bombarded by meteorites until all traces of early land forms were obliterated, petrologists have been able to deduce the form and composition of the Moon's ancient surface. It is divided into two distinct parts, the highlands, consisting of very plagioclase-rich rocks called *anorthosites*, and the maria, which were vast expanses of basaltic lava. The scale of magmatism responsible for this division was so great that many lunar geologists believe that during the early accretionary stage of its history, the surface of the Moon had an ocean of basaltic magma some 400 km deep, in which great blocks of anorthosite floated like icebergs in the sea. The ages of the anorthosite bodies (about 4.4 billion years) is considerably older than that of the basalts in which they are immersed (from 3.9 to 2.5 billion years), and thus, the anorthosite probably represent a very early crust that formed by extraction of a felsic melt from the Moon's interior.

It is likely that the Earth underwent a similar episode of extensive impact melting about 4.45 billion years ago. Although no visible record of it has been preserved, we have a more recent example that illustrates how dramatic the effects of a meteorite impact can be. The Sudbury Complex of southeastern Ontario was formed 1850 million years ago when a meteorite about 12 kilometers in diameter struck the Earth at a velocity of about 25 kilometers a second. It created a crater 200 kilometers in diameter and generated 30,000 cubic kilometers of impact melt with temperatures in the range of 1700°C. Temperatures at the point of impact reached 4000°C, and the momentary pressure was equivalent of that at the center of the Earth. The crustal rocks, along with the meteorite itself, were completely vaporized. Great as it was, this event was relatively mild compared with those to which the Earth and Moon were subjected.

Formation of the Earth's core probably occurred during this early episode of massive melting when droplets of molten nickel-iron and iron sulfides began to gravitate toward the center of the Earth. One can calculate the composition that the mantle would have after this event by subtracting a nickel-iron core with a mass of 32 percent from an Earth with an original composition close to that of the average chondritic meteorite. The resulting values, shown in Table 1-1, no. 2, fall within the range of certain types of exotic rocks brought up from below the crust in deep-focused, explosive eruptions (Table 1-1, no. 3 and no. 4). Although the exact proportions of some components remain uncertain, this may provide a crude estimate of the overall composition of the source region from which most early igneous rocks were derived.

Formation of the Crust

The Earth's crust began to form as early as 4.5 billion years ago. It was probably similar in composition to the modern oceanic crust, but we can only infer this because it has been swept away and renewed many times over. It was intruded by a more silica-rich continental crust that began to appear about 200 million years later and grew at an accelerating rate to reach a peak at the end of the Archean, about 2.7 billion years ago. By that time, the crust had about 80 percent of its present volume. Thereafter it grew more slowly in widely spaced episodes until it gradually approached a steady state in which gains and losses were nearly in balance. Although it has been fragmented and redistributed, the total volume of the continental crust does not seem to have changed much in the last 2 billion years. In fact, some would argue that the continents are now becoming slightly smaller because more crust may be returned to the mantle by subduction than is being added in the form of new igneous rocks.

As far as we can tell, the Earth is the only body in the solar system that has gone through this stage of evolution. Notably absent from the thousands of meteorites that have been studied are any specimens resembling terrestrial crustal rocks, such as shale, granite, rhyolite, or schist. Mercury and the asteroids were too small and lost heat too quickly to segregate a crust. Mars is closer to the Earth in this respect and even seems to have had substantial amounts of water at an early time in its history, but its crust has a basaltic character like that of the Earth's oceans, and there seems to be nothing resembling our continental crust.

Formation of the continental crust was essentially a process of segregation in which the first partial melts produced during the initial heating of the mantle formed buoyant masses of magma that rose to the surface in great plumes. Chief among the components of these early melts were the large, oxygen-seeking lithophile elements, silicon and aluminum, along with lesser amounts of iron, magnesium, calcium, and alkalies. The elements segregated in this way are not necessarily light ones. Many, like uranium, lead, and gold, are heavy metals, but because their compounds have low melting temperatures or high solubilities in silica-rich melts, they were scavenged from the mantle and concentrated in the continental crust. So effective was this process that even though the crust accounts for less than one percent of the total mass of the Earth it contains most of the Earth's lithophile elements. Potassium, for example, has an average abundance in the mantle of only 7 or 8 parts per million, but in the continental crust, it accounts for about 25,900 ppm.

This efficient concentration of lithophile components into the crust has had enormous consequences. Without it, extraction of metals and industrial minerals from ore deposits would be virtually impossible. Even more important, this same group of lithophile elements includes the heat-producing radioactive isotopes of potassium, uranium, and thorium, and their segregation into the crust has greatly reduced the rate of heat production in the mantle (**Table 1-2**). This in turn reduces the intensity of magmatic activity to a level that makes our planet habitable.

Table 1-2 Estimated contributions of the principal radioactive elements to the heat production of the mantle.

	Half-life (10⁹ years)	Heat production (cal g⁻¹ yr⁻¹)	Abundance in Mantle (ppm)	Heat production per gram of mantle (10⁻⁶ cal/g/yr)
²³⁸U	4.5	0.706	0.08 ± 0.05	0.05 ± 0.04
²³²Th	13.9	0.202	0.27 ± 0.15	0.03 ± 0.01
⁴⁰K	1.3	0.211	0.13 ± 0.08	0.03 ± 0.02
				0.11 ± 0.07

■ Isotopic Evolution of the Crust and Mantle

The decay of the radioactive isotopes provides an invaluable tool for tracing the temporal and compositional evolution of the Earth. The more important of these—rubidium, uranium, and samarium—decay at known rates to a daughter element that can be identified by its distinctive isotopic nature. Because these elements have differing geochemical properties, they have been partitioned in different ways during development of the core, mantle, and crust so that isotopic evidence of their redistribution provides a way of inferring the timing and effects of major geochemical processes.

As rubidium-87 decays to strontium-87, the ratio of the latter to the nonradiogenic isotope strontium-86 increases with time. In a similar way, as the radioactive isotope of uranium, ²³⁸U, decays to lead-206, and samarium-147 decays to neodymium-143; the ratios of these decay products to their nonradiogenic counterparts, ²⁰⁴Pb and ¹⁴⁴Nd, increase with time. Details of the isotopic systems are outlined in Appendix D, but for our present purposes, it will suffice to note how the rates of growth of the radiogenic isotopes enable us to trace major geochemical changes far back in time. All three systems have decay constants long enough to produce measurable effects on the time scale of the Earth's history.

Figure 1-4 shows how the isotopic ratios of Sr, Pb, and Nd are thought to have evolved since the planet first began to take its present form. Consider first the case of strontium (Fig. 1-4a). The Rb/Sr ratio of the Earth as a whole was determined at the time of accretion from the solar nebula, but the original concentrations and ratio of the parent and daughter elements have not remained uniform. Both elements were strongly excluded from the Ni-Fe melt forming the core. As a result, both were residually enriched in the mantle, and because their relative proportions were not affected, the rate of increase of

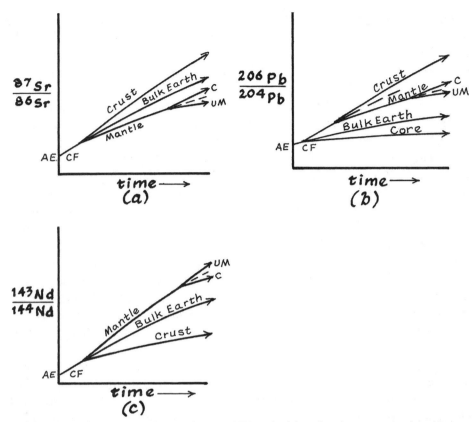

Figure 1-4 Each of the isotopic systems, Sr, Pb, and Nd, evolved through various stages starting with accretion of the Earth (AE) and continuing at differing rates according to the partitioning of the isotopes at the time of formation of the core (CF) and separation of the continental crust (C) from the upper mantle (UM) (see text for explanation).

$^{87}Sr/^{86}Sr$ remained constant. Later events, such as those leading to formation of the continental crust, had a quite different effect. Because Rb had a greater affinity than Sr for the first liquids produced by melting of the primitive mantle, it was more strongly partitioned into the crust, and the Rb/Sr ratio of the remaining mantle was significantly reduced. The subsequent increase of radiogenic ^{87}Sr and its ratio to the nonradiogenic isotope, ^{86}Sr, was therefore slower in the mantle than it was in the Rb-rich continental crust.

In a similar way, the present distribution of U and Pb in the Earth has resulted from a series of events beginning with accretion of the planet and continuing through the various stages of evolution of the core, mantle, and crust. The mobility of the individual isotopes of U and Pb during the earliest stages is not well understood, but it is thought that formation of the core could have resulted in separation of Pb from U with much of the Pb going into the core, probably as a sulfide mineral. Uranium remained in the mantle to be later concentrated in molten silicates that were segregated to form the upper continental crust. The slope of the growth curve of the radiogenic isotope ^{206}Pb relative to the nonradiogenic form, ^{204}Pb, changed with each of these events so that the present ratios of the isotopes in the crust and mantle differ, as shown schematically in Figure 1-4b.

Because of the marked enrichment of U in the upper crust, continental Pb tends to be much more radiogenic than that of mantle-derived basalts, but neither is strictly uniform. Pb ratios have regional patterns that reflect the age and local history of the crust and mantle. It is thought, for example, that much of the lead concentrated in the continents has been carried into the oceans by rivers and then cycled back to the continents by subduction. This Pb has a fairly uniform isotopic ratio because of global mixing in the oceans, and it is so rich in radiogenic ^{206}Pb that when it is subducted and incorporated into the magmas of island arcs, it overwhelms the much smaller contribution of the lead-depleted mantle and in this way provides a way of distinguishing the origins of the various components of subduction-related magmas.

The development of the rare-earth system, Sm-Nd, resembles that of Sr and Pb but has important differences. The parent–daughter ratio, Sm/Nd, was not greatly altered by formation of the core because all the rare-earth elements were equally excluded from the metallic melt. During subsequent evolution of the mantle, the ratio of ^{143}Nd to ^{144}Nd continued to increase as Sm decayed. The rate of increase in the mantle, being a direct function of the relative abundances of Sm and Nd, corresponded to a parent–daughter ratio of about 0.31, about the same as that of chondritic meteorites. This ratio is altered in magmas produced by partial melting of mantle rocks because the parent isotope, Sm, is partitioned less strongly into a silicate melt than is its daughter product, Nd. Thus, unlike Rb/Sr, the parent–daughter ratio, Sm/Nd, is reduced in magmas relative to the mantle from which they are derived.

Although the rare-earth elements, including both Sm and Nd, are strongly concentrated in the crust, their partitioning is not identical. The abundance of Nd in the crust is about 25 times that of the mantle, whereas that of Sm is somewhat less, about 16. Because of this difference in the parent–daughter ratios, the rate of increase of ^{143}Nd/^{144}Nd has been slower in the continental crust than in the mantle from which crustal rocks are derived (Fig. 1-4c). Thus, even though the absolute abundances of the rare-earth elements are greater in the crust than in the mantle, the Nd isotopic ratio evolves more rapidly in the mantle than it does in the crust—just the opposite of

$^{87}Sr/^{86}Sr$. The Nd ratio differs according to the local age of the crust, but it is always less than would be expected for a chondritic parent–daughter ratio, whereas that of the mantle is greater by a complementary amount.

Assuming these differences are the result of differing degrees of separation of Sm and Nd from a mantle that initially had chondritic abundances of the rare earths, it is possible to use a simple mass balance to estimate the extent to which these lithophile elements have been depleted from the mantle by formation of the continental crust. If the mass of the continents is estimated from their areal extent and thickness, and the average Sm and Nd abundances are known for both the upper mantle and crust, one can easily calculate the mass of the depleted mantle. When this is done, one finds that the depleted part of the mantle would extend to a depth of no more than 600 to 700 km. If, as seems probable, the entire mantle is not mixed by throughgoing convection, depletion of these and other lithophile elements from the upper levels of the mantle would cause it to be compositionally layered on a global scale. We shall see some of the consequences of this later when we consider the origins of mantle-derived magmas.

■ Compositional Evolution of the Crust and Mantle

After the early events that established the basic division of the Earth into a core, mantle, and primitive crust, a much longer period elapsed before the continents took on a form similar to what they have today. Apart from a few remnants in places like Wyoming, Northern Canada, Greenland, and western Australia, the early crust has been almost totally buried or lost to erosion. The record in the isolated remnants is therefore fragmentary and may be far from complete, but their remarkable similarity wherever they can be seen suggests that the conditions they record were broadly uniform throughout the early continents.

Early Evolution of the Continental Crust

After their initial formation, the continents continued to grow and evolve as magmatism added to their mass and surficial processes of weathering and erosion brought their composition into closer equilibrium with the atmosphere and hydrosphere. The dominant character of volcanic and plutonic activity during this early period was quite unlike that associated with modern tectonic regimes. Even though the amount of heat generated in the mantle was greatly reduced by the partitioning of radioactive, heat-producing elements into the crust, the thermal gradient of the mantle and crust must have remained much steeper than it is today. Metamorphic gradients in many Archean rocks reflect unusually high temperatures at shallow depths, and it appears that even crustal rocks were subject to partial melting when buried at relatively shallow depths. The earliest Precambrian igneous rocks include an unusual type of ultramafic lavas, *komatiites*, that must have been erupted at temperatures at least 200 to 400 degrees above those observed in modern volcanoes. Other types of igneous rocks, such as *trondhjemite*, a sodic variety of granite, were not totally confined to the earlier periods but were proportionately much more common than they are today.

The record of this period, although incomplete and difficult to decipher, is preserved in two types of terranes, one dominated by high-grade metamorphic and plutonic rocks and the other by more weakly metamorphosed volcanic rocks and sediments of *greenstone belts* (**Fig. 1-5**). The former consist of circular or oval bodies, up to 200 or 300 kilometers in diameter, formed by diapiric bodies of trondhjemites, tonalites, and granodiorites that rose as plume-like bodies into the greenstones. Most of these rocks are multiply deformed, and many show signs of having been partly remelted to the extent that the distinctions between igneous and metamorphic rocks are blurred. The deformation is most intense around the domes of gneiss and plutonic rocks that compressed the surrounding greenstones into tight, complex fold belts.

The more weakly metamorphosed rocks of greenstone belts represent shallower parts of the crust that are thought to have developed as accumulations of sediments and volcanic rocks between 2800 and 2500 million years ago. The stratigraphic sequences differ from place to place, but most have a thick, lower section consisting of komatiites, peridotites, and pillow basalts with only minor sedimentary interbeds. These mafic and ultramafic rocks grade into a sequence of more intermediate compositions—andesites, dacites, and rhyodacites—that tend to become more felsic and pyroclastic upward. The volcanic rocks give way upward to cherts, greywackes, shales, and conglomerates laid down shortly before the entire mass was strongly compressed and folded.

The Archean was clearly a time of intense tectonic and magmatic activity. Plate movements driven by strong, shallow convection of the mantle must have been very rapid. Rates of spreading at divergent plate boundaries are estimated to have been about six times faster than they are today. Subduction, on the other hand, was much more limited. Despite the presence of andesites and other rocks common to modern island arcs, the geologic evidence for structures of the kind associated with modern subduction is hard to find. Plate motion was probably driven by the viscous drag of the rapidly convecting mantle, and it had little, if any, contribution from the gravitational pull of subducted plates. Most of the driving force of modern subduction is thought to be the negative buoyancy of descending slabs of

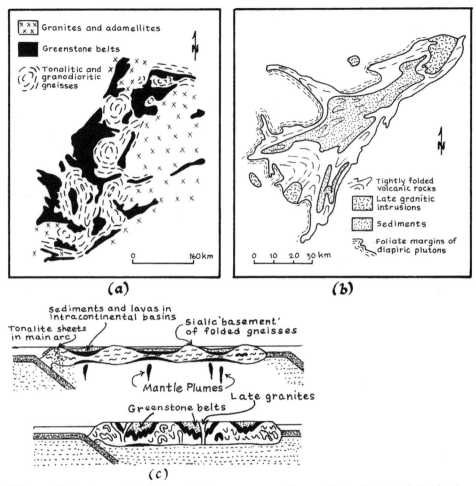

Figure 1-5 The characteristic features of Archean terranes are illustrated by the Barberton Belt of South Africa (a). It consists of islands of foliated gneiss and plutonic rocks separated by tightly folded greenstone belts. The latter (b) consist of mafic lavas and sediments that have been compressed into steep-sided synclines and intruded by later granites. (c) The entire assemblage was compressed by the strong motion of the underlying mantle. (a and b are adapted from C. R. Anhaeusser, 1971, *Geol. Soc. Australia Sp. Pub. No. 3*, 103–120, and M. J. and R. P. Viljoen, 1970, in Clifford, T.N. and Gass, I.G., eds., *African Magmatism and Tectonics*, Edinburgh: Oliver and Boyd, 27–49. c is adapted from G. C. Brown and A. E. Mussett, 1993, *The Inaccessible Earth*, London: Chapman & Hall.)

basalt that is converted to *eclogite*, a garnet-pyroxene rock with a density greater than that of the average mantle. If the descending basalts were remelted before reaching the depths where this transition to eclogite occurs, they would not have the high density needed to pull the slab farther into the mantle. It was not until the Proterozoic, about 2 billion years ago, that the thermal gradient declined enough for the basalt-eclogite transition to become possible. By that time, the continents had become thicker and stronger, and plate tectonics was taking on a more modern aspect.

Meanwhile, the continents continued to grow, both by marginal accretion and by addition of mantle-derived magmas. With time, they became richer in SiO_2, Al_2O_3, and K_2O as weathering and erosion concentrated these components in increasing volumes of sediments. Magmatic rocks of relatively low density were added to the upper crust, while denser, more mafic magmas tended to intrude at greater depths, and in this way, the crust became compositionally zoned with the proportion of dense, mafic rocks increasing with depth (**Table 1-3**).

Table 1-3	Average composition of the continental and oceanic crusts. Major elements are in weight percent, trace elements in part per million (ppm). All iron is shown as FeO*.

	1.	*2.*	*3.*	*4.*	*5.*
		Continental			*Oceanic*
	Upper	Middle	Lower	Total	Crust
SiO_2	66.3	60.6	52.3	59.7	50.5
TiO_2	0.7	0.8	0.54	0.7	1.6
Al_2O_3	14.9	15.5	16.6	15.7	15.3
FeO*	4.68	6.4	8.4	6.5	10.4
MnO	0.07	0.1	0.1	0.1	0.2
MgO	2.46	3.4	7.1	4.3	7.6
CaO	3.55	5.1	9.4	6.0	11.3
Na_2O	3.43	3.2	2.6	3.1	2.7
K_2O	2.85	2.0	0.6	1.8	0.2
P_2O_5	0.12	0.1	0.1	0.11	0.2
Rb	87	62	11	53	1
Sr	269	281	348	299	90
Ba	626	402	259	429	7
U	2.4	1.6	0.2	1.4	0.05
La	29	17	8	18	2.5
Sm	4.83	4.4	2.8	4.0	2.6
Ni	60	70	88	73	150

Source: From a compilation by K. C. Condie, 2005, *Earth as an Evolving Planetary System*, Amsterdam and Boston: Elsevier.

Post-Archean Magmatism

A marked change in the character of magmatism came at the end of the Archean around 2.7 billion years ago. Thereafter, the ultramafic komatiites have been very rare, but plagioclase-rich anorthosites became increasingly common. The character of basalts also changed. Archean basalts resemble those of the modern ocean floor in most respects, but they have about twice as much potassium—0.14 to 0.26 percent K_2O compared with less than 0.10 percent for the basalts erupted at modern oceanic ridges. Differences of this kind are interpreted as evidence that the upper mantle from which these magmas have been derived was becoming more depleted with time.

It was also about this time that plate tectonics took on a more modern form, and the continental crust attained a mass and average composition close to what it has today. As tectonic and magmatic activity declined, the ancient parts of the continental crust became increasingly stable. Although individual continents have changed size and shape, the simple fact that they have survived prolonged periods of erosion that should long ago have stripped away the upper crust suggests that the crust has been steadily renewed. Without some sort of replenishment, nothing but the roots of continents would remain, and because these deeper levels tend to be more mafic, the average compositions of sediments should have changed with time. The fact that the modern products of erosion differ little from those laid down a billion years ago means that gains and losses are approximately in balance.

Part of the explanation for this apparent stability is that almost 60 percent of the sediments produced by erosion come from older sediments rather than from metamorphic and igneous rocks of the basement. When this material is deposited on the sea floor, most of it is scraped from the oceanic plates and accreted to the continental shelves at convergent plate boundaries. Nevertheless, about 1.3 to 1.8 km^3 of sediments are returned to the mantle by subduction each year, while this much or more is added to the crust in the form of subduction-related magmas.

The most conspicuous additions to the continents from the mantle are the huge volumes of basaltic magma that have been erupted at odd intervals throughout Phanerozoic time. These lavas, however, are more mafic than the upper crust, and unless their contribution is balanced by corresponding amounts of felsic material, the result should be a steady dilution of the lithophile elements and a progressive basification of the continents. There may be a weak trend in this direction, but the fact that no major change is seen must mean that some other input of a complementary composition helps to maintain a rough balance of mafic and felsic components. Moreover, the proportions of these components must have remained approximately constant for much of geologic time.

The huge volumes in granitic batholiths are obviously a major contribution of felsic components. Another is the great sheets of rhyolite and rhyodacite that cover large parts of the continents and rival the flood basalts, both in thickness and extent. Much of this granitic and rhyolitic material is derived from older crustal rocks and contributes no net addition to the crust, but at least part of the siliceous magma added to the continents is derived from deeper sources. Simple melting of mantle peridotite cannot yield any important amount of magma of this kind, but we shall see later (in Chapter 10) that amphibolitic rocks in the lower lithosphere could provide a more fertile source. If this is indeed what happens, one must conclude that the process of separation of lithophile components from the mantle did not end abruptly with formation of the primitive continents but has continued, although at a much slower rate, to the present day.

Spatial and Temporal Distribution of Igneous Rocks

Although the nearly constant average composition of continental sediments is well established, it is more difficult to say whether igneous rocks have also remained unchanged through Phanerozoic time. Although stratigraphers have determined the proportions of different types of sediments during earlier geologic periods, no comparable data have been compiled on the volumes or average compositions of igneous rocks. Without this information it is impossible to quantify long-term trends in the geochemical exchange between the mantle and continents. Until someone takes up this formidable task, we can only speculate on broad, highly subjective impressions.

There is little doubt that global rates of magmatism have varied widely through time. One of the best records, oddly enough, is found in the rise and fall of sea level that accompanies increases and decreases of spreading rates on oceanic ridges. Because the elevations and volumes of the ridges vary directly with the rate of sea-floor spreading and hence the rate at which basalts are added to the crust, periods of high sea level are also times of increased magmatic activity in the oceans. We can infer that increased spreading rates entail a comparable increase of subduction-related magmatic activity as well.

The distribution of global magmatism today is largely confined to a small number of geologic settings where it is closely related to the motion of lithospheric plates (**Fig. 1-6**). This pattern provides a convenient way of dividing this activity into a small number of distinct types, each with its own compositional and eruptive characteristics.

By far the most important focus of Cenozoic magmatism has been at divergent plate boundaries, principally oceanic spreading axes (**Table 1-4**). Extensional regions of the continental crust, such as the great rift of East

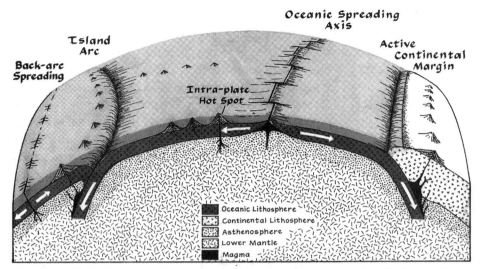

Figure 1-6 Modern igneous activity has been concentrated in three tectonic settings. The largest volume (about 21 cubic km per year) is produced at spreading axes (center). Lesser amounts (between 1 and 10 cubic km per year) are associated with converging plate boundaries in island arcs (left), where the overlying plate is oceanic, and continental margins (right), where it is continental. Subordinate amounts of magmatism are found in intraplate regions (left center) where moving plates pass over a persistent thermal anomaly in the mantle.

Africa, also have considerable volcanism but far less than in the oceans. We can observe the oceanic type of activity in Iceland, where the Mid-Atlantic Ridge can be seen emerging from the sea, crossing the island, and descending again into the Arctic Ocean. At intervals of a few decades, eruptions of fluid basaltic lava pour from fissures to flood wide areas of the surface,

Table 1-4   Global rates of Cenozoic magmatism.

Location	Rate (km³/yr)	
	Volcanic Rocks	Plutonic Rocks
Oceanic ridges	3	18
Convergent plate boundaries	0.4–0.6	2.5–8.0
Continental intraplate regions	0.03–0.1	0.1–1.5
Oceanic intraplate regions	0.3–0.4	1.5–2.0
Global total	3.7–4.1	22.1–29.5

Source: From estimates of J. A. Crisp, 1983, *Jour. Volc. Geoth. Res.*

while the feeder-dikes of these same eruptions solidify at shallow depths and add as much as a meter at a time to the lithosphere along the central axis of the island. In the same way, the ocean floor must be spreading laterally as new lava and dikes are emplaced along the oceanic ridges.

The greatest number of the world's active land volcanoes is concentrated along the lines of plate convergence where the oceanic lithosphere descends into the mantle. They form a "chain of fire" that, with few interruptions, follows the entire circumference of the Pacific Ocean. Other volcanic belts, such as those of the Antilles Arc and Indonesia, are in similar settings. Volcanism is notably rare at convergent boundaries where both plates are continental. A few plutons have been intruded into these collisional zones, but only under very exceptional circumstances.

Many of the great belts of batholiths were probably roots of volcanic chains like those seen today at convergent plate boundaries, but others are well within the interiors of continents in quite different tectonic settings. Some of the latter may have been related to great outpourings of ignimbrites, such as the siliceous pyroclastic flows that cover wide areas in western North America. They are commonly associated with zones of continental rifting, but a few of the large centers, such as the caldera complex of Yellowstone or the granitic ring-complexes of New England, may be continental manifestations of mantle hotspots equivalent to those forming long chains of volcanoes in intraplate regions of the oceans.

Some of the great outpourings of flood basalts that have covered thousands of square kilometers of the continental interiors have a similar origin. These *large igneous provinces* are thought to result from giant "superplumes" that produce huge amounts of magma over relative brief periods every 10 to 20 million years. The Permo-Triassic basalts of Siberia, for example, cover 2.5 million square kilometers, and the Jurassic Karoo basalts of South Africa and the Cretaceous Paraná basalts of Brazil are almost as great. Flood basalts such as these are not confined to the continents. The Ontang-Java Plateau of the southwestern Pacific was produced by similar outpourings of basalt during the Cretaceous.

■ The Nature of the Mantle

Our knowledge of the mantle comes from a variety of evidence, including geophysical measurements of its physical properties, samples brought up from deep sources in volcanic eruptions, and experimental studies of rocks at very high pressures. Here we look briefly at the geophysical evidence and what can be inferred from it. The chemical and mineralogical composition of the mantle is discussed in more detail in Chapters 8 and 9.

Geophysical Properties of the Mantle

The velocities of seismic waves transmitted through the crust and mantle vary with the mineralogical character, density, and temperature of the rocks through which they travel. In this way, they place broad limits on the temperatures and compositions of the crust, mantle, and core.

The sharp, step-like *Mohorovicic discontinuity*, or *Moho*, that marks the base of the crust lies about 5 to 10 km below the oceans but ranges from 25 to 150 km or more below the continents. Below that level, the compressional wave velocities (>7 km/sec) and densities (>3.2 g/cm^3) correspond to rocks composed entirely of dense, mafic minerals, such as olivine and pyroxene. Except for a sharp decrease caused by partial melting in the *low-velocity zone* at a depth of about 100 kilometers, the velocity increases steadily with depth.

The *lithosphere* includes both the crust and that part of the upper mantle that behaves like an elastic solid, whereas the *asthenosphere* on which it floats has a viscosity low enough to permit isostatic adjustment and motion of the drifting lithospheric plates. The lithosphere is somewhat thicker under the continents than under the oceans, and in both places, it is slowly thickening as it cools. The lithospheric mantle is also thought to change composition with time as it gains mobile components from the underlying asthenosphere and gives them up to the crust.

The thermal gradient of the lithosphere is governed primarily by the rate at which heat is transferred by conduction. At greater depths, where heat is transferred more rapidly by convection, the thermal gradient is less steep. The depth at which the conductive gradient of the lithosphere intersects the convective gradient of the asthenosphere corresponds fairly well to the low-velocity zone where the mantle approaches temperatures at which it begins to melt. This cannot be a layer of totally molten rock, however. Because shear waves, which are not normally transmitted through liquids, are only slightly retarded as they pass through this zone, the molten fraction cannot be more than a percent or two.

Two worldwide seismic discontinuities at depths of about 400 and 670 km define the top and bottom of a *transition zone* separating the upper and lower mantle (**Fig. 1-7**). They are thought to mark pressure-induced changes in the mineral assemblages of the mantle. At pressures equivalent to about 400 km, olivine takes on a denser spinel-like structure, and at those corresponding to the 670-km discontinuity, it changes to an even denser form with the structure of perovskite. The mantle is therefore stratified by a density barrier that inhibits the exchange of chemical components from one level to the other. As a result, the upper part of the mantle has evolved more or less independently of the lower part. The lower mantle may not be totally

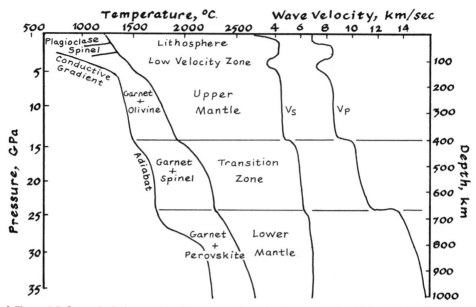

Figure 1-7 Some physical properties of the crust and mantle. The compressional (P) and shear (S) wave velocities on the right are the principal evidence for interpreting the temperature and density of the mantle. The geothermal gradient has two parts, a conductive section above and a convective one below. The latter is divided into two parts by a transition zone in which olivine changes to denser forms in response to the pressure at depths of 400 and 660 to 670 kilometers.

isolated, however. Some of the mantle plumes that are thought to be responsible for intraplate hot spots and large igneous provinces may originate as deep as the core–mantle boundary.

Seismic data, if combined with precise satellite measurements of the configuration of the Earth and measurements of its surface heat flow, provide a way to analyze the thermal regime of the mantle and identify broad zones of deep-seated magmatic activity. Beyond that, however, it is difficult to define the precise temperature and density of the mantle, even at shallow levels. Only about half of the heat coming from the Earth's interior could be generated by decay of the present abundances of radioactive isotopes. The other part must come from heat inherited from the early history of the planet when thermal energy was generated by the impact of large meteorites and by decay of radioactive isotopes that were more abundant than they are today.

These conditions could be determined more precisely if we had reliable determinations of the thermal gradient of the lithosphere, but this is one of the most difficult physical properties to measure. Direct measurements of the Earth's thermal gradient in deep mines and bore holes indicate that temperatures near the surface increase with depth at rates of about 15°C per km. It is clear, however, that these measured gradients cannot extend to much deeper levels, for if they are projected downward, they rapidly reach and exceed temperatures at which any plausible mantle rocks would be largely molten. Hence, the rate at which temperature increases must decline rapidly with depth.

It might seem that a better way of estimating the thermal gradient in the lithosphere would be to measure the amount of heat reaching the surface from the Earth's interior (between 1 and 2×10^{-6} cal cm^{-2} sec^{-1}) and, making due allowance for the distribution of radioactive heat sources, to calculate the thermal gradient that would be required to produce the observed flux of heat through rocks of known thermal conductivity. Unfortunately, as we shall see in the next chapter, thermal conductivity is strongly dependent on temperature, and we find that in order to estimate its value, we must first know the temperature of the rocks, which, of course, is what we are trying to determine from the conductivity! Better estimates of the temperature gradient are obtained by correlating the seismic velocities with experimental measurements of the effects of temperature and pressure on the properties of mantle rocks.

Convection in the Mantle

Any estimate of the thermal gradient depends on whether the heat is transferred purely by conduction or by some combination of conduction and convection. It is probably safe to assume that conduction is dominant in the lithosphere, especially in regions remote from spreading axes and zones of subduction, but this is less likely to be true for the mantle below the lithosphere.

Figure 1-8 illustrates a conception of the mantle that is consistent with what we know today. It depicts the upper and lower mantle as essentially independent systems separated by a transition zone in which, as we noted earlier, olivine changes to successively denser forms. The energy exchange required for minerals to adjust their crystalline structure to a higher or lower pressure tends to prevent the entire mantle from convecting as a unit.

The temperature within any convecting horizon is probably close to an *adiabatic* gradient. That is to say, the heat content of the rocks remains nearly constant while temperature adjusts to the changing pressure, much as it does in the Earth's atmosphere, where the temperature decreases as air rises and expands but increases as it descends and is compressed. Just as the temperature of air is buffered by the heat absorbed or released when water vaporizes or condenses, the temperature of the mantle must be moderated by the heat absorbed or released when minerals melt, crystallize, or change their crystalline structure in response to changes of pressure.

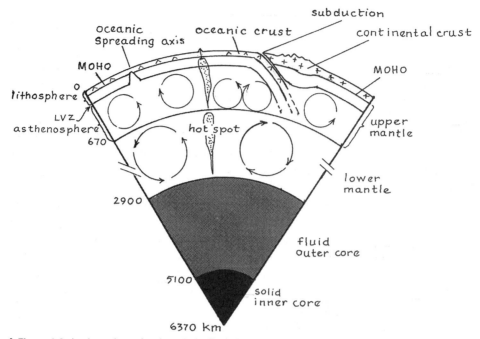

Figure 1-8 A schematic section through the Earth. For clarity, the thicknesses of shallow units have been exaggerated.

The amount of this adiabatic cooling resulting from relief of pressure on a rising magma can be estimated from the relationship

$$\frac{dT}{dP} = \frac{T\alpha}{\rho C_p} \tag{1-1a}$$

where α is the coefficient of thermal expansion (about 3×10^{-5} deg^{-1}), C_p is heat capacity (about 0.3 cal g^{-1} deg^{-1}), and the mechanical equivalent of heat is 41.84 cal^{-1}. Thus, for mantle temperatures and densities, the adiabatic effect would be about

$$\frac{(1200 + 273) \times 3 \times 10^{-5}}{2.7 \times 0.3 \times 41.84} = 1^{\circ}\text{kbar}^{-1} \tag{1-1b}$$

or between 0.2 and 0.3 degrees for each kilometer that the convecting mantle rises or descends. To compare this with the conductive gradient, we can estimate the latter from the Fourier equation for conductive heat transfer:

$$q = -K\frac{dT}{dZ} \tag{1-2}$$

and calculating the thermal gradient necessary to produce the observed heat flux through the lithosphere (about 1×10^{-6} cal cm^{-2} sec^{-1}). Assuming the conductivity of mantle peridotite is about 8×10^{-3} cal cm^{-1} sec^{-1} deg^{-1}, we find that the gradient, dT/dZ, would be about 12.5 degrees per kilometer. This value is probably too high because it does not take into account the fact that much of the surface heat flow comes from radioactive heat sources in the lithosphere, but in any case, the conductive gradient is at least an order of magnitude greater that the adiabatic gradient.

■ Mechanisms of Magma Generation

In considering the manner in which magmas are generated in the mantle, we should note at the outset that it is unlikely that melting is simply a spontaneous result of an accumulation of heat produced by decay of its radioactive elements. Mantle concentrations of U, Th, and K (Table 1-2) are so small that even if all their heat were to contribute to melting, times of the order of 10 million years would be required to yield just one weight percent of melt. Moreover, these elements are so effectively partitioned into the first fraction of melt that separation of this liquid would leave the mantle so depleted of heat-producing elements that it would immediately lose its capacity to generate further magma. The prolonged magmatism we see in many parts of the Earth would be impossible.

The largest and most readily available source of energy for melting is the heat in the rocks themselves. The problem of magma generation is primarily one of finding mechanisms that draw on the stored heat of large masses of rock to produce relatively small amounts of melt. A number of mechanical and thermal schemes can be visualized to explain how this might be accomplished.

Consider a temperature profile in the lithosphere and a convecting asthenosphere, such as the one shown schematically in **Figure 1-9**. Several changes could lead to melting under these conditions: (a) an influx of heat and increase of temperature at constant pressure and composition, (b) a decrease of pressure at constant heat content and composition, and (c) a change of composition at constant heat content and pressure. Certain physical limits can be placed on each of these.

(a) Assume, first, that an influx of heat from some unspecified deeper source enters the horizon where rocks are already close to the solidus temperature at which melting begins (Fig. 1-9a). A critical thermal property of silicates is their large latent heat of fusion, ΔH_f (about 65 to 100 cal g^{-1}), relative to their heat capacity, C_p (about 0.2 to 0.3 cal g^{-1} deg^{-1}). Each calorie of added heat produces about 0.01 gram of melt, but the same amount of heat can raise the temperature of a gram of rock (while still below its melting temperature)

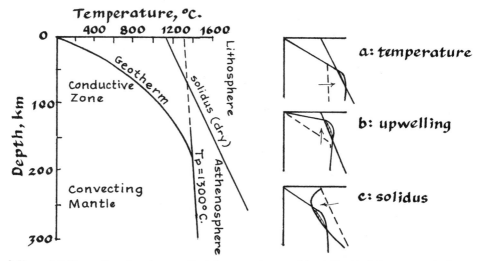

Figure 1-9 The configuration shown on the left represents a possible relationship between a conductive thermal gradient in the lithosphere and an adiabatic gradient in a convecting asthenosphere having a potential temperature of 1300°C. These two gradients are separated by a thermal boundary layer near the boundary between the lithosphere and asthenosphere. The solidus temperature for the mantle is shown for dry conditions. The small diagrams on the right illustrate the effects of (a) raising temperature by an influx of heat, (b) lowering pressure by upwelling of the asthenosphere, and (c) lowering the solidus curve by introducing another component, such as a volatile. In all three cases, the temperature of the cross-hatched zone of melting is less than it would be if there had been no melting because the heat absorbed on melting lowers the ambient temperature. (Adapted from D. P. McKenzie and M. J. Bickle, 1988, *Jour. Petrol.* 29:625–679; D. M. Latin, J. E. Dixon, J. G. Fitton, and N. While, 1990, *Geol. Soc. Spec. Publ.* 55:207–227.)

by more than three degrees. Only a relatively small amount of heat is needed to raise the temperature of a rock to the point where it begins to melt, but any subsequent melting absorbs such large amounts of heat that it impedes any further rise of temperature. The same effect works in reverse as well. The amount of heat that must be removed to crystallize a gram of melt, being the same as that required for melting, is so much greater than the heat capacity that any decline of temperature requires much more heat loss than simple cooling of a solid rock. In this way, melting or crystallization has the effect of stabilizing temperature gradients and smoothing the effect of any heat flux.

Another interesting property of silicate melts is the way their thermal conductivity varies with temperature. Conductivity declines with rising temperature until radiative heat transfer becomes increasingly important as the opacity of the rock or melt declines. It is interesting that the temperature at which this happens in mafic rocks is around 1200° to 1400°—the melting range at mantle pressures. Thus, an influx of heat into rocks at temperatures below this conductivity minimum tends to lower conductivity and retard the heat transfer, which in turn raises the temperature even further until melting begins. This "run-away effect" has been suggested as a possible melting mechanism for rocks that have an influx of heat from below.

Other mechanisms have been proposed by which melting could come from the conversion of mechanical energy to heat. We have already noted the example of meteorite impact. Another possibility is tidal friction, which could account for the strong volcanism on the satellites of large planets like Jupiter or Saturn.

(b) Suppose that, instead of an influx of heat, the rocks are raised to a level of lower pressure, possibly as a result of convective upwelling of the mantle (Fig. 1-9b). A rock that was already close to the temperature at which it would begin to melt would now be in a realm where it should be partly liquid. Even though no heat is added, crystals will begin to melt by drawing on their stored heat. Because the curve for the solidus temperature (i.e., the temperature at which melting begins) is so much steeper than that for the adiabatic temperature, an upwelling mantle is almost certain to reach the solidus, and as it continues to rise, increasing amounts of melt will be produced in response to the declining pressure. The heat required to achieve this melting is drawn from the stored heat of the rocks and is a direct function of the difference between the temperature of the mantle solidus, T_s, and that of the adiabat, T_a. The fraction of melting, F, for a given temperature difference is

$$F = (T_s - T_a) C_p / H_f \tag{1-3a}$$

where H_f is the heat of fusion and C_p is the heat capacity. The amount of melting for a given decrement of pressure, dP, would therefore be

$$\frac{dF}{dP} = \frac{(dT_s/dP - dT_a/dP) C_p}{H_f} \tag{1-3b}$$

The solidus temperature increases at a rate of about $10°$ kbar^{-1}, whereas we have seen that the adiabatic gradient is of the order of $1°$ kbar^{-1}; thus, for typical values for H_f (about 100 cal g^{-1}) and C_p (about 0.3 cal g^{-1}deg^{-1}), it turns out that about 2.7 percent of the mantle could melt for each kilobar or about 0.8 percent per kilometer of rise (**Fig. 1-10**).

(c) The third melting mechanism requires neither an influx of heat nor a change of pressure but an addition of components, such as H_2O or alkalies, that lower the melting temperature (Fig. 1-9c). This fluxing action might result from an introduction of volatiles, either from a source at greater depth or from breakdown of a volatile-bearing mineral in the rock itself.

Although the combined volatile contents of H_2O and CO_2 in basalts are probably less than 1 weight percent, even these small amounts could have a

Temperature

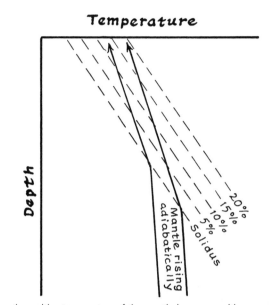

Figure 1-10 Because the melting temperature of the mantle increases with pressure at a greater rate than the adiabatic gradient, relief of pressure in upwelling mantle will cause melting to begin when the temperature of mantle rocks reaches that of the solidus. As the mantle continues to rise, the amount of melt will increase, as shown schematically by the dashed lines for percent melt. Because this melting draws on the stored heat in the rocks, the temperature declines at a greater rate than it would without melting. The amount of melt also depends on the initial temperature of the rising mantle.

large effect on melting. Melting temperatures of basalt are lowered as much as 100°C by only 0.1-percent water, and other volatiles, as well as the components they are likely to carry in solution, have comparable effects. We shall see in Chapter 9 that this form of melting is probably responsible for the magmas of subduction-related volcanoes.

The ability of magma to rise through the lithosphere is strongly dependent on its volume and physical properties, such as heat content, viscosity, and density. It seems unlikely, therefore, that the lavas reaching the surface are an accurate sampling of all the magmas the mantle is capable of producing. One can think of the light, relatively cool rocks of the lithosphere as a crude barrier that tends to filter out magmas that are much denser than the crust or too viscous to rise through narrow channels. It also impedes the rise of magmas that contain too little heat to survive strong cooling at shallow levels. The large increase of viscosity that comes with advanced degrees of crystallization severely limits the ability of crystal-rich magmas to rise through the crust. When the proportion of crystals reaches 40 or 50 percent, the rate of flow becomes so slow that the magma loses heat to the crust and solidifies before it reaches the surface.

In the next chapter, the physical properties that govern these processes are examined in greater detail, and in subsequent chapters, the mechanisms of magma generation are considered in the context of the tectonic settings in which the major types of igneous rocks are found.

Selected References

Condie, K. C., 2005, *Earth as an Evolving Planetary System*, Amsterdam and Boston: Elsevier, 447 p. A very comprehensive, up-to-date survey of the evolution of the Earth.

Eriksson, P. G., *et al.*, editors, 2004, *Precambrian Earth: Tempos and Events*, Elsevier Precambrian Research Series, 941 p. A collection of outstanding papers on all aspects of the Earth's early history.

Jackson, I., editor, 1998, *The Earth's Mantle: Composition, Structure, and Evolution*. New York: Cambridge University Press, 566 p. A very comprehensive series of papers dealing with the mantle.

Murray, B., M. C. Malin, and R. Greeley, 1981, *Earthlike Planets*, New York: W. H. Freeman & Co., 387 p. An excellent summary of recently acquired knowledge of the solar system.

Norton, O. R., 1994, *Rocks from Space*, Missoula, MT: Mountain Press, 446 p. A good semipopular review of meteorites.

2 Magmas and Igneous Rocks

Basic to all the diverse aspects of igneous geology is the nature of magma itself—its physical and chemical properties and its relationships to its geologic environment. Magma can be defined as a completely or partly molten natural substance, which on cooling, solidifies as a crystalline or glassy igneous rock. Most magmas are rich in silica and are capable of deforming by viscous flow under moderate stresses. They may have crystals in suspension, and most have volatile components that tend to exsolve when confining pressure is relieved.

The differences between magmas and crystalline rocks are not as sharply defined as are those between simpler liquids and solids. The temperatures through which solidification takes place may span as much as 200°C, and within this range, the physical properties of magma gradually change in response to the increasing proportions of crystals and the changing characteristics of the liquid in which they grow.

■ Composition

Igneous rocks are often divided into two broad groups according to their proportions of light or dark minerals. Granites, for example, are said to be *felsic* or *leucocratic* because they consist almost entirely of the light-colored minerals, quartz and feldspar, whereas gabbros, which contain abundant, dark, ferromagnesium minerals, are referred to as *mafic* or *melanocratic*. Rocks totally devoid of light minerals are said to be *ultramafic*. The corresponding terms, *acidic* and *basic*, often found in older literature, were based on a long since discredited concept. Although petrologists shun them, their use is so entrenched that they are difficult to avoid. No sharp boundary can be drawn between mafic and felsic rocks because the terms are loosely qualitative. A more precise measure is *color index*, which is simply the volumetric proportion of ferromagnesium minerals.

Igneous rocks are either *intrusive* or *extrusive* and, depending on their depth of emplacement and solidification, are divided into three broad groups: *plutonic* rocks that have cooled slowly at moderate to great depths in the crust or mantle, *hypabyssal* rocks emplaced in shallow, subvolcanic conditions, and *volcanic* rocks erupted on the surface. Each plutonic rock has a volcanic equivalent, but on a broad scale, the proportions of different chemical compositions in the two groups are not the same. Plutonic suites tend to have volumetrically greater proportions of felsic rocks, whereas volcanic suites tend to be more mafic, but like most generalities, this one has many exceptions.

The chemical compositions of igneous rocks are usually considered in terms of two groups of elements—a fixed number of *major elements* that normally account for 98 to 99 percent of the total and a variety of *trace elements* with much lower concentrations. Standard analyses, such as those in **Tables 2-1 and 2-2**, report the 13 major components as weight percentages of oxides listed in a conventional order. The trace elements reported in an analysis depend on the nature of the rock and the purpose of the analysis. They are reported not as oxides but as single elements in parts per million (ppm) and may be listed in any convenient order.

Petrologists attach so much importance to the compositional diversity of igneous rocks that they tend to overlook the basic regularities that characterize the abundances of their chemical constituents. Inspection of any compilation of chemical analyses, such as those in Tables 2-1 and 2-2, shows that compositions are neither limitless nor random. Each individual component accounts for a characteristic proportion of the rock. Silica is by far the dominant oxide, normally making up between half and three quarters of the total, whereas alumina is usually second in abundance and accounts for 10 to 20 percent of the total. The oxides of iron, magnesium, and calcium have the greatest variability. They commonly range from 10 or 15 percent to essentially zero. Apart from a few exotic alkaline rocks, the alkalies, Na_2O and K_2O, rarely total more than 10 percent and normally are less than half that amount. These variations are very systematic in the sense that the abundance of each one is usually linked either positively or negatively to that of the others. One would be justifiably suspicious of an analysis of an igneous rock with 70 percent SiO_2 in which CaO is more abundant than Na_2O or MgO more abundant than the iron oxides, even though the absolute abundance of each individual component is within its normal range for igneous rocks as a whole.

The fact that we recognize certain abundances or ratios of components as typically igneous and instinctively question compositions that violate these patterns shows that natural magmas are not randomly mixed solutions but the products of systematic processes controlled by strict physical–chemical laws.

Table 2-1 Typical chemical and normative compositions of volcanic rocks in Figure 2-4.

	Subalkaline Rocks							
	Tholeiitic Series				High Alumina Basalt	Calcalkaline Series		
	Picrite Basalt	Olivine Tholeiite	Tholeiite	Icelandite		Andesite	Dacite	Rhyolite
SiO_2	46.4	49.2	53.8	61.8	49.2	60.0	69.7	73.2
TiO_2	2.0	2.3	2.0	1.3	1.5	1.0	0.4	0.2
Al_2O_3	8.5	13.3	13.9	15.4	17.7	16.0	15.2	14.0
Fe_2O_3	2.5	1.3	2.6	2.4	2.8	1.9	1.1	0.6
FeO	9.8	9.7	9.3	5.8	7.2	6.2	1.9	1.7
MgO	20.8	10.4	4.1	1.8	6.9	3.9	0.9	0.4
CaO	7.4	10.9	7.9	5.0	9.9	5.9	2.7	1.3
Na_2O	1.6	2.2	3.0	4.4	2.9	3.9	4.5	3.9
K_2O	0.3	0.5	1.5	1.6	0.7	0.9	3.0	4.1
P_2O_5	0.2	0.2	0.4	0.4	0.3	0.2	0.1	0.1
Q	–	–	6.7	14.9	–	12.3	22.8	28.8
Il	2.7	3.2	2.9	1.8	2.1	1.4	0.6	0.3
Mt	2.6	1.4	2.8	2.5	3.0	2.0	1.2	0.3
Or	1.7	3.0	9.2	9.6	4.2	5.4	17.9	24.5
Ab	14.0	19.8	28.0	40.0	26.3	35.3	40.7	35.5
An	14.8	25.0	20.8	17.8	33.5	23.7	12.5	5.9
Di	15.8	22.3	13.7	3.8	11.3	3.6	0.3	–
Hy	16.2	14.8	15.1	8.7	13.9	16.0	4.0	3.1
Ol	31.8	10.3	–	–	5.1	–	–	–
Ap	0.4	0.4	0.9	0.9	0.6	0.4	0.2	0.2
C	–	–	–	–	–	–	–	1.1

SiO_2	44.1	44.2	45.4	47.9	49.7	55.6	60.7	60.6
TiO_2	2.7	1.6	3.0	3.4	2.1	0.9	0.5	0.1
Al_2O_3	12.1	15.6	14.7	15.9	17.0	16.4	20.5	18.3
Fe_2O_3	3.2	4.4	4.1	4.9	3.5	3.1	2.3	2.8
FeO	9.6	6.1	9.2	7.6	9.0	4.9	0.4	1.2
MgO	13.0	8.9	7.8	4.8	2.8	1.1	0.2	0.1
CaO	11.5	9.7	10.5	8.0	5.5	2.9	1.4	0.8
Na_2O	1.9	4.0	3.0	4.2	5.8	6.1	6.2	8.9
K_2O	0.7	1.8	1.0	1.5	1.9	3.5	6.7	5.1
P_2O_5	0.3	0.7	0.4	0.7	0.5	0.7	0.1	–
Il	3.8	2.3	4.3	4.8	3.0	1.3	0.7	0.1
Mt	3.4	4.7	4.4	5.2	3.7	3.9	2.4	–
Or	4.2	10.8	6.0	9.1	11.5	21.5	38.9	22.1
Ab	12.5	13.2	22.2	37.7	41.2	57.0	47.0	44.4
An	22.6	19.6	24.1	20.6	15.1	7.3	6.2	3.9
Di	26.3	19.7	21.0	12.3	7.6	2.3	–	–
Hy	–	–	–	–	–	3.3	–	–
Ol	23.9	14.5	14.1	8.3	9.7	2.3	–	1.5
Ne	2.8	13.9	3.1	0.5	7.2	–	4.7	14.4
Lc	–	–	–	–	–	–	–	5.9
Ap	0.6	1.5	0.9	1.5	1.1	1.5	0.2	–
Ac	–	–	–	–	–	–	–	7.7

Table 2-2   Average chemical and mineral compositions of selected plutonic rocks.

| | Chemical Composition | | | | | | | | |
	Granite	Syenite	Granodiorite	Quartz Diorite	Diorite	Gabbro	Anorthosite	Dunite	Lherzolite (Peridotite)
SiO_2	70.18	60.19	65.01	61.59	56.77	48.24	54.44	40.49	43.95
TiO_2	0.39	0.67	0.57	0.66	0.84	0.97	0.67	0.02	0.10
Al_2O_3	14.47	16.28	15.94	16.21	16.67	17.88	25.61	0.86	4.82
Fe_2O_3	1.57	2.74	1.74	2.54	3.26	3.16	0.93	2.84	2.20
FeO	1.78	3.28	2.65	3.77	4.40	5.95	1.26	5.54	6.34
MnO	0.12	0.14	0.07	0.10	0.13	0.13	0.07	0.16	0.19
MgO	0.88	2.49	1.91	2.80	4.17	7.51	0.93	46.32	36.81
CaO	1.99	4.30	4.42	5.38	6.74	10.99	9.92	0.70	3.57
Na_2O	3.48	3.98	3.70	3.37	3.39	2.55	4.58	0.10	0.71
K_2O	4.11	4.49	2.75	2.10	2.12	0.89	1.01	0.04	0.21
H_2O	0.84	1.16	1.04	1.22	1.36	1.45	0.35	2.88	1.08
P_2O_5	0.19	0.28	0.20	0.26	0.25	0.28	0.25	0.05	0.02
Density	2.67	2.76	2.72	2.81	2.84	2.98	2.97	3.29	3.33

Mineral Composition

Mineral									
Quartz	25	–	21	20	2	–	2	–	–
K feldspar	40	72	15	6	3	–	–	–	–
Oligoclase	26	12	–	–	–	–	–	–	–
Andesine	–	–	46	56	64	–	–	–	–
Labradorite	–	–	–	–	–	65	91	–	–
Biotite	5	2	3	4	5	1	–	–	–
Amphibole	1	7	13	8	12	3	–	2	–
Hypersthene	–	–	–	1	3	6	–	2	15
Augite	–	4	–	8	14	21	–	–	–
Olivine	–	–	–	–	–	7	–	10	71
Magnetite	2	2	1	2	2	2	1	2	1
Ilmenite	1	1	–	–	2	2	2	–	–
Apatite	0.2	0.5	0.4	0.5	0.5	0.6	0.5	–	–
Spinel	–	–	–	–	–	–	–	1	3

Source: After Daly and Larsen, *Geol. Soc. Am. Spec. Paper 36*, 1942, with modifications and additions.

Equally important is the regularity with which certain rock types are found together in characteristic geologic settings. The lavas erupted in different tectonic environments are often distinctive and, in some instances, unique. The fact that granitic plutons, for example, are nowhere found within the deep oceanic basins but only on continents places severe limits on processes that could account for their formation. Throughout most of geologic time, certain types of rocks have formed distinctive suites, and members of one of these suites rarely occur in others. The regularity of these natural groupings of rocks must reflect fundamental differences in their origins.

Nomenclature

The ostensible aim of igneous petrology is to understand these relationships, and in order to do this, one needs an orderly way of referring to igneous rocks and identifying their salient features. Unfortunately, igneous petrologists have been slow in reaching this simple goal. Until recently, they were burdened with hundreds of rock names, many of which were used in a different sense from one country or institution to another. Many of these names have now been abandoned, and the remaining ones are more rigorously defined.

Although petrology is still plagued with conflicting systems of classification and nomenclature, a consensus is beginning to emerge. The one for plutonic rocks outlined in **Figures 2-1 and 2-2** and explained in **Table 2-3** is as widely accepted

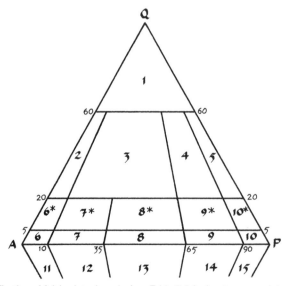

Figure 2-1 Classification of felsic plutonic rocks (see Table 2-3 for key to nomenclature). (Adapted from IUGS Subcommission on the Systematics of Igneous Rocks, *Geotimes*, October, 1973.)

Table 2-3	Determinative key for plutonic rocks with less than 90 percent mafic minerals and less than 60-percent feldspathoids (Fig. 2-1).

Field	Rock Name
I. Q > 60% of light minerals	(1) Quartz-rich granitoids
II. Q = 20–60% of light minerals	
a. P = 0–10% of feldspar	(2) Alkali-feldspar granite
b. P = 10–65% of feldspar	(3) Granite
c. P = 65–90% of feldspar	(4) Granodiorite
d. P = 90–100% of feldspar	(5) Tonalite
(trondhjemites are leuco-tonalites (M = 0–10%) containing oligoclase or andesine	
III. Q = 5–20% of light minerals	
a. P = 0–10% of feldspar	(6*) Alkali quartz syenite
b. P = 10–35% of feldspar	(7*) Quartz syenite
c. P = 35–65% of feldspar	(8*) Quartz monzonite
d. P = 65–90% of feldspar	(9*)
An < 50	Quartz monzodiorite
An > 50	Quartz monzogabbro
e. P = 90–100% of feldspar	(10*)
An < 50	Quartz diorite
An > 50	Quartz gabbro
M = 0–10%	Quartz anorthosite
IV. Q = 0–5% of light minerals	
a. P = 0–10% of feldspar	(6) Alkali syenite
b. P = 10–35% of feldspar	(7) Syenite
c. P = 35–65% of feldspar	(8) Monzonite
d. P = 65–90% of feldspar	(9)
An < 50	Monzodiorite
An > 50	Monzogabbro
e. P = 90–100% of feldspar	(10)
An < 50	Diorite
An > 50	Gabbro
M = 0–10%	Anorthosite
V. F = 0–60% of light minerals	
a. P = 0–10% of feldspar	(11) Feldspathoidal alkali syenite
b. P = 10–35% of feldspar	(12) Feldspathoidal syenite
c. P = 35–65% of feldspar	(13) Feldspathoidal monzonite
d. P = 65–90% of feldspar	(14)
An < 50	Feldspathoidal monzodiorite
An > 50	Feldspathoidal monzogabbro
e. P = 90–100% of feldspar	(15)
An < 50	Feldspathoidal diorite
An > 50	Feldspathoidal gabbro

Source: Adapted from IUGS Subcommission on the Systematics of Igneous Rocks, *Geotimes,* 1973.

A is alkali feldspar, P plagioclase, Q quartz, F feldspathoids, and M mafic minerals. The divisions on the A–P axis are at 10, 35, 65, and 90 percent plagioclase.

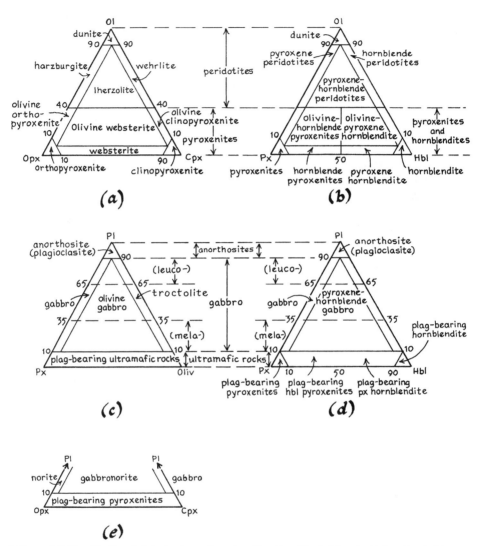

Figure 2-2 Classification of plutonic ultramafic rocks without hornblende (a) and with hornblende (b). Gabbroic rocks without hornblende are given in (c) and those with hornblende in (d). Gabbroic rocks with orthopyroxene are shown in (e).

as any is likely to be. The same is not true, unfortunately, for volcanic rocks. Volcanic rocks present a special problem because two magmas of identical chemical composition can crystallize to very different mineral assemblages, depending on their rates and conditions of crystallization. The term *heteromorphism* is used to describe this phenomenon. Thus, a basaltic magma may form many heteromorphic rocks, including gabbro, dolerite (a medium-grained hypabyssal rock), various types of basalt, and sideromelane (a clear, brown basaltic glass). Because of the extreme heteromorphism of volcanic rocks, they are more difficult to define mineralogically than are their plutonic counterparts.

In order to surmount this problem, modern petrologists tend to rely more on chemical than on petrographic criteria to characterize volcanic rocks. An important step was made in this direction when, in 1971, two Canadian geologists, Neil Irvine and Robert Baragar, published a statistical study of chemical analyses of large numbers of volcanic rocks and established compositional limits for the major types of natural associations. Their classification scheme **(Figs. 2-3 and 2-4 and Table 2-4)** corresponds closely to current usage, particularly in the English-speaking world. It has the virtue of not depending on any particular genetic interpretation; it merely groups rocks according to their observed compositions and geological associations.

Although one might wish that classification systems were based on sound genetic relationships and that names could be made to fit neat pigeon

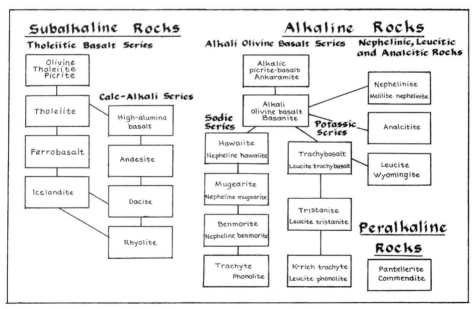

Figure 2-3 General classification of the common volcanic rocks. The lines joining boxes serve to outline common associations. The order of the columns is one of increasing differentiation downward. The tholeiitic and calc-alkaline divisions of the subalkaline rocks are closely related, and several rock names, such as dacite and rhyolite, are common to both series. Rock names in small print are variants of the main rock name in the box. (Adapted with minor modifications from Irvine and Baragar, 1971, *Can. J. Earth. Sci.* 8:523–548.)

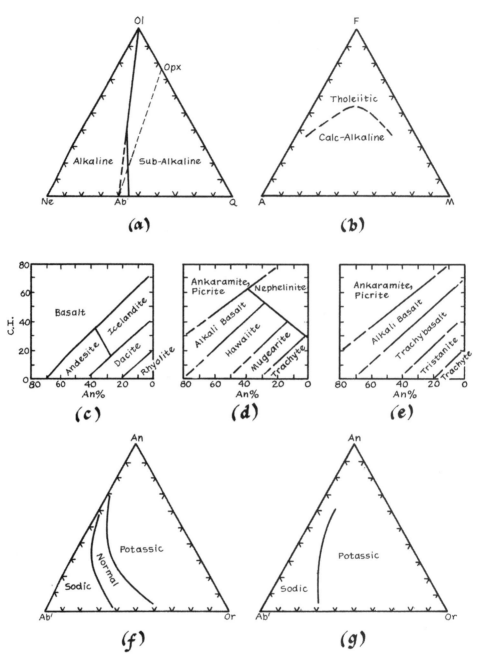

Figure 2-4 Compositional limits of the common volcanic rocks according to the classification system of Table 4.

Table 2-4	A volcanic rock may be classified on the basis of its chemical and normative composition according to the system of Irvine and Baragar (1971), depicted in Figs. 2-3 and 2-4. The procedure is as follows.

1. If the rock has normative acmite (Ac), it is probably peralkaline. Otherwise go to Step 2.

2. Classify as subalkaline or alkaline according to the boundary line in Fig. 2-4a and the criteria given in Chapter 7. If the rock is subalkaline, go to Step 3; if alkaline, go to Step 4.

3. (a) Classify as calc-alkaline or tholeiitic using the boundary lines in Fig. 2-4b. See Chapter 9 for details.

   (b) If olivine (Ol) is greater than 25%, the rock is a tholeiitic picrite. Otherwise, classify as basalt, andesite, icelandite, dacite, or rhyolite according to Fig. 2-4c.

   (c) Classify as potassic, "normal," or sodic according to Fig. 2-4f. Go to Step 5.

4. (a) Decide whether the rock is a nephelinite, leucitite, or analcitite on the basis of its petrographic character and the criteria in Chapter 11. Otherwise continue here.

   (b) If olivine (Ol) is greater than 25%, the rock is an alkaline picrite. If modal olivine plus augite exceed 25% it is an ankaramite. Assign the rock to one of the two alkaline basalt series using Fig. 2-4g. Classify it further according to Figs. 2-4d or 2-4c as appropriate.

5. Check the result against analyses in Table 2-1 and elsewhere in the text. Most important, check the result against the rock's petrographic characteristics. If the rock has been significantly altered or has an unusual composition, this step will help to avoid errors.

holes, experience shows that most rock names are too deeply entrenched to be abandoned or redefined, no matter how logical the revision might seem to those who propose it. The ingenious system designed to classify rocks on the basis of "normative" minerals is an excellent example of this problem.

Around the beginning of the previous century, four petrologists, Cross, Iddings, Pirsson, and Washington, devised a thoroughly quantitative and logical scheme for the classification of volcanic rocks. They discarded all previous nomenclature and set up a new set of names based entirely on parameters that could be calculated from a chemical analysis. The nomenclature and classification system was received with admiration and acclaim but was used by scarcely anyone except its authors and was promptly abandoned.

Normative Composition and Classification of Igneous Rocks

In contrast to the resounding failure of their classification scheme, the techniques that Cross and his co-workers developed for relating chemical analyses of volcanic rocks to mineralogical assemblages proved to have more lasting value. They provide a convenient way of comparing diverse chemical compositions by recasting the analysis of a rock into a standard set of idealized, or "normative," minerals. They do this by transforming the weight percentages of the major components into minerals that might form with complete equilibrium crystallization under plutonic conditions in the upper crust. In this way, the nature of a rock can be defined independently of differing conditions of crystallization and equilibration. The calculation is designed to yield a mineral assemblage, known as a *norm*, which corresponds remarkably well to the actual mineral assemblage, or *mode*, of a totally crystalline rock formed at low pressure under conditions of ideal equilibration.

A number of variants of the basic norm calculation have been devised. The one used here and explained in Appendix A is referred to as the Barth-Niggli or Molecular Norm. It has an advantage over the older CIPW norm (named from the initials of the four petrologists who devised the original system) in that the resulting mineral assemblage is closer to the volumetric proportions observed in the mode. The CIPW norm yields an assemblage based on weight percentages and overemphasizes iron and titanium oxides. Moreover, the simple molecular proportions are easier to convert to alternative minerals in reaction equations.

Genetic Classifications

The ideal classification system would, of course, be one that groups igneous rocks genetically according to the physical–chemical or tectonic factors governing their composition and mode of occurrence. Although far from achieving this ideal, some of the broad categories to which rocks are commonly assigned have developed because petrologists perceived a genetic link in suites of rocks that share particular chemical and mineralogical attributes in distinct geological settings. The most notable example is the division between alkaline and subalkaline rocks. The distinction between these two major groups can be traced back at least as far as the "Atlantic" and "Pacific" types proposed by Alfred Harker at the beginning of the last century. Their basic compositional difference is mainly in their relative concentrations of silica and alkalies. Subalkaline rocks are "critically saturated" with SiO_2. That is to say, they have enough of that component to form a set of modal or normative minerals that does not include an alumina silicate with less silica than feldspar. The alkaline rocks are deficient in silica relative to other components, especially Na_2O and K_2O, and for this reason contain a normative feldspathoid, such as nepheline (Ne) or leucite (Lc). Alkaline rocks can be subdivided on the basis of how strongly they are undersaturated with silica and whether potassium or sodium is the dominant alkali. Rocks in which the molecular proportion of $Na_2O + K_2O$ exceeds that of Al_2O_3 are called peralkaline, and because all the alkalies cannot enter alumina silicates, they contain the sodic pyroxene acmite (Ac) in their norm. Their mode may include any of a number of sodic pyroxenes or amphiboles.

The subalkaline group has two major subdivisions, a *tholeiitic* and a *calc-alkaline* series, which differ mainly in their relative concentrations of iron and alumina. The mafic and felsic end-members are basalts and rhyolites that differ in only minor ways between the two series, but intermediate tholeiitic magmas are richer in iron, whereas calc-alkaline rocks tend to have more Al_2O_3 and, hence, more plagioclase.

Some rocks have more Al_2O_3 than can be incorporated in feldspar. In the norm, this results in corundum (C), and in plutonic rocks, the excess Al_2O_3 normally enters muscovite. Such rocks are called *peraluminous*.

These major divisions correspond rather well to the tectonic association mentioned in the preceding chapter, and for that reason, they form a logical basis for the system illustrated in Figure 2-3. Tholeiitic basalts and their differentiates are by far the most voluminous and ubiquitous group. Although they occur in almost every geological environment, they are most characteristic of the ocean floor and divergent plate boundaries; calc-alkaline rocks are associated almost exclusively with orogenic belts and convergent plate boundaries. Alkaline rocks are found in intraplate regions of both the oceans and continents and only rarely at plate boundaries. Most oceanic rocks are sodic, in contrast to their continental counterparts, which tend to be more potassic. Although they seldom account for large volumes, the alkaline rocks are exceedingly diverse and include as many petrographic varieties as all other rocks series combined.

Volatile Components

Even though volatiles constitute only a small weight fraction of magmas, their molecular weights are low so that their mole fraction can be large. Moreover, they have disproportionately large effects on physical properties, such as viscosity, and on the composition, order of appearance, and form of crystallizing minerals. Although much of the volatile content of volcanic rocks is lost before minerals crystallize at low pressures, plutonic rocks retain more of their volatiles and incorporate them into stable minerals. Water, for example, enters the hydrous minerals, amphibole and mica, and sulfur combines with iron or copper to form sulfides. Few igneous rocks contain carbonates or other CO_2-bearing minerals, but some, notably carbonatites, contain major amounts. Halogens are essential constituents of apatite, tourmaline, topaz, and some amphiboles.

All natural magmas contain dissolved water, carbon dioxide, and other volatile components that may be absorbed and exsolved in complex ways as magmas rise from their source to cool and crystallize at shallower depths. It is difficult to know what proportions of these gases were constituents of the original magma at its source. Recent studies of continental volcanoes indicate that CO_2 may be the dominant gas at depth and that much of the water released during eruptions is absorbed from the crust. The most reliable data on mantle-derived volatile components come from glassy lavas dredged from the deep ocean floor, where pressures are so great that little exsolution takes place before the lava is quenched. CO_2, H_2O, and lesser amounts of sulfur gases account for by far the greater part of the total volatile content and together amount to less than one percent by weight. Chlorine and fluorine are present in smaller amounts.

The solubilities of gases in magmas vary with pressure, temperature, and with the compositions of both the gas and the liquid (**Fig. 2-5**). Every student is aware (or should be) that warm beer foams more than cold beer and that uncapping a bottle results in an escape of bubbles. These simple observations illustrate the fact that solubility decreases when confining pressure on the gas is reduced and when temperature increases. The same principles hold for magmas. In addition, if the pressure of the gas alone is constant and less than the total load pressure on the liquid, solubility decreases as load pressure increases. The effect can be thought of as one of squeezing the gas from the liquid because the volume of the liquid with dissolved volatiles is greater than the volume of the liquid alone. For this reason, the water content of columns of magma can be greater at the top than at depth.

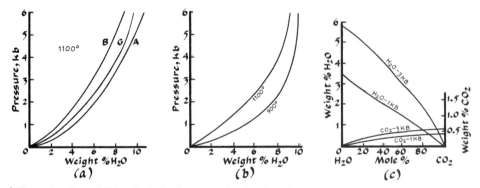

Figure 2-5 The solubility of volatiles in magmas is a function of pressure, temperature, and the compositions of the liquids and gases. (a) shows the solubility of H_2O in basalt (B), granite (G), and andesite (A), all at 1100°C, as a function of pressure of H_2O. Although the three compositions have about the same solubilities at the same temperature, the three types of magmas are liquid at very different temperatures, and their solubilities under natural conditions will differ accordingly. As shown in (b), the solubility in granite increases with falling temperature; basalt and andesite show similar effects. Hence, natural basalts have the lowest solubility as natural liquids and granites the highest. The pressure of another gas, such as CO_2, decreases the solubility of H_2O. (c) shows solubilities in a basaltic melt at 1200°C and two different total pressures, one and three kilobars. (Data from Hamilton and Anderson, 1967, *Basalts* 1:445–482; Goranson, 1931, *Amer. J. Sci.* 22:481–502; and Kadik et al., 1972, *Geoch. Internat.* 9:1041–1050.)

Most silica-rich magmas can hold more water in solution than can more mafic magmas (Fig. 2-5a), and this effect is augmented by the lower temperatures of the former (Fig. 2-5b). If another gas, such as CO_2, is present, it reduces the solubility of water, as shown in Figure 2-5c.

■ Physical Properties of Magmas

The wealth of data on the chemical and mineralogical nature of igneous rocks stands out in striking contrast to the meager information available on the physical properties of the liquids from which these rocks crystallized. Until recently, this deficiency has been a serious handicap for petrologic interpretations, but more studies are now being focused on physical aspects of magmas, and our knowledge of their behavior is rapidly increasing.

Temperatures and Other Thermal Properties

Temperatures recorded for erupting lavas rarely fall outside the range between 800°C and 1200°C. The lower values are observed in partly crystalline lavas and probably correspond to the limiting conditions under which magmas can flow. Magmas containing more than about 40 to 50 percent crystals are so viscous that they rarely erupt at the surface. The fact that low temperatures are observed in lavas with differentiated or felsic compositions and high-temperature eruptions are confined to basalts implies that few magmas have a wide enough range of crystallization to remain mobile at temperatures much below that at which crystallization begins.

Very few lavas are hotter than the temperature at which their first minerals crystallize. Basalts, for example, are rarely observed at temperatures above 1200°C, and even at their highest temperatures, they have little or no superheat. That is to say they can cool through only a small temperature interval before crystals begin to nucleate and grow. This is a significant observation, for it tells us much about the manner in which magmas reach the surface.

When we consider that the melting temperatures of common igneous rocks increase with pressure at a rate of roughly 3°C for each kilometer of depth, it is apparent that at depths where magmas are generated (at least 50 km) their melting temperatures must be at least 150 degrees higher than those observed at the surface. The fact that erupting lavas are close to their crystallization temperatures at atmospheric pressure indicates that they must lose heat as they rise and that the temperatures observed at the surface cannot be those at which the magma formed at depth. Some of the consequences of these relationships are discussed in later chapters.

Some rhyolites may have been exceptions to the rule that magmas erupt with little or no superheat. Estimates of temperatures of obsidian flows based on the viscosity they seem to have had when erupted suggest that they may have been as much as 100 or 200 degrees above their crystallization temperatures, but these estimates are subject to so many unknown factors that they are far from reliable. The absence of crystals in very siliceous glasses may simply be due to the difficulty of nucleating crystals in such viscous magmas. It is unfortunate that not a single rhyolitic eruption has been observed in which temperatures of the magma could be measured or even estimated.

One of the most distinctive thermal properties of magmas and igneous rocks is the great contrast between their heat capacity and heat of fusion or crystallization. *Heat capacity*, C_p, is defined by the relationship

$$C_p = \frac{\Delta H}{\Delta T} \qquad (2\text{-}1)$$

in which ΔH is the amount of heat that must be added or removed to raise or lower the temperature of 1 gram by 1 degree centigrade. (The term *specific heat* refers to the heat capacity of a substance relative to water at the same conditions.) *Heat of fusion*, ΔH_F, is the heat that must be added or removed to melt or crystallize one gram that is already at the temperature where the liquid and solid co-exist. For most rocks, C_p is about 0.3 cal g^{-1} deg^{-1}, whereas typical values for ΔH_f are around 65 to 100 cal g^{-1}. As we noted in the preceding chapter, roughly the same amount of heat is involved in melting or crystallization of a rock as in raising or lowering the temperature of the rock or melt by 200 to 300 degrees.

Igneous rocks and melts are exceptionally poor conductors of heat. Their thermal conductivity (**Fig. 2-6**) is a function of two mechanisms of heat transfer, ordinary lattice or phonon conduction and radiative or photon conduction. The first reflects the thermal vibration of the crystal lattice, whereas the second is a form of electromagnetic radiation. Because the lattice structure expands and begins to break down with increasing temperature, phonon conductivity declines at a decreasing rate up to temperatures of about 1200°C. Beyond that temperature, the opacity of the rocks declines rapidly so that radiative heat transfer increases and the total conductivity becomes much greater. Siliceous rocks, such as andesite and rhyolite, have lower opacities and hence greater radiative conductivity at lower temperatures than do mafic rocks.

The Physical–Chemical Nature of Silicate Liquids

Several lines of evidence show that silicate liquids are not simple ionic solutions but have a certain degree of ordering that approaches that of minerals (**Fig. 2-7**). The small differences of densities and refractive indices between crystalline rocks and their glasses (rarely more than a few percent) indicate that

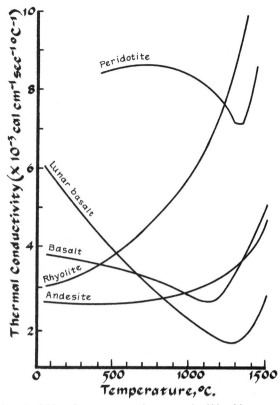

Figure 2-6 Thermal conductivities of some common igneous rocks. (After Murase and McBirney, 1973, *Geol. Soc. Amer. Bull.* 84:3563–3592.)

the structural arrangement of atoms does not differ greatly in the two states. Moreover, the amount of heat involved in changing from the crystalline solid to the liquid state is small compared with that of vaporization and indicates that the crystalline structure undergoes relatively little disruption in passing from the solid to the liquid. The difference between the structure of the liquid and crystalline forms lies mainly in the extent of linkage of the atoms. The atoms in silicate melts seem to be bonded by forces similar to those between the atoms of crystals but with less long-range order, probably on a scale of about 20 Angstrom units (Å) or 20×10^{-8} cm.

The nature of the silicon ion itself is responsible for many of the overall properties of magmas. Because of its high charge (+4), small radius (0.39 Å), and its coordination number with oxygen (4), its bonding with oxygen is much stronger than that of other common cations. Al, although not so strongly bonded to oxygen as Si, is nevertheless stronger than Ca, Mg, or

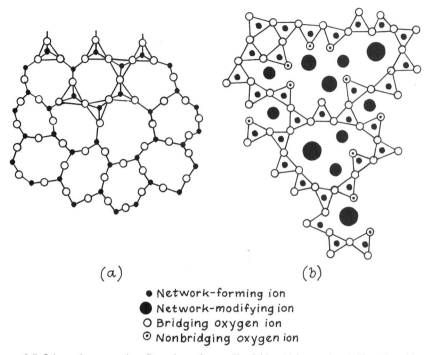

(a)   (b)

● Network-forming ion
● Network-modifying ion
○ Bridging oxygen ion
⊙ Nonbridging oxygen ion

Figure 2-7 Schematic structural configurations of pure silica (a) in which an ordered silica mineral is shown in the upper part of the field and a less ordered liquid or glass below. A silicate liquid containing network-modifying ions is shown in (b). (After Carmichael et al., 1974, *Igneous Petrology*, McGraw-Hill.)

Na, and in some respects, Si and Al play similar roles in both liquids and crystalline solids.

In pure silica melts, silica tetrahedra are linked in three-dimensional networks with oxygen atoms, most of which are shared by two or more tetrahedra. Other cations can enter the melt in limited amounts as independent ions occupying spaces between the tetrahedra and thereby modifying the basic framework (Fig. 2-7b).

In this sense, Si and Al are often referred to as *framework-forming* elements. *Framework modifiers*, the ions occupying spaces within the basic framework structure, are Mg, Ca, Fe, Na, and most other metals. These components can be accommodated in amounts up to about 20 cation percent before the basic framework begins to break down into smaller geometric units. The basic network of tetrahedra in which Si is bonded to O atoms in a ratio of 1 to 2 is reduced to less extensive units with smaller Si/O ratios as other cations take the place of Si in sharing O atoms. In breaking the continuity of the structure into smaller units, the framework changes from extensive three-dimensional networks of tetrahedra, almost all linked by shared oxygens, to configurations with more nonbridging oxygens. When more than 66 percent of the cations are framework modifiers, the liquid takes the form of orthosilicates in which each Si atom forms a separate tetrahedron.

Viscosity

The effect of the proportion of Si atoms on the framework linkage of silicates is seen most clearly in the viscosity of their liquids. Viscosity, η, is a measure of the resistance to flow of any fluid substance. More strictly, it is defined as the ratio of the applied shear stress to the rate of deformation or shear strain (**Fig. 2-8**) and is expressed by the general relationship

$$\sigma = \sigma_0 + \eta \left(\frac{du}{dz}\right)^n \tag{2-2}$$

where σ is the total shear stress, applied parallel to the direction of deformation, σ_0 is the stress required to initiate flow, and du/dz is the gradient of velocity, dx/dt, over a distance z normal to the direction of shear, x. The exponent n has a value of 1.0 or less, depending on the nature of the specific fluid and the velocity gradient produced by shear strain. For so-called Newtonian fluids (curve A in **Fig. 2-9**), n is 1.0, and σ_0 is zero, but many "pseudoplastic" fluids and liquids with suspended solid particles have a nonlinear relationship of shear stress to strain rate, such as that depicted by curves B and C in Figure 2-9, for which the value of n is less or more than 1.0. Other substances, including many highly polymerized materials, have a finite yield strength that must be overcome before they deform permanently (curve D,

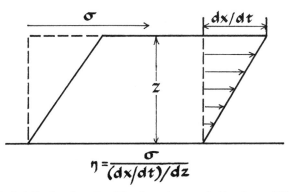

Figure 2-8 Viscosity is defined as the ratio of the shear stress applied to a layer of thickness z to the rate at which it is permanently deformed in a direction x parallel to the stress. See the text for explanation.

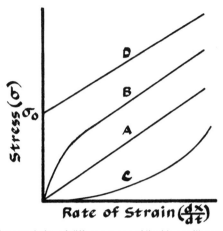

Figure 2-9 The physical characteristics of different types of liquids are illustrated by the relationships between the rate at which they are permanently deformed by shear (dx/dt) and the stress (σ) parallel to the direction of shear. Newtonian fluids (A) have a linear relation with an origin at zero on both axes. Some non-Newtonian fluids have nonlinear relations, such as those with shear thinning (B) or shear thickening (C), whereas others (D) have a finite yield strength (σ_0) that must be exceeded before they are permanently deformed by viscous flow.

Fig. 2-9). At stresses below σ_0, they have an elastic behavior, but at higher stresses, they are viscoelastic.

Although the viscosities of most silicate melts can be explained, at least crudely, in terms of their proportions of Si/O bonds, we lack a comprehensive theory to account for the individual effects of all the major cations. It is possible to estimate viscosities of superheated liquids from their chemical composition (Appendix B), but the relationships in the range of temperatures of natural magmas are more complex. **Figure 2-10** shows the measured viscosities of some common igneous rocks in and above their melting ranges and illustrates the principal effects of composition, temperature, and crystal content. As one would expect, viscosities drop with increasing temperature and decreasing silica content. The effect of pressure is to reduce viscosity because it causes ionic complexes to take on more compact structural configurations. Water and most other volatile components tend to reduce viscosity because they break the continuity of framework structures in the melt. A small increase in H_2O results in a large decrease of viscosity, as shown for a granitic composition in Figure 2-10c. The effect on more mafic magmas is less dramatic. Suspended crystals have the opposite effect; viscosity rises rapidly as crystals nucleate and grow.

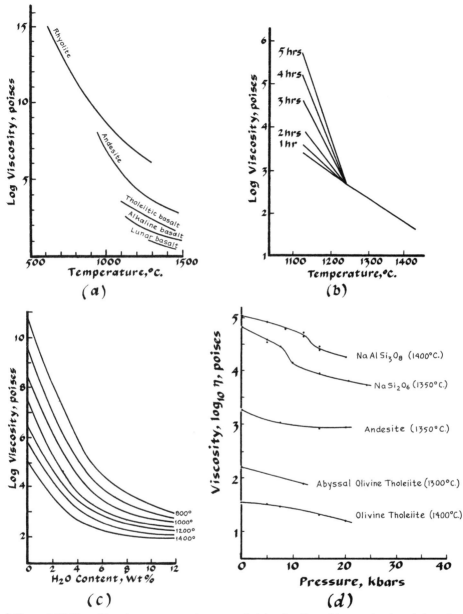

Figure 2-10 Viscosities of some common igneous melts (a) as functions of temperature in and above the melting range. (b) shows the increase in viscosity of a basalt after cooling for different periods of time. The inflection point is close to the temperature at which the basalt begins to crystallize. The effect of water on viscosity is illustrated by the example of a granitic liquid shown in (c). The effect of pressure is to reduce the viscosity of most silicate melts (d). (a, b, and c are from Murase and McBirney, 1973, *Geol. Soc. Amer. Bull.* 84:3563–3592. d is from Scarfe, Mysen, and Virgo, 1987, *Geoch. Soc. Spec. Publ. No. 1*, 59–67.)

The term *effective viscosity* is used to designate the viscosity of an inhomogeneous fluid, such as a magma with suspended crystals or bubbles. Its value can be estimated from the relationship:

$$\eta_{\text{eff}} = \eta_0 (1 - \phi/\phi_c)^{-2.5} \qquad (2\text{-}3)$$

in which η_{eff} is the effective viscosity of a fluid having a crystal-free viscosity of η_0 and a fractional volume of solids, ϕ. The term ϕ_c is the critical packing of crystals and has a value of 0.74 for spheres of uniform size or about 0.6 for solids with the shapes and size distribution of crystals in magmas. The equation shows how the effective viscosity of a magma may increase by an order of magnitude or more when its crystal content approaches 50 percent. At the same time, of course, the remaining liquid changes composition so that the viscosity also changes by virtue of that factor as well. (Methods of estimating viscosities of such magmas are given in Appendix B.)

Density

Next to viscosity, density is probably the most important property affecting the physical behavior and chemical differentiation of magmas. The densities of most magmas range between about 2.2 and 3.1 g cm^{-3}. They vary with temperature, pressure, and composition.

The effect of temperature is expressed by the volumetric coefficient of thermal expansion, α, the proportional change of volume, V, per degree at constant pressure:

$$\alpha = \frac{1}{V} \left(\frac{\partial V}{\partial T} \right)_P \qquad (2\text{-}4)$$

α normally has a value for silicate melts of about 3.0×10^{-5} per degree. The corresponding effect of pressure, isothermal compressibility, often designated β, is defined as

$$\beta = \frac{1}{V} \left(\frac{\partial V}{\partial P} \right)_T \qquad (2\text{-}5)$$

and has a value of about 7.0×10^{-6} bar^{-1} for most silicate liquids. Thus, the effect of a one-degree increase of temperature on a totally liquid magma with an original density of 2.7 g cm^{-3} would be a decrease of density of about 0.001 percent. This is equivalent to a difference of pressure of 3.9 bars or the weight of about 14.6 m of the same magma. The magnitudes of both thermal expansion and compressibility are lower for crystalline rocks, the respective values being about 2×10^{-5} deg^{-1} and 1.3×10^{-6} bar^{-1}. The values for partly crystalline magmas are, of course, intermediate between those of the liquid and solid.

The effect of composition on density is large relative to that of temperature and pressure. **Figure 2-11** shows the densities of liquids with the compositions of common igneous rocks over a broad range of temperature. Note that the decrease of density due to thermal expansion is much less than that of changing composition. The main component responsible for this large compositional effect is iron. In a given series of differentiated magmas, density may increase or decrease, depending on whether iron is enriched or depleted in the liquid as it evolves in composition from basaltic to more felsic magmas, such as rhyolite.

■ Effects of Cooling and Crystallization

It can be seen from relationships such as these that the properties of magmas differ from those of simple solutions, in that no sharp boundary can be drawn between the liquid and the solid. Consider first what happens when a magma that cools and crystallizes without changing its bulk composition. As it solidifies, it passes gradually through a succession of stages in which its density, viscosity, and yield strength increase with falling temperature. At the highest temperatures, where only a few crystals are held in suspension, the magma can be permanently deformed by viscous flow, even under small stresses. Its behavior approaches that of a true Newtonian fluid, but with falling temperature and increasing proportions of crystals, viscosity increases abruptly.

When crystals make up as much as half the volume, the magma can still be deformed by viscous flow, but the stresses required for a given shear rate are higher, and an increasing yield strength must be exceeded to produce permanent strain. Short-term stresses lower than the yield strength produce recoverable elastic strain, whereas small stresses of long duration may be absorbed by various forms of creep in the crystal–liquid mush. When the proportion of crystals becomes great enough for grain-to-grain contact, the partly molten assemblage behaves like an elastic solid when subjected to small short-term stresses. Very slow deformation can take place at low stresses by diffusion through the liquid fraction, but this mechanism decreases in importance with further cooling so that when the last fluid disappears, the rock behaves as a brittle solid for short-term stresses.

Consider next a magma that evolves in composition as it cools and differentiates. Because changing composition can affect the physical properties in different ways, depending on how the magma evolves, the variations of viscosity and density may be quite different from one igneous series to another (**Fig. 2-11b**). In calc-alkaline magmas, the effect of declining temperature is augmented by that of increasing silica content to cause a steady increase of viscosity throughout the entire range, from basalt to rhyolite.

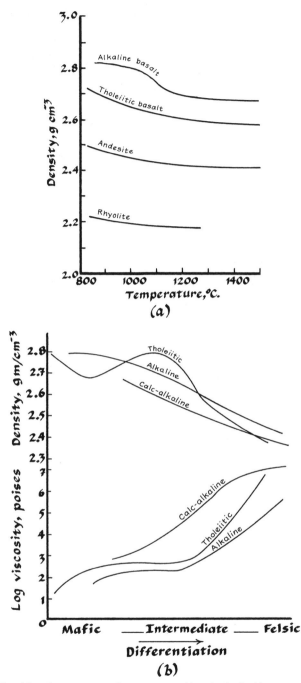

Figure 2-11 (a) Densities of some common igneous compositions in the liquid state as functions of temperature. (b) Variations of density (above) and viscosity (below) of the three main types of magmas as they evolve toward more felsic compositions. The differences result mainly from their differing rates of enrichment of silica, iron, and alkalies. The curves for tholeiites are based on lavas of Hawaii and the Galapagos Islands. The calc-alkaline curves are for rocks of the Cascade Range, and the alkaline ones are for the phonolitic series of Tahiti.

This increase would be even greater, of course, if the effect of phenocrysts, which are plentiful in these magmas, were also taken into account. Tholeiitic magmas show little increase of viscosity until the composition passes the point of maximum iron enrichment. The effect of the increased concentration of iron, a network modifier, offsets that of falling temperature, and viscosity remains nearly constant through the early and middle stages of differentiation. An intermediate type of variation is seen in some alkaline magmas in which alkalies play a strong role as network modifiers, especially in very silica-poor liquids.

Similar differences are found in the variations of density. Density declines throughout the entire calc-alkaline series because of the steady decrease in iron and increase in silica that characterizes the entire range from basalt to rhyolite in magmas of this type. Because tholeiitic and most alkaline magmas become progressively enriched in iron through most of their mafic and intermediate range of compositions, the effect of thermal contraction is reinforced by that of composition, and the liquids increase in density as they cool and differentiate. Beyond a certain point, however, iron enrichment reaches a maximum, and as silica and alkalies begin to increase more rapidly, density declines abruptly until it reaches a minimum in the most felsic end-members of the series.

■ Transfer of Mass and Energy by Diffusion

Nearly all processes of igneous geology involve a transfer of mass or energy to approach a state of equilibrium at a lower energy level. When lava is discharged at the summit of a volcano and descends to lower elevations, it loses potential energy and at the same time gives up heat to its surroundings. Both forms of energy are transferred by virtue of a difference or gradient of energy level, in one case elevation and the other temperature, and the direction of transfer is always from the high to the low energy level. This is the essence of the Second Law of Thermodynamics, which simply states that heat flows from a hot body to a cold one or, in a more general sense, that energy is always transferred from a higher to a lower level. In magmatic processes, we are concerned mainly with the ways in which this principle governs processes of heat and mass transfer leading to different types of igneous rocks.

In the case of a descending lava flow, three types of processes are taking place simultaneously—transfer of heat, mass, and momentum—but the laws governing the rates of these processes have almost identical forms. Just as the rate of heat loss is related to the rate of change of internal energy, the rate of chemical diffusion between crystals and liquid is related to the rate of

change of concentration of chemical components, and the rate of accelera-
tion of velocity is related to the transfer of momentum within a viscous fluid.

We can understand the basic equations governing these processes if we
consider those for heat transfer and bear in mind that the same principles
govern chemical diffusion and transfer of momentum. The Fourier equation
for heat conduction:

$$q = -K \frac{dT}{dZ} \tag{2-6}$$

describes the rate of heat transfer, q, by conduction through a unit cross-
sectional area of a layer of thickness Z when the two sides of the layer are
held at constant temperatures, such that $T_1 > T_2$. K is the coefficient of thermal conductivity in units, such as calories cm^{-1} deg^{-1} sec^{-1}, and the negative
sign reflects the fact that heat flows in the direction of declining temperature.
An identical relationship, known as Fick's Law, describes the rate of transfer
of mass by chemical diffusion, and the equation for viscous flow of a
Newtonian fluid (Eq. 2-2) shows the corresponding relations for transfer of
momentum.

For most geological purposes, we are not interested in the steady-state
condition so much as the changes that take place with time, as, for example,
when hot magma intrudes a cooler horizon or blocks of roof rocks are incorporated into a pluton and equilibrate chemically and thermally with the
magma. Under conditions such as these, the heat removed or added to a volume of rock, V, causes temperatures to change at a rate that is governed by
the rate of change of the heat flux, $\partial q/\partial z$, with distance z, so that

$$\rho C_p \frac{\partial T}{\partial t} V = -\frac{\partial q}{\partial z} V \tag{2-7}$$

Canceling V and substituting the Fourier equation (2-6)

$$\frac{\partial T}{\partial t} = \frac{K}{\rho C_p} \frac{\partial^2 T}{\partial z^2} \tag{2-8}$$

From the definition of thermal diffusivity,

$$\kappa = \frac{K}{\rho C_p} \tag{2-9}$$

we can substitute that term and get

$$\frac{\partial T}{\partial t} = \kappa \frac{\partial^2 T}{\partial Z^2} \tag{2-10a}$$

which is the basic diffusion equation for the rate of change of temperature in a nonuniform thermal gradient. The corresponding equations for chemical diffusion and transfer of momentum are

$$\frac{\partial C}{\partial t} = D \frac{\partial^2 C}{\partial Z^2} \tag{2-10b}$$

and

$$\frac{\partial V}{\partial t} = \nu \frac{\partial^2 V}{\partial Z^2} \tag{2-10c}$$

in which C = concentration,
      V = velocity,
      D = chemical diffusivity, and
      ν = kinematic viscosity (η/ρ).

Typical values of D, ν, and κ are given in **Table 2-5**. Note that the units of all three are identical: cm^2 sec^{-1}.

When solved for different limiting conditions, these equations can be applied to specific problems of thermal, chemical, or mechanical diffusion. Here we need note only a few examples using the thermal case as illustrative of the general behavior. (The derivation of the equations given here, as well as others for different geometric configurations, can be found in references at the end of this chapter. Many solutions to these equations involve the so-called error function and its complement [erf and erfc]. The table in Appendix C will facilitate their numerical evaluation.)

The most simple and useful form (**Fig. 2-12a**) applies to the case of two semi-infinite bodies with the same thermal diffusivity, κ, and each with an initially uniform temperature such that $T_1 > T_2$. The distribution of temperature, T, of either body relative to its initial temperature, T_0, is a function of

|Table 2-5   Some typical physical constants for igneous rocks and melts

Symbol	Property	Typical Values
C_p	Heat capacity	0.25 to 3.0 cal g^{-1}
ΔH_F	Heat of fusion	65 to 100 cal g^{-1}
α	Coefficient of thermal expansion	3×10^{-5} deg^{-1}
η	Viscosity (poise)	10^2 to 10^5 g cm^{-1} sec^{-1}
κ	Thermal diffusivity	4×10^{-3} cm^2 sec^{-1}
ν	Kinematic viscosity (stokes)	10^2 to 10^4 cm^2 sec^{-1}
D	Chemical diffusivity	10^{-5} to 10^{-8} cm^2 sec^{-1}

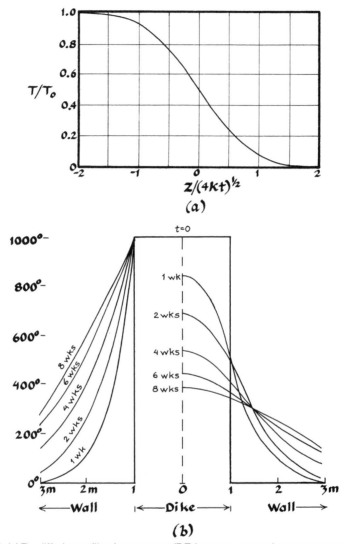

Figure 2-12 (a) The diffusion profile of temperature (T/T_0) across a contact between two semi-infinite bodies with the same thermal properties is a function of the square root of time and diffusivity, k. The distance Z is measured from the contact at Z = 0. The profile for concentration of a chemical component would differ only in that k would be replaced by chemical diffusivity, D (after J. Crank, 1956, *The Mathematics of Diffusion*, Oxford Press, 347 p.). (b) If one of the bodies has a finite dimension normal to the contact, as would an igneous intrusion, the temperature would decline as shown by this example based on a suddenly emplaced dike. The left side of the section shows the temperature in the walls after various times if the magma continues to flow through the dike, as it would if it reached the surface to feed a large lava flow. In this case, the dike acts as a heat source of constant temperature, and the temperature in the walls can be calculated according to Equation 2-18. The right side shows the temperature distribution if the magma is instantly emplaced and immediately ceases to flow. The temperature profiles are calculated from Equation 2-16 or from the graph above.

the square root of time, t, and distance, Z, from the contact at Z = 0. It is given by

$$T(Z,t) = \frac{1}{2} T_0 \, \text{erfc} \, \frac{Z}{2\sqrt{\kappa, t}} \qquad (2\text{-}11)$$

An example of a situation to which this equation might be applied would be that of two large bodies of rock with differing temperatures suddenly brought into close contact by faulting so that heat flows from one to the other. It can also be applied to chemical diffusion between two bodies of different compositions. The general solution illustrated in Figure 2-12a shows the characteristic form of the temperature distribution with distance scaled to the square root of time and diffusivity. If the diffusivities of the two bodies differ, the scaling factors for each side of the contact will also differ accordingly. **Figure 2-12b** shows how the temperature distribution in and around a dike would change with time as heat is transferred from the dike into its walls.

The distance that the effects of diffusion extend beyond the boundary is a function of only time and diffusivity and is independent of the magnitude of the initial temperature contrast. A small contrast of temperature, composition, or momentum propagates its effect just as fast as a large one in materials of the same physical properties, the only difference being the amount that is transferred.

The assumption that diffusivity is the same on both sides of the boundary does not apply to the case of a crystallizing magma. As we have seen, the heat of crystallization of magmas is large compared with the heat capacity of either the liquid or solid, and the effect of this difference is to reduce the rate of cooling of the magma as it passes through the crystallization interval. For most purposes, the heat of crystallization, ΔH_F, given up during crystallization through a temperature interval, $T_1 - T_2$, can be allowed for by adding the proportional heat per degree to the heat capacity to obtain an effective heat capacity:

$$h = C_P + \Delta H_F / (T_1 - T_2) \qquad (2\text{-}12)$$

which can be substituted for C_p in Equation 2-9 to obtain an appropriate value for thermal diffusivity.

In Equation 2-11, transfer is assumed to be entirely by diffusion. If heat is being removed by mass transfer, as it would be, for example, at the top of a lava flow emplaced under rapidly convecting water, then the interface may be maintained close to a constant temperature, T_1, and the equation for the distribution of temperature in the semi-infinite source becomes

$$\frac{T - T_1}{T_0 - T_1} = \text{erf} \, \frac{X}{2\sqrt{\kappa t}} \qquad (2\text{-}13)$$

An example of a geological application of these principles is shown in **Figure 2-13,** which shows the predicted and measured temperature distributions in a cooling lava lake in Hawaii. The base of the lava cooled by conduction to the ground, while the upper part lost heat to the air and to rainwater that fell on the surface and seeped down cracks in the crust. When the water vaporized, it absorbed heat and tended to hold the temperature close to the boiling point of water. As long as the interior of the magma was

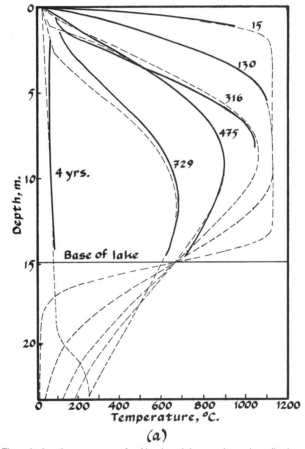

(a)

Figure 2-13 (a) The calculated temperatures for Alae lava lake are shown here (broken lines) together with temperature profiles measured in bore holes drilled after the number of days indicated by the numbers on each curve. The upper contact is at a nearly constant temperature of the atmosphere, while the lower one is governed by conduction to the underlying rocks. The average difference between the calculated and measured temperatures is less than two degrees. (Based on work of the U.S. Geological Survey.) Note that the temperature of the interior was held nearly constant by the relatively large heat of crystallization then declined more rapidly after almost all the liquid had crystallized.

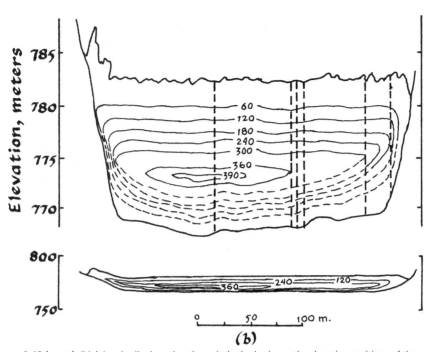

Figure 2-13 (cont.) (b) A longitudinal section through the body shows the changing positions of the 1000°C isotherm as the lens of melt became smaller with time. The numbers on the contours give the number of days elapsed since the lake was formed. The contours are dashed where they are based on extrapolations. In the upper figure, the vertical scale is 10 times greater than the horizontal, whereas in the lower figure they are equal. (Modified from Peck, 1978, *U.S. Geol. Surv. Prof. Paper* 935B.)

still at least partly molten, the temperature of the molten zone remained nearly constant. In the solidified zones, the temperature distribution was governed by conduction to a zone of constant temperature above and to an essentially infinite body of rock below. The importance of the heat of crystallization of the magma can be seen by comparing the slow rate of cooling while the magma was still above its solidus temperature (slightly above 1100°C) with the more rapid rate after it had crystallized.

Although hydrothermal convection is an effective way of transferring heat away from an intrusion, the internal cooling is governed primarily by the rate of conductive heat transfer, and because this is slow compared with the rate of convective heat transfer, the former is the limiting factor for all thermally driven processes. The "thermos bottle effect" of the conductive layer around an intrusion insulates the interior so that the rate of conductive heat loss governs the rate of heat transfer by convection, not the reverse. We saw an excellent example of this in the Hawaiian lava lake where the cooling history could be explained quite well solely in terms of conduction. As a result, most of the heat loss is taken up in crystallization and is confined to a narrow thermal boundary layer.

■ Flow of Magma in the Mantle and Crust

Under the conditions of high temperature and pressure that prevail in the mantle, small amounts of liquid are probably able to infiltrate through intracrystalline spaces, much as water flows through an aquifer. Judging from experimental studies of partly melted rocks at high pressure, it seems unlikely that a liquid fraction of more than 2 or 3 percent can be formed before the melt begins to migrate upward. Although the nature of this flow cannot be observed directly in the mantle, it can be inferred from the laws governing flow through porous media.

Flow Through Porous Rocks

When rocks begin to melt at high pressures, the first liquid tends to be concentrated where the edges of three mineral grains come together. Tube-like channels along these intersections form a network of interconnected passageways, and if there is a pressure gradient on the liquid, it can migrate along these channels. As long as the dimensions of the channels are small compared with those of the system as a whole, the flow can be treated in terms of Darcy's Law for flow through a uniform porous medium:

$$q = \frac{k}{\eta} \frac{dP}{dz} \qquad \qquad (2\text{-}14)$$

where q is the volume flowing through a unit area, dP/dz is the pressure gradient in the direction z, k is the permeability, and η is viscosity. (If there is a temperature or compositional difference that affects the density of the fluid, dP should also include a term $gd\rho$, because flow can also be driven by density differences.) Thus, the rate of flow varies directly with the permeability and pressure gradient and inversely with viscosity. The units of k are cm^2 if η is in poise, and typical values range from about 10^{-6}, the permeability of sand, to 10^{-16} for a solid coarse-grained plutonic rock. The variation is particularly large where the liquid fraction, ϕ, is small.

Because the value of q is in units such as cm^3/cm^2 sec, it has the same dimensions as velocity, cm/sec, but because the area of the total cross-section is much larger than that of the channels, its average velocity is much less than the actual velocity of the fluid in the channels. The permeability of partially molten rocks varies with the volumetric fraction of the pore space, also known as porosity, and with the geometry and degree to which the spaces are interconnected in pathways through the solid. Pumice, for example, has a large percentage of void space but low permeability, because few of the voids are connected, whereas fractured basalt may have little porosity but great permeability.

The pressure gradient, dP/dZ, can result from stress on the deformable crystalline matrix or from differences of density between one part of the liquid and another. As we saw earlier, the densities of silicate liquids can vary greatly as their compositions evolve, and this can result in a condition in which a light liquid is overlain by a relatively heavy one. In most instances, the liquid fraction will be less dense than the crystals, and at high temperatures and pressures, the latter can be deformed by solid-state creep at rates that are rapid enough to cause the liquid to migrate upward under the force of gravity. At the same time, porosity decreases as crystallization advances and the viscosity of the liquid changes as its composition evolves. The net effect is for the term κ/η in Equation 2-14 to vary by large factors with depth in the zone of crystallization.

Diapiric Rise

Density and viscosity also affect the manner in which magmas rise from their source and pass through the lithosphere (**Fig. 2-14**). The mechanism of intrusion of a particular magma depends mainly on its physical properties and those of the rocks through which it passes. At deep levels where rocks can be deformed plastically, many magmas are thought to rise as plumes or diapirs. The velocity of ascent of these buoyant masses is determined, at least approximately, by the general form of Stokes' Law

$$U = \frac{g\Delta\rho r^2}{3\eta_1}\left(\frac{\eta_2 + \eta_1}{\eta_2 + \frac{3}{2}\eta_1}\right) \tag{2-15}$$

in which the velocity, U, of a spherical body is governed by its radius, r, density contrast, $\Delta\rho$, viscosity, η_2, and the viscosity, η_1, of the material through which it moves. Thus, the most rapidly ascending magmas would be those of large size and density contrast rising through rocks of low viscosity. Note that in the two limiting cases where the body has a viscosity, η_2, of zero or infinity, the numerical factor in Equation 2-15 changes only from 2/9 to 1/3 so that the viscosity of the body has almost no effect on its velocity of ascent. In a few rare instances, earthquake epicenters have been observed to rise from mantle depths at rates of 500 to 2000 meters per day before eruption of magma at the surface. This rate has been interpreted as that of a rising body of magma fracturing and wedging aside the rocks through which it rises.

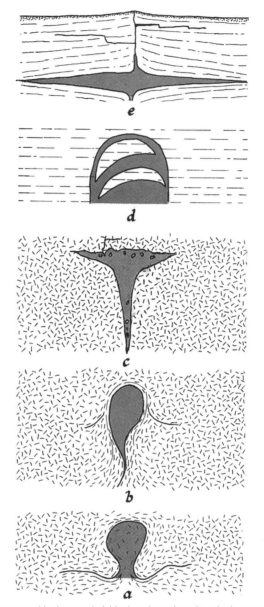

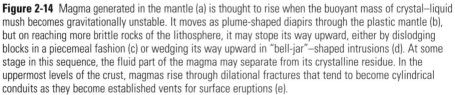

Figure 2-14 Magma generated in the mantle (a) is thought to rise when the buoyant mass of crystal–liquid mush becomes gravitationally unstable. It moves as plume-shaped diapirs through the plastic mantle (b), but on reaching more brittle rocks of the lithosphere, it may stope its way upward, either by dislodging blocks in a piecemeal fashion (c) or wedging its way upward in "bell-jar"–shaped intrusions (d). At some stage in this sequence, the fluid part of the magma may separate from its crystalline residue. In the uppermost levels of the crust, magmas rise through dilational fractures that tend to become cylindrical conduits as they become established vents for surface eruptions (e).

In the upper levels of the brittle lithosphere, magmas rise by a variety of mechanisms, such as flow through fractures and stoping. The latter is a mechanism by which fracturing and wedging aside of roof rocks provides the space into which magma rises. Blocks stoped from the roof founder in the magma and may be assimilated or otherwise incorporated into the rising mass. In so doing, they alter the temperature and composition and can have an important effect on the evolution of magmas in the crust.

Flow through fractures is probably confined to the upper few kilometers of the crust, where the strength of rocks can support channels with rigid walls. The average velocity of flow, U, varies with the gradient of pressure, dP/dh, and with the width of the fracture, W, or radius, R, of a cylindrical conduit and inversely with the viscosity of the magma, η, according to the relationships

$$U = \frac{R^2}{12\eta}\left(\frac{dP}{dz}\right) \tag{2-16a}$$

$$U = \frac{W^2}{8\eta}\left(\frac{dP}{dz}\right) \tag{2-16b}$$

Because magmas passing through narrow fractures in cool rocks are so drastically cooled, there is a critical balance between the velocity of ascent and the rate of heat loss that determines how far a magma can rise.

The most efficient mechanism of fracturing brittle rock is by propagation of fissures because the broad area of a planar surface results in the greatest total force for a given hydraulic pressure. After magma begins to flow through such a channel, however, the shape tends to become less planar and more cylindrical. The reason for this change is that a cylindrical conduit is thermally and mechanically the most efficient channel for fluid flow. Because a smaller surface to volume ratio reduces heat losses and frictional drag, magma remains hotter and flows most rapidly in the widest part of a fissure so that, with time, it erodes its walls and establishes more equidimensional conduits. In thinner, less efficient sections, it loses more heat, flows more slowly, and eventually freezes. For this reason, shallow conduits tend to become more pipe-like the longer flow continues.

Depending on the properties of the magma and the width of its channel, flow may be either *laminar* or *turbulent*. In laminar flow the stream lines or tracks of particles are smooth and nearly parallel, as they are, for example, in glaciers. Turbulent flow is characterized by random swirls, just as one sees in a rapidly flowing river in flood. For most fluids, the transition to turbulence occurs when the dimensionless Reynolds number

$$Re = \frac{\rho UL}{\eta} \tag{2-17}$$

reaches a critical value of about 2000. In this expression, ρ is density, U is velocity, L is a characteristic dimension, and η is viscosity. The Reynolds number is the ratio of the inertial to the viscous or frictional forces in a moving fluid. In laminar flow, viscosity dominates, whereas in turbulent flow, viscosity is less important than inertial forces.

The petrologic importance of this distinction lies in the fact that in turbulent flow intimate mixing can take place between magmas of differing composition, and processes involving transfer of heat and mass are more efficient. In laminar flow, these processes are much slower because they are more dependent on diffusion.

■ Convection

There have been widely divergent opinions among petrologists over the question of how certain bodies of magma convect or, in some instances, whether they even convect at all. The problem lies in differing interpretations of exactly what happens at the margins of a cooling intrusion. When magma gives up heat to its surroundings and a solidification front advances toward the interior, the effects on the stability of the magma can take a variety of forms that are subject to different interpretations.

Cooling and crystallization produce gradients of temperature and composition across a *boundary layer* between the zone of crystallization and fluid interior **(Fig. 2-15)**. The forms that these gradients take are governed by the respective rates of diffusion of heat and chemical components, and their effects on density combine to produce a third gradient, namely that of velocity and momentum. The thermal boundary layer tends to be wider than the compositional one because, as we have seen, thermal diffusivity is much greater than chemical diffusivity, but the two do not vary independently. They are closely linked through the crystal–liquid relationships.

A declining temperature produces a positive density contrast in the thermal boundary layer, whereas the density of the compositional layer can either increase or decrease depending on the net effect of the various components that are enriched or depleted in the liquid. In addition, crystals suspended in the liquid can also augment or reduce density, depending on their compositions. The effects on the stability of the magma are best seen in the case of crystallization on a steep wall **(Fig. 2-16)**. When the thermal and compositional effects reinforce one another, the flow is uniformly downward, but if they have opposite signs, a compositional layer may rise close to the wall, while the outer part of the thermal layer descends.

The effects of cooling and crystallization under a roof are more complex. Unlike the conditions at a steep wall where any horizontal density gradient

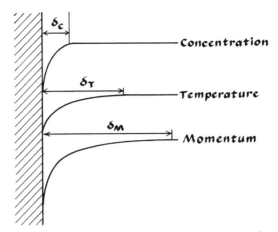

Figure 2-15 Three different types of boundary layers may form at the margins of a cooling body of magma. Their relative widths depend on their respective diffusivities. Because chemical diffusivity is very slow, the compositional boundary layer, shown by the concentration profile at the top, is normally narrower than the thermal boundary layer. Both temperature and composition affect the density of the fluid and cause it to sink or rise. The velocity imparted by gravitational forces is transferred farther into the mass of the fluid in the interior to form a broad momentum boundary layer. The wall depicted here is highly schematic; it is not necessarily the same for all three boundary layers but may extend different distances into a partly crystallized mushy zone.

can drive convection, the stability of a solidification zone cooled from above is governed by the magnitudes of the forces tending to drive convection relative to those opposing it. The former is a function of the temperature-related density contrast, $g\alpha\Delta T$, over a dimension L; the opposing factors are the viscous resistance to flow, ν, and conductive heat transfer, κ. The more viscous the magma is the more force is required to move it, and the more heat that is transferred by conduction the weaker the temperature gradient.

The relationship of these factors is expressed by the *thermal Rayleigh number*:

$$Ra = \frac{g\alpha\Delta TL^3}{\nu\kappa} \qquad\qquad (2\text{-}18a)$$

A very similar *compositional Rayleigh number* describes the corresponding effect of density changes due to a compositional gradient.

Recent work has raised the question of whether it is appropriate to use this relationship to evaluate the convective behavior of sheet-like magma bodies. The thermal Rayleigh number was not designed for a condition in which a melt is crystallizing as it cools but for one in which all the heat loss is taken up by cooling of a crystal-free Newtonian liquid. When we examine the crystallization of thick sills and lava flows in Chapter 6, we shall see

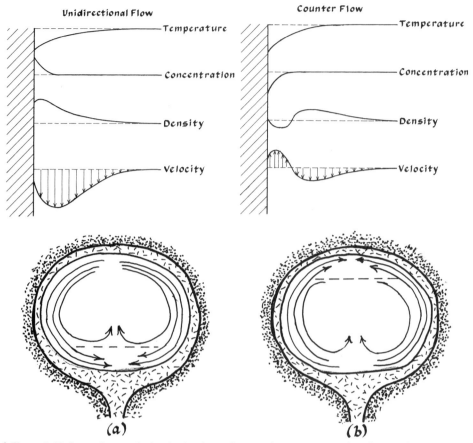

Figure 2-16 Depending on whether the fractionated magma becomes more or less dense as its composition evolves, the convective flow due to crystallization on the steep side wall and intrusion may be unidirectional (a) or counterflow (b). In the former the compositional and thermal effects on density reinforce one another and both parts of the boundary layer descend. In the latter, the compositional boundary layer close to the wall rises, while the thermal layer descends. In both cases, the compositional layer tends to collect on the floor or under the roof where it can evolve independently of the main mass.

strong evidence that they remained essentially stable until they were completely solidified. For example, compositional inhomogeneities introduced by successive injections of magma are preserved throughout the entire course of subsequent cooling and solidification. If there had been any through-going convection, it would have mixed the different pulses of magma and destroyed the stratification.

This is not to say that liquids of differing compositions do not migrate from one level of a sill to another. Interstitial liquids can drain from the crystal mush under the roof, or they can be expelled by compaction on the floor. Discrete masses of highly buoyant felsic differentiates can rise to collect under the roof. These are independent mechanisms that have no direct relation to convection.

It is probably safe to assume, therefore, that the principal driving force of convection in magmas is cooling and crystallization at a steep side wall. It can take two general forms depending of the sign of the compositional effect. If the two components of density reinforce one another, the entire boundary layer descends as a unit. If they are opposed, the result is counterflow with a rising compositional boundary close to the wall and a descending thermal one towards the interior. In either case, if the flow is laminar, the compositional and thermal segments preserve their separate identities and may separate on reaching the floor or roof. The body then becomes compositionally and thermally stratified much as a room does when hot air rises to the ceiling and cool air accumulates near the floor. A separate zone that develops under the roof can then evolve independently of the main body below. We shall see in Chapter 5 (Fig. 5-5) that this may be an effective way of fractionating liquids of differing compositions and densities.

Slumping of crystals from the wall could cause the flow to become turbulent, in which case mechanical mixing rather than diffusion governs the rates of transfer of heat and chemical components, and the widths of the thermal, compositional, and momentum boundary layers are more nearly equal. The crystal-laden magma can flow across the floor of the intrusion much as sediment-loaded density currents do on the sea floor.

In the coming chapters, we shall see examples in which the different forms of cooling and crystallization have had distinctive effects on the trends and extent of magmatic differentiation.

Selected References

Irvine, T. N., and W. R. A. Baragar, 1971, A guide to the chemical classification of the common volcanic rocks, *Can. Jour. Earth Sci.*, 8:523–548. A standard reference for the nomenclature of volcanic rocks based on their chemical compositions.

Jaeger, J. C., 1968, Cooling and solidification of igneous rocks, in H. H. Hess and A. Poldervaart, eds., *Basalts: The Poldervaart Treatise on Rocks of Basaltic Composition, Vol. 2*, Elmsford, NY: Pergamon, 503–571 pp. A simplified but very useful treatment of heat and mass transfer during cooling and solidification of magmas.

Turcotte, D. L., and G. Schubert, 1982, *Geodynamics: Applications of Continuum Physics to Geological Problems*, Wiley, 450 pp. A comprehensive treatment of physical processes with special attention to the mechanisms of plate tectonics.

3

Crystal–Liquid Relations

Two recent eruptions of Kilauea volcano, one in 1959 and the other in 1965, poured basaltic lava into pit-craters where it ponded, formed a solid crust, and slowly cooled. The rates of solidification, although rapid compared with those of plutonic rocks, were slow enough to permit geologists to drill through the crust and follow the crystallization of the lava as its temperature gradually declined. **Figure 3-1** shows a series of samples recovered from the drill holes and quenched to preserve the liquid (now glass) and minerals that were growing from it at successive stages of cooling.

Solidification of this slowly cooled basalt illustrates how different the crystallization of natural magmas is from more familiar types of less complex liquids. First, the minerals did not crystallize at a single temperature, as ice does when it freezes at 0°C; instead, the course of solidification extended through at least 200 degrees of cooling. Second, the individual minerals began to nucleate and grow at temperatures well below those at which the same crystals would form from a liquid of their own composition. Augite, for example, began to crystallize between 1180 and 1160°C, but the melting temperature of this mineral (in the absence of other components) is about 1250°C. Labradorite appeared at around 1160°C, but in a liquid of its own composition, this mineral starts to crystallize near 1350°C. Once formed, these minerals continued to grow as the liquid slowly cooled (**Fig. 3-2**). Olivine, on the other hand, crystallized early but later ceased to grow and was partly resorbed in the cooler evolving liquid. Finally, and most important, the liquid changed composition throughout the course of crystallization. It first became more iron rich until iron-oxide minerals began to crystallize. It then became poorer in iron but richer in silica and other components so that the last liquid differed markedly from the initial lava. In contrast to the original basaltic composition with about 50 percent SiO_2, the last liquid to crystallized contained more than 75 percent SiO_2 and had a composition close to that of rhyolite. Had the liquid fraction been separated at any stage of this process, it could have formed rocks with compositions very different from that of the original magma.

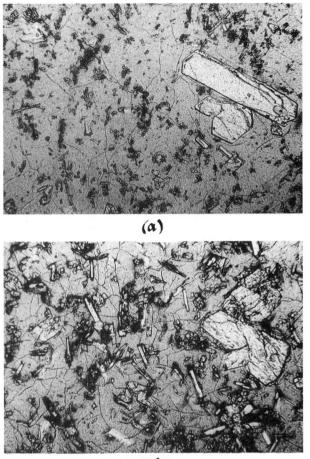

Figure 3-1 Photomicrographs of samples recovered from different depths in a crystallizing lava lake on Kilauea Volcano, Hawaii. The width of field in each photograph is 1.2 mm (T. L. Wright, U.S. Geological Survey). (a) A sample from the glassy crust quenched from a temperature of about 1170°C. Euhedral phenocrysts of olivine grew at greater depths before the basalt was erupted. Pyroxene has just begun to nucleate from the matrix of clear basaltic liquid, now glass. (b) At 1130°C, the pyroxene is larger and plagioclase has appeared.

It is these characteristics of magmatic crystallization that account for much of the compositional diversity of natural igneous rocks. To understand these processes, we should first consider a few simple theoretical aspects of crystallization and then see how these principles apply to the phenomena we saw in the lava lake.

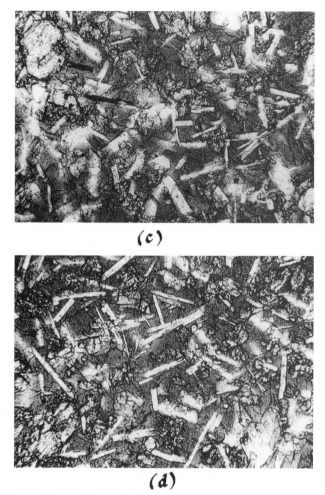

Figure 3-1 (Cont.) (c) At 1075°C, crystallization is well advanced. The dark reddish brown color of the glass reflects its enrichment in iron and titanium. (d) At 1065°C, opaque minerals (magnetite and ilmenite) have nucleated when the liquid became saturated with these minerals. Note the rounded and embayed form of the olivine at the lower right. It shows signs of reaction with the liquid.

■ Some Basic Thermodynamic Relationships

If we consider the behavior of a body in isolation from its surroundings, we speak of it as a *system*. Most of the systems we discuss are simpler than the lava lake, and they are studied under artificially controlled conditions that permit more precise observations. A part of such a system is referred to as a *phase*, if it has a particular set of homogeneous properties that make it

Figure 3-1 (Cont.) (e) At 1020°C, the lava is about 80-percent crystals. Small pools of iron-poor and silica-rich liquid remain in the interstices of the crystalline mush. Note the small needles of apatite that began to grow when the residual liquid became enriched in phosphorus. Crystallization of magnetite and ilmenite has reduced the iron and titanium contents of the remaining liquid to low levels. (f) Below 990°C, the rock is almost totally crystalline. Note the rounded grains of resorbed olivine.

mechanically separable from the rest of the system. A homogeneous liquid is an example of a single phase, and an assemblage of olivine crystals is another; however, a mixture of olivine and pyroxene would constitute two phases because these minerals have different physical properties, such as density or magnetic susceptibility, that can be used to separate them.

A crystallizing silicate liquid, like any other system capable of a physical or chemical transformation, has certain properties, such as volume, pressure,

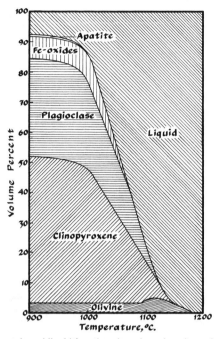

Figure 3-2 Proportions of crystals and liquid for a basalt such as that shown in Figure 3-1. Olivine was present as phenocrysts at the time of eruption and later was partly resorbed in the cooling magma. With falling temperature, olivine was followed by clinopyroxene, plagioclase, magnetite, and ilmenite, and finally, apatite, and the proportion of liquid declined to less than 10 percent. The liquid indicated in the low-temperature range would be in the form of glass, as it is in the quenched samples recovered by drilling. If cooling were slower, the basalt would be entirely crystalline at all temperatures below about 1000 degrees.

temperature, and composition, that define its chemical and physical nature. These properties, often referred to as *parameters of state*, are of two kinds, *extensive* and *intensive*. As their name implies, extensive parameters are those whose magnitude depends on the extent or amount of the system or phase. Intensive properties, on the other hand, have constant values in any part of a system, regardless of the mass or volume. For example, pressure and temperature are independent of the mass of the substance they describe. The ratio of any two extensive parameters is also independent of the total mass and has a constant value in a homogenous system. Thus, the ratio of weight or mass to volume is an intensive property, density, as are specific volume (the reciprocal of density) or molar volume (the volume per unit molar weight).

The properties of most value in describing an igneous phase or system are those that define energy states because these determine whether a geological system is stable and how it tends to change. The total energy, E, of a geological system has three main components, thermal, mechanical, and chemical, each of which is the product of an intensive and an extensive property.

Thermal energy (Q) = temperature (T) × entropy (S)
Mechanical energy (W) = pressure (P) × volume (V)
Chemical energy (U) = chemical potential (μ) × mass (M)

so that

$$E = TS + PV + \mu M \qquad (3\text{-}1)$$

The terms *entropy* and *chemical potential* are less familiar than temperature, pressure, volume, and mass and may require some explanation. Entropy, denoted S, is a measure of the amount of energy represented by the randomness or degree of disorder of the arrangement of atoms. As more energy is added to a substance, it becomes progressively less ordered. Thus, a gas has higher entropy than a liquid of the same composition and temperature, and a liquid has a higher entropy than the crystalline form of the same substance. When a system undergoes a change from one level of entropy to another, as from a liquid to a solid, the amount of that entropy change is measured by the amount of energy given up or absorbed, dQ, divided by the absolute temperature.

$$dS = dQ/T \qquad (3\text{-}2)$$

With falling temperature and energy content, the structure of crystals becomes increasingly well ordered. Fewer impurities and less solid solution can be accommodated, and at a temperature of absolute zero, an ideal crystal would be pure, its atoms would be at rest and perfectly ordered, and its entropy would be zero.

Chemical potential, μ, can be defined as the amount by which the energy of a system is increased when a given infinitesimal amount of an element or compound is added in a way that does not alter the thermal or mechanical energies. The total chemical energy of a system is equal to the sum of the chemical energies of the components present or

$$\text{Total chemical energy} = \mu_1 n_1 + \mu_2 n_2 + \cdots = \mu n \qquad (3\text{-}3)$$

where n is usually expressed in moles.

Because all systems tend to seek their lowest energy level, the driving forces of magmatic processes are energy differences between one condition and another, and the change of energy is equal to the algebraic sum of the individual components.

$$dE = dQ + dW + dU \tag{3-4}$$

In a given reaction or change, the signs for individual types of energy may be positive or negative. Positive values indicate that the system gains energy; negative values indicate energy that is given up to the surroundings. Thus, in a process involving heat energy, if dQ is negative, the result is a liberation of heat, and the process is said to be *exothermic*, whereas a change that brings an increase of heat energy to the system is *endothermic*.

There is no way of establishing an absolute measure of energy; only relative energy levels can be measured, but for most practical purposes, this poses no problem. The change of energy, which is our main concern, can be determined by measuring the amount of heat a system absorbs or the work performed as the system goes from one state to another. These changes can usually be measured or calculated with enough precision to serve our purposes.

The general equation expressing the change of energy level for a change occurring at a given temperature in a system at constant pressure (the condition under which most geological processes take place) can be written

$$dE = dQ - PdV + \mu dn \tag{3-5}$$

The negative sign of the second term results from the fact that when volume increases work is expended by the system on its surroundings as it expands against the prevailing lithostatic pressure. Thus, an increase of volume entails a decrease of mechanical energy.

The third term represents any chemical energy absorbed or lost. Most processes of petrologic interest involve only thermal and mechanical energy, and the concentrations of the chemical components are constant. If so, Equation 3-5 reduces to

$$dE = dQ - PdV \tag{3-6}$$

Enthalpy

In the processes of igneous petrology, the energy levels of systems of constant composition are most conveniently defined in terms of heat content and mechanical energy, PV. The term *enthalpy*, H, used for this purpose, is defined as

$$H = E + PV \tag{3-7}$$

Differentiating and combining with Equation 3-6 gives

$$dH = dQ + VdP \tag{3-8}$$

and for isobaric conditions (i.e., $dP = 0$),

$$dH = (dQ)_P \tag{3-9}$$

which simply states that, for a process at constant pressure, the change of heat content or enthalpy is equal to the heat exchanged between the system and its surroundings.

This may seem like a devious way of stating the obvious, but dH is not just a measure of how much hotter or cooler a substance has become. Its magnitude is a function of the change of state, which entails more than just temperature, pressure, and volume. Consider, for example, the changes that take place as heat is added to ice at atmospheric pressure (**Fig. 3-3**). If the ice is initially at some temperature T_1, below 0°C, the added heat will raise the temperature of the ice at a rate that is governed by the *heat capacity* at constant pressure, C_p. Because the change of enthalpy, ΔH, is equal to the amount of heat absorbed at constant pressure (Equation 3-9), then

$$\Delta H = \Delta T C_p \tag{3-10}$$

and the value of C_p determines the slope of the curve between T_1 and T_2.

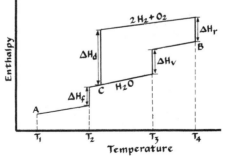

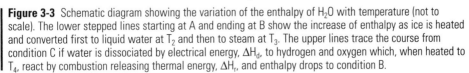

Figure 3-3 Schematic diagram showing the variation of the enthalpy of H_2O with temperature (not to scale). The lower stepped lines starting at A and ending at B show the increase of enthalpy as ice is heated and converted first to liquid water at T_2 and then to steam at T_3. The upper lines trace the course from condition C if water is dissociated by electrical energy, ΔH_d, to hydrogen and oxygen which, when heated to T_4, react by combustion releasing thermal energy, ΔH_r, and enthalpy drops to condition B.

On reaching T_2, the ice must change from the solid to liquid state before any further change of temperature can take place. The amount of heat required to do this, ΔH_f, is known as the *latent heat of fusion*. Hence, the enthalpy of the system increases at constant T and P, and after the ice is melted, the next increment of heat absorbed, dQ, will again result in a rise of temperature at a rate governed by the heat capacity of the liquid. A similar step in enthalpy occurs at 100°C when the liquid passes to the gaseous state, and the heat of vaporization, ΔH_v, is added to effect this transition.

The value of enthalpy at any stage includes all the heat that has been added to bring the system to its present state, regardless of how that state was reached. For example, if liquid water at condition "C" were dissociated by means of an electrical current to equivalent amounts of hydrogen and oxygen at the same temperature, the combined enthalpies of the two gases would be greater than that of water (for a reason that we note shortly). Adding heat raises the temperature of the gases to T_4, as shown in the upper curve of Figure 3-3. If the two gases are now combined to form H_2O, they will liberate heat during the combustion reaction. If this process takes place under isothermal conditions (i.e., at constant temperature), the gases will arrive at the same state as was reached in the first example and have the same enthalpy, despite the fact that the way heat was exchanged was very different.

Enthalpy is usually expressed in calories per mole and, for the uncombined elements, is taken as zero at an arbitrary reference point—atmospheric pressure and room temperature (298° Kelvin), usually referred to as the *standard state*. The enthalpy of a compound is measured with reference to the same standard state but includes its *heat of formation*, ΔH_0, from the elements, a quantity that can be determined by measuring the amount of heat absorbed or given off when the compound is formed at the standard state. For example, when one mole of quartz is formed from silicon and oxygen, 205 calories are given off so that $\Delta H_0 = -205$ cal/mole.

■ Equilibrium in Reversible or Irreversible Processes

The exchanges of energy during the changes of H_2O from one state to another require interaction with the surroundings of the system. For work to be expended, there must be a pressure difference across the boundaries of the system, for if there were not, the system would be in mechanical equilibrium. Likewise, the system can absorb heat from its surroundings only if it is at a lower temperature. If there is no temperature difference, the system is in thermal equilibrium. In the same way, if the chemical potentials of all components in all phases are equal, the components have no tendency to react, and the system is in chemical equilibrium. When a system has no differential pressure, temperature, or chemical potential, it is stable and said to be in thermodynamic equilibrium with its surroundings.

The term *reversible* is used for a process or change resulting from an infinitesimal displacement of the conditions of equilibrium. The direction of the process can be reversed so that it goes in one direction or the other depending on the nature of the change of conditions. Thus, a reversible process can be visualized as the passage of a system through a series of equilibrium states. An example would be the change of the amount of vapor that accompanies a gain or loss of heat by a liquid-vapor combination at its boiling temperature. The process will go in either direction as heat is added or removed. This is in contrast to processes that are irreversible, such as flow of heat from a hot to a cold body or mixing of gases or liquids of differing compositions; these cannot reverse their direction any more than water can flow uphill.

Free Energy

The step-like changes of enthalpy that accompany a reversible process at constant temperature (Fig. 3-3) stem from a change of energy at constant temperature, $T\Delta S$, that represents the increase or decrease of the degree of internal ordering of the substance. It is sometimes referred to as bound energy because it is that part of the heat content that is not available to perform work. Because the change of volume involved in the phase change is a necessary consequence of that change, it cannot perform work on the surroundings without reducing or reversing the process. The same holds for any change of state; part of the energy change is absorbed in establishing the new level of internal order and becomes a specific property of each substance at a given condition.

In a reversible process of this kind, the heat gained or lost is

$$dQ = d(TS) = TdS + SdT \tag{3-11}$$

and if, as in the previous example, the process is isothermal, $dT = O$, this reduces to

$$dQ = TdS \tag{3-12}$$

At constant pressure, from Equation 3-9,

$$dH = dQ_P = TdS \tag{3-13}$$

which is the basic equation for equilibrium of a reversible isothermal process at constant pressure.

For practical purposes, the energy relations that are of most interest are those that express the net amount of heat and work associated with a particular process. The term "free energy," which serves this function, is the difference between the combined heat and work that a process involves and that part of the energy that is in the form of "unavailable work" (TS). Free energy is usually expressed as *Gibbs Free Energy*, designated G and defined as

$$G = H - TS \tag{3-14}$$

This quantity is useful for determining the equilibrium conditions of a system. At equilibrium, $\Delta G = 0$, and for isothermal changes at constant pressure, the magnitude of ΔG is a measure of the divergence of a system from the equilibrium state.

Clapeyron's Equation

It is often important to be able to predict how changes of conditions will affect equilibrium at different depths in the Earth. The Clapeyron equation is very useful in this respect because it defines the relationship between pressure and temperature at which two phases are in equilibrium.

If the system is at equilibrium, dQ = TdS (Equation 3-12), and we can write Equation 3-6 as

$$dE = TdS - PdV \tag{3-15}$$

By definition, H = E + PV, so that

$$dH = dE + PdV + VdP \tag{3-16}$$

Combining Equations 3-15 and 3-16, we obtain

$$dH = TdS + VdP \tag{3-17}$$

By definition, Equation 3-14, G = H − TS, and

$$dG = dH - (TdS + SdT) \tag{3-18}$$

Substituting the value for dH in Equation 3-17, we obtain

$$dG = VdP - SdT \tag{3-19}$$

This equation holds for both of the phases that are in equilibrium, 1 and 2, so

$dG_1 = V_1 dP - S_1 dT$ and

$dG_2 = V_2 dP - S_2 dT$

Combining these two equations and making $V_2 - V_1 = \Delta V$ and $S_2 - S_1 = \Delta S$, and recalling that at equilibrium conditions, $\Delta G = 0$, we have

$$dP\Delta V = dT\Delta S \tag{3-20}$$

Because, from equation 3-13, $\Delta H = T\Delta S$, at equilibrium, we can write Equation 3-20 as

$$\frac{dT}{dP} = \frac{T\Delta V}{\Delta H} \tag{3-21}$$

In the sections to follow, we see how these relationships can be applied to a great variety of important geologic problems.

■ One-Component Systems

Figure 3-4 is a pressure–temperature (P–T) diagram showing the stability ranges of the three principal phases of H_2O: steam, water, and ice. It can be thought of as a map showing the realms of pressure and temperature within

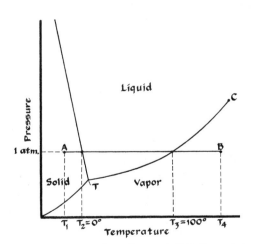

Figure 3-4 Schematic pressure–temperature phase diagram for H_2O. The scale has been distorted, and the various forms of ice have been omitted in the interest of clarity.

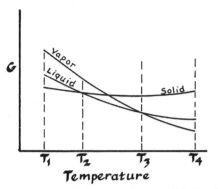

Figure 3-5 Free-energy versus temperature for the three phases of H_2O. The slopes are a function of the differing entropies of the solid, liquid, and vapor. It can be seen from Equation 3-19 that at constant pressure the slopes of the curves, dG/dT, are equal to –S.

which only a single phase is stable, and the boundaries between the three fields delineate the temperatures and pressures at which the two adjacent phases can coexist in equilibrium.

The changes we considered earlier can be traced along the path for atmospheric pressure, A–B, and show how, with increasing temperature, the system passes from one phase to another. These changes can be related to the free energies, G, of the three phases, which are shown schematically in **Figure 3-5** as a function of temperature at constant pressure. Because G = H – TS, the slopes of these curves reflect the differing entropies of the three phases. Ice, having the lowest entropy, has the lowest slope, and vapor, with the highest entropy, has the steepest slope. At temperatures above T_3, the vapor phase has the lowest free energy and is therefore the stable phase. With falling temperature, first the liquid and then the solid has the lowest free energy and becomes the stable phase.

In real systems, phases do not always change spontaneously on crossing their boundaries but may persist for some distance into an adjacent field. At temperatures slightly above or below the intersections in Figure 3-5, the energy differences may be smaller than those required to form stable nuclei of the low-energy phase, and another with a slightly higher free energy may persist as a meta-stable phase. This is why very pure water can be cooled several degrees below 0°C before freezing, and obsidian persists for thousands of years without crystallizing, even though it is far below the temperatures at which its minerals should nucleate and grow. Even in slowly cooled lavas, plagioclase may not begin to crystallize until undercooled by as much as 25°C. This is an important factor in crystallization of igneous rocks, especially rapidly cooled lavas and pyroclastic ejecta.

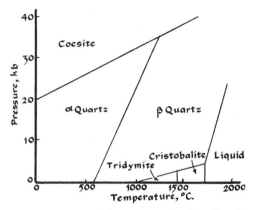

I **Figure 3-6** Part of the phase diagram for SiO_2. Several crystalline forms have been omitted.

Figure 3-6 is a phase diagram for SiO_2. It resembles that of water, except for the absence of a vapor field that is at such a high temperature that it has been omitted. Most diagrams like Figure 3-6 have been constructed by conducting a series of experiments at different temperatures and pressures to determine the stable phase under given conditions. Proceeding one point at a time, the limits of the stability fields of the various phases are delineated.

Experimental construction of phase diagrams for natural minerals requires much time and care, mainly because of the large numbers of experiments that are needed and the difficulty of controlling conditions with the required precision, especially at very high temperatures and pressures. Because of the problems of nucleation mentioned earlier, it may be difficult to determine the exact positions of the phase boundaries. This uncertainty can be avoided by "seeding" the sample with small crystals and seeing whether they increase or decrease in size. Another technique is to "reverse" the phase changes by testing transitions in both directions across phase boundaries. To determine a melting curve, for example, two starting materials, crystals and glass (or liquid), are held at the same temperature and pressure. If, at the end of the experiment, both take on the same form, there can be little doubt that the observed phase is the stable one.

A second method of constructing phase diagrams, although less laborious, has even more serious limitations. If the thermodynamic properties of the substance are well known, the positions and slopes of the phase boundaries can be determined by the Clapeyron equation (3-21). The diagram in Figure 3-4 could be constructed in this way. Ice, as we have seen, melts at 273°K with a 10-percent decrease of specific volume and a release of latent heat, ΔH, of about 79.7 cal g^{-1} (which, in units of mechanical energy, is equivalent to $79.7 \times 41.3 = 3292$ atm cm g^{-1}). The slope of the equilibrium curve is therefore

dT/dP = $273 \times (-.1)/3292 = -.008°$ atm^{-1}

(Ice is unlike most common substances in that its volume decreases on melting, and for this reason, the melting curve has a negative slope.) Using this method, each of the phase boundaries can be drawn to complete the phase diagram, provided, of course, that at least one point is known on each curve.

Until recently, thermodynamic properties were too poorly known to calculate phase diagrams with sufficient accuracy to define the field boundaries precisely. As more data have been obtained, the situation has greatly improved, but only after most of the commonly used phase relationships had already been worked out experimentally. For conditions that are difficult to reproduce in the laboratory, the theoretical approach may be the only feasible way of determining phase relationships. In later chapters, we shall see examples where this method has been of great value, especially when dealing with very high temperatures and pressures.

■ The Principle of Le Chatelier

If we examine the behavior of systems in which two or more phases coexist in equilibrium, we find a simple basic rule that can be very useful in predicting the response of a system to changes of temperature or pressure. Consider, for example, a system consisting of a liquid and vapor initially in equilibrium, to which heat is added at constant volume. Some of the liquid will be vaporized, absorbing heat until equilibrium conditions are restored at a higher temperature. As long as the liquid phase has not been entirely vaporized, the increased temperature and greater proportion of vapor cause an increase in pressure, and the effect of this is to raise the temperature at which the two phases remain in equilibrium. In other words, the effect of increasing temperature is absorbed by the system in such a way that equilibrium is restored at a new condition, in this case at an elevated pressure. In a similar fashion, if the pressure on the system is lowered, liquid will be vaporized to occupy the enlarged volume. Heat is absorbed, the temperature falls, and the system moves to a lower position on the liquid-vapor equilibrium curve.

These relationships are expressed in a more general way by the *Principle of Le Chatelier*, which simply states that

Any external change that affects a system of two or more phases at equilibrium will cause a reaction to proceed in the direction of least resistance. By absorbing the change, the reaction tends to restore equilibrium by adjusting to the new condition.

In each of the examples just considered, the tendency of the two-phase system is to absorb and moderate the changes imposed on it. It acts much as a thermostat does, in that it stabilizes the system and prevents it from leaving the conditions defined by the two-phase equilibrium curve. This effect is often referred to as a *buffer*. For example, the climates of regions near the sea are more moderated than those of inland areas because their temperatures are buffered by the nearby water. When conditions are altered by an influx of warm air, water evaporates, absorbs heat, and cools the air; an influx of cold air has the opposite result. Thus, by Le Chatelier's principle, the system responds in a way that absorbs the change and buffers the system at equilibrium.

■ The Phase Rule

Another fundamental relationship seen in stability diagrams, such as Figure 3-4, is expressed by the Phase Rule of J. Willard Gibbs, which is usually written

$$F = C - P + 2 \qquad\qquad (3\text{-}22)$$

F is the number of degrees of freedom (sometimes called the "variance") of a system and is the number of independent intensive variables, normally temperature, pressure, or the concentration of a component. C is the minimum number of components by which the compositions of the phases can be expressed, and P is the number of phases that can coexist under equilibrium conditions.

The examples in Figures 3-4 and 3-6 have only a single component, H_2O or SiO_2, but they have several phases including liquid, vapor, and one or more crystalline polymorphs. In the diagram for the one-component system H_2O we see that a single phase—solid, liquid, or vapor—can exist within a two-dimensional field representing ranges of the two intensive variables, temperature and pressure. Within these areas T and P can vary independently. Thus, because

$$F = C - P + 2 = 1 - 1 + 2 = 2$$

the system has two degrees of freedom, temperature and pressure and is said to be *divariant*.

Liquid and solid, liquid and gas, or solid and gas can coexist under the conditions defined along the boundaries of the individual stability fields, but on these boundaries pressure and temperature must vary sympathetically according to the slopes of the P–T curves if the two phases are to remain in equilibrium. Only one factor, temperature or pressure, can vary independently. Hence,

$$F = 1 - 2 + 2 = 1$$

and the system is *univariant*. The "triple point," where three phases coexist, has a unique temperature and pressure. Hence

$$F = 1 - 3 + 2 = 0$$

and the system has no degrees of freedom. It is said to be *invariant*. Neither temperature nor pressure can vary without at least one phase becoming unstable.

In the chapters to follow, we shall see examples in which the Phase Rule can be used to predict the behavior and stable mineral assemblages in geological systems, but first we must examine the effects of adding additional components to the simple one-component systems considered thus far.

■ Two-Component Systems

Systems with two or more components in varying concentrations have much in common with simple aqueous solutions. A solution containing NaCl becomes saturated and precipitates halite when the chemical potential, μ, of the crystalline phase falls below that of NaCl in the liquid, and once saturated, the equilibrium is independent of the number or volume of crystals. As long as a single crystal is present, solid NaCl can be added or removed without upsetting the equilibrium. Changing the amount of NaCl in the liquid is a different matter. If the liquid is diluted, the chemical potential of dissolved NaCl is decreased, equilibrium is upset, and solid NaCl will dissolve to restore the equilibrium. This basic difference reflects the fact that the amount of crystalline material is an extensive property of the system, whereas the concentration or chemical potential of the liquid is an intensive property, and as the phase rule indicates, only the latter affects the number of stable phases.

These relationships can be better understood if we consider the relationship between chemical potential and temperature. It can be shown that at constant pressure these two intensive properties are linked in a very simple way:

$$d\mu = -SdT \tag{3-23}$$

At the saturation or melting temperature of a pure substance, A, the chemical potentials of the liquid A and solid A are equal, and the two phases are in equilibrium. If temperature falls below the equilibrium value, Equation 3-23 shows that the chemical potentials of both phases increase (because dT is negative), but that of the solid increases less than that of the liquid because its entropy, S, is smaller. Thus, the solid, having the lower chemical potential, becomes the stable form. The chemical potential of liquid A can also be lowered by diluting it with another component, B, and in this way, the liquid could still be in equilibrium with the solid at a temperature below the melting temperature of pure A. This effect, known as "melting-point lowering," explains why the freezing temperature of water is lowered by an addition of NaCl or antifreeze.

The chemical potential of any component i at a given temperature in an ideal solution is

$$\mu_i = \mu_0 + RT \ln x_i \qquad (3\text{-}24)$$

where

μ_0 = the chemical potential at a standard reference state
R = the gas constant (1.9872 cal mole^{-1} deg^{-1}) and
x_i = the mole fraction of i

Most real solutions deviate from the ideal logarithmic relation, and a factor γ_i, known as the *activity coefficient*, takes this difference into account. (The quantity $\gamma_i x_i$ is known as the *activity*, a_i, of component i.) Thus, a more general form of (3-24) would be

$$\mu = \mu_0 \, RT \ln \gamma x = \mu_0 + RT \ln a \qquad (3\text{-}25)$$

The relationship between the melting temperature of a component i and its concentration in the liquid can be obtained by integrating Equation 3-23 from T_i to T, substituting the equivalent value, $\Delta H_i / T_i$, for ΔS (assumed to be constant in the range between T and T_i), and combining the result with Equation 3-25 to obtain

$$\ln a_i = \ln \gamma X_i = \frac{\Delta H_i}{R}\left(\frac{1}{T_i} - \frac{1}{T}\right) \qquad (3\text{-}26)$$

where T_i is the melting temperature of pure i and ΔH_i, the heat of fusion, is positive and assumed to be constant. This is the basic equation defining the temperature at which crystals of A are in equilibrium with a liquid having varying concentrations, x, of that component.

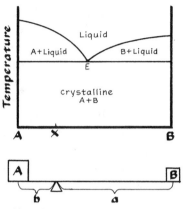

Figure 3-7 The principal fields and boundary curves for a simple two-component system with a eutectic. The proportions of components or phases at any particular temperature can be calculated by the lever rule using the proportional lengths of constant-temperature lines parallel to the base. The composition X, consisting of A and B in the proportions of a and b, can be thought of in terms of a lever with its fulcrum at X.

These relationships are depicted in **Figure 3-7**. Each phase, A and B, has its own melting or saturation curve, known as a *liquidus*, which defines the compositions and temperatures of liquids in equilibrium with the corresponding solids. The intersection of these curves at point E is referred to as a *eutectic*.

Theoretically, the form of the liquidus curves can be calculated from Equation 3-26, if all the thermodynamic functions are known and, in many cases, the calculated curves correspond closely to those determined experimentally. For some systems, however, the fit is poor, especially in the middle compositional range. Much of this discrepancy can be attributed to the heat of mixing of intermediate liquids, which has not been taken into account in Equation 3-26.

Free-Energy Relationships

The stability relations of a two-component system can also be depicted in terms of free energy, G, much as it was for a single component in Figure 3-5, but with addition of another intensive variable, concentration, the system cannot be plotted in a two-dimensional diagram as it was in the earlier example; a series of temperature sections is required (**Fig. 3-8**).

As we saw earlier, the free energy of crystals of fixed composition is determined by

$$G = H - TS \qquad\qquad (3\text{-}14)$$

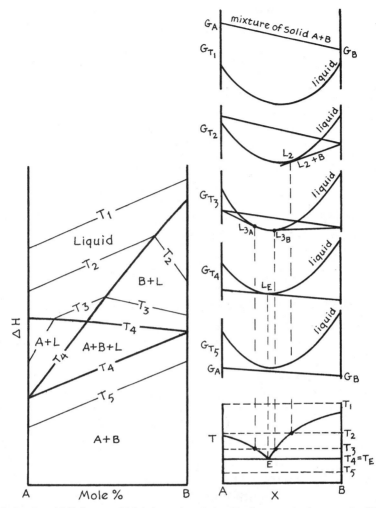

Figure 3-8 A series of G-X diagrams (right) shows the relationships between the free energies (G) of the liquid and solid phases at different temperatures. X is in mole proportions (adapted from S. A. Morse, 1980, *Basalts and Phase Diagrams*, p. 59). The same system can be shown in an H-X diagram (left) in which the stability fields are defined in terms of enthalpy and composition, and temperature is shown by contours. The advantage of such a diagram is that the proportions of phases can be determined in terms of the extensive variable, enthalpy, which, for most geological conditions, is of more interest than temperature. Note that the temperature in the triangular area (A + B + L) is constant at $T_4 = T_E$, but the proportion of phases varies with composition and enthalpy. The diagram assumes that changes of entropy with temperature are proportionately small. (Adapted from H. S. Yoder, Jr., 1976, Generation of Basaltic Magma. *Nat. Acad. Sci.*, p. 265.)

and the free energy of any mechanical mixture of two crystalline phases, A and B, is a simple linear function of the free energies of the two phases. This is not true of a solution, however. Mixture of two end-member components of the liquid results in a spectrum of intermediate liquids with chemical potentials that are a function of the proportions of their components. Unlike mechanical mixtures of solids, intermediate liquids must have higher entropies than their pure end-members, and by Equation 3-14, their free energies must be lower. The G–X curves for the liquids must be concave upward.

Consider the relationships between the molar free energy curves for the solid and liquid phases in the two-component system shown schematically in the right-hand part of Figure 3-8. Free energy has been plotted for the solid and liquid at five temperatures corresponding to T_1 through T_5, and the resulting stability relations are summarized in the T–X diagram at the base of the figure.

At the highest temperature, T_1, the entire range of liquid compositions has a lower free energy than does any solid, and the liquid is the stable phase for all values of X. With falling temperature, the free energy of the solid rises less rapidly than that of the liquid (because of the higher entropy of the liquid, as illustrated in Figure 3-5), and as it does so, the value for B falls below that of the liquid. Crystalline B is therefore stable, whereas A is not. The composition of the liquid in equilibrium with crystals of B at T_2 is given by the point of tangency of a line from solid B to the G–X curve for the liquid. All compositions between these two points form stable mixtures of crystalline B and this liquid along the straight line between B and the point L_2. (The liquid cannot have any composition other than that of L_2 because any other lines representing B and coexisting liquids would intersect the liquid curve at two points with higher values of G.) For all compositions richer in A than L_2, the liquid has a lower free energy and is therefore the only stable phase.

At a lower temperature, T_3, A is now stable and coexists with a liquid defined by the tangent at L_{3A}, and two crystal-liquid assemblages are possible, one for compositions between A and L_{3A} and another between B and L_{3B}. Intermediate compositions are still all liquid. At T_4, the temperature of the eutectic E, the straight line between A and B is exactly tangent to the liquid curve at the composition of the eutectic so that A, B, and the eutectic liquid co-exist in equilibrium. These three coexisting phases constitute an invariant assemblage (at constant pressure). At any lower temperature, such as T_5, the solid is stable for all compositions.

The two-component system (Fig. 3-8) can also be depicted in terms of heat content (enthalpy, H) just as the one-component system was in Figure 3-3.

Because the two end-members have different thermal properties (C_p, S, and ΔH_f), the rates of temperature change and crystallization or melting for a given gain or loss of heat will depend on the rate of change of enthalpy and the proportions of the two components.

Crystallization, Melting, and Assimilation

Crystallization and melting in a simple two-component system with a eutectic are illustrated by **Figure 3-9**. In tracing the following examples, it will be helpful to refer back to Figure 3-8 and observe how the behavior is expressed in terms of free energy and enthalpy.

Any intermediate liquid, such as X, at an initial temperature T_1, will not begin to crystallize until it cools to T_2 at which point its chemical potential is equal to that of crystalline A and it begins to precipitate that phase. With removal of heat, crystals of A are formed, and the residual liquid is increasingly enriched in B. The composition of the cooling liquid follows a course down the liquidus until it reaches the eutectic at E. At that point, the liquid has been so enriched in B that it is saturated with both A and B, and if more heat is removed, the liquid will crystallize those two minerals without changing temperature or composition until the last drop of liquid is gone. Only then will the temperature of the system decline again below T_E. A similar

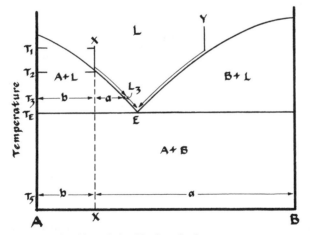

Figure 3-9 Crystallization and melting relationships in a simple two-component system with a eutectic. The proportions of phases at different temperatures can be measured by the lever rule (Fig. 3-7). For example, at T_3, the bulk composition, X, consists of crystals of A and liquid L_3 in the proportions of a to b. At T_5, the bulk composition consists entirely of crystals of A and B in the proportions a to b.

course would be followed for a liquid Y except that it would crystallize B first, and A would only appear at the eutectic.

In systems of this kind, the course of melting is simply the reverse of crystallization. If heat is added to a mixture of crystals of A and B with the proportions of X, the temperature rises until melting begins at T_E, and an initial liquid forms with eutectic composition E. Melting continues at constant temperature (and pressure) until the last crystal of B disappears, at which time crystalline A is still present. With further additions of heat, the temperature rises. More of A melts, and the liquid is enriched in A as it follows the path of the liquidus upward. When the last crystal of A disappears at T_2, the liquid has the composition of the original crystalline mixture, and any additional heat merely raises the temperature of the liquid.

The effect of contamination can be seen if we consider two simple cases. Suppose first that a very small crystal of B, already at temperature T_3, is added to a large volume of liquid L_3 that has evolved by cooling and crystallization of the original liquid X. Because the liquid is undersaturated with B, the crystal will dissolve. This tends to shift the liquid toward B, but this is impossible as long as the liquid is in contact with crystals of A. If the process is isothermal, a small amount of A will dissolve and by contributing that component to the liquid, maintain the original composition L_3. Solution of the crystalline phases is endothermic, however, and would tend to lower the temperature. Using what you now know, can you deduce how the system would respond in that case?

Suppose, now, that the added crystals were solid A, initially at a temperature lower than T_3. The crystal would be chemically stable because the liquid is saturated with that phase, but it absorbs heat from the rest of the system and lowers the temperature of the entire system. This causes more A to crystallize from the saturated liquid, which then becomes poorer in A, the component that was added to it. The bulk composition, of course, must shift slightly toward A.

The effect of pressure is illustrated in **Figure 3-10**. The liquidus curves in the two-dimensional diagram become curved surfaces in a three-dimensional diagram for T, X, and P, and the point E becomes the line of intersection of the two liquidus surfaces. The effect of pressure may differ for the two components and cause the eutectic to migrate toward one or the other as it rises in temperature.

The Phase Rule can be applied to such a two-component system, but we now have a third intensive variable, X, the composition of the liquid, which adds another possible degree of freedom. For example, in Figure 3-10, the three-dimensional space above the two liquidus surfaces represents the conditions under which a single liquid phase can vary independently in temperature, pressure, and composition, so that

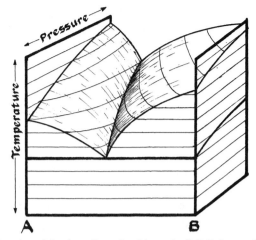

Figure 3-10 Perspective view of the three dimensional form of a T–P–X diagram for a two-component system with a eutectic.

$$F = C - P + 2 = 2 - 1 + 2 = 3$$

On the liquidus surfaces where crystals coexist with a liquid, temperature, pressure, or composition can all vary, but fixing two of these determines the third. The assemblage has two degrees of freedom, and the liquidus surface is said to be divariant. At the intersection of the two surfaces, three phases, A, B, and liquid, coexist, but if any of the three variables T, P, or X, changes, the other two must also change in a proscribed way to remain in the linear axis of the trough. The system is therefore univariant. An invariant condition, which would be represented by a point, is possible if a fourth phase, such as a vapor is in equilibrium with the liquid and crystalline A and B.

In the region below the eutectic temperature where only crystals of A and B are present, it would appear that there are three degrees of freedom, T, P, and X, but for two components and two phases the phase rule indicates that there should be only two ($F = 2 - 2 + 2 = 2$). The explanation of this apparent anomaly lies in the fact that in this region there is no liquid, and X, the ratio of two crystalline phases, is not an intensive variable but merely an extensive one. With only two degrees of freedom, T and P, and two phases, A and B, the phase rule is still valid. Similar reasoning applies to the volumes representing A + liquid and B + liquid under the liquidus surfaces and above the temperature of the eutectic. Having three dimensions, these volumes appear to have three degrees of freedom, T, P, and X, so that, again, the phase rule seems to give $3 = 2 - 2 + 2$. But, no liquids can have a composition and temperature within the limits defining this space; it is occupied only by mixtures of crystals and liquids, and because a variation of the proportions of these phases is not an intensive variable, F is only 2 (T and P).

Systems with an Intermediate Compound

If two components have an intermediate crystalline form, the relationships resemble those of two individual eutectic systems. For example, if two end-members, A and B, have an intermediate compound, AB, each adjacent pair can be treated as a binary system (**Fig. 3-11**).

If the melting temperature of one of the end-members is very high, its liquidus may override part of the liquidus of the intermediate compound. This condition is illustrated in **Figure 3-12a**. On heating, AB cannot melt directly to a liquid of its own composition because its melting temperature is below the liquidus of A. Instead, it melts *incongruently* to a liquid, R, and crystals, A, neither of which has the same composition as the original material.

What happens on melting is this: as heat is added to crystals of AB, they reach the temperature T_R and begin to form a liquid, not of composition AB but of R (Fig. 3-12b). At the same time, the crystals are converted to the form and composition of crystalline A. This process continues as more heat is added, and when the last crystal of AB disappears, the assemblage, still at temperature T_R, consists of liquid of composition R and crystals of A. Then as more heat is added, the crystals of A begin to melt, the liquid becomes richer in A, and the temperature rises. At T_2, the last crystal of A disappears, and the liquid has the original composition of AB. The process on cooling is the reverse.

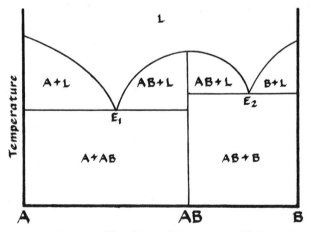

Figure 3-11 A two-component system with an intermediate compound, AB, that melts congruently to a liquid of its own composition.

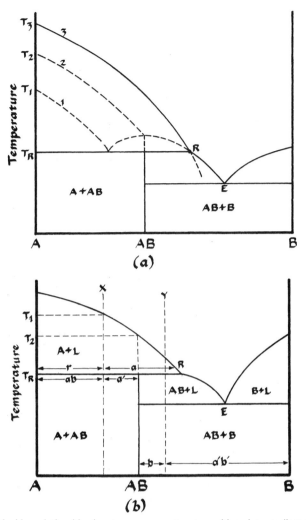

Figure 3-12 (a) Liquidus relationships in a two-component system with an intermediate compound, AB, that melts incongruently. As the liquidi of increasingly refractory forms of A rise in temperature from condition 1 to 3, their intersections with the liquidus of AB change from a eutectic to a reaction point, R. When AB rises in temperature under condition 3, the first liquid forms at R where the liquidus of A intersects that of AB. The temperature cannot rise above T_R as long as AB is still present because the liquidus of AB passes below the liquidus of A, making AB unstable at higher temperatures. AB must dissolve at the same time that A crystallizes before the temperature can rise. (b) The reverse holds for cooling and crystallization. A liquid with the composition of AB will first crystallize A and continue to do so until it intersects the liquidus of AB at R. At any lower temperature, the liquidus of A falls below that of AB so that only the latter is stable and earlier-formed crystals of A must dissolve as AB crystallizes.

The most interesting effects in such a system are those of compositions between A and AB. A liquid X, for example, crystallizes A on cooling and is enriched in B as it does so. Although the low-temperature crystalline assemblage should consist of A and AB, the liquid that evolves during crystallization of A must eventually reach a composition R between AB and B before AB can start to form. If liquid R were to crystallize independently, it would first precipitate AB and finally AB + B. If earlier formed crystals of A are present, however, this would result in a violation of the phase rule. Although the system has three crystalline phases, the number of components is only two because that is the minimum number necessary to define all compositions in the system. Hence, the assemblage of three crystalline phases, A, AB, and B, is incompatible with a two-component system with two degrees of freedom, T and P, in the low-temperature region. In other words, A and B cannot coexist in equilibrium without forming AB.

The liquid R cannot continue to crystallize down to the eutectic E under equilibrium conditions so long as crystals of A are present. Instead, it reacts with some of the earlier formed crystals, dissolving them while forming new crystals of AB. Even though the heat consumed in dissolving A is gained by crystallizing AB, and the temperature remains constant, heat must be removed from the system as a whole because the amount of AB that crystallizes is more than equivalent to the amount of A that is dissolved. If it were not, the composition of the liquid would become richer in A than the composition of R, an obvious impossibility that is prevented by crystallization of additional AB.

The proportions of phases involved in this reaction can be determined by means of the lever rule. When the liquid first reaches R, it contains liquid and crystals in the proportions r and a, respectively. As reaction proceeds, liquid is consumed in forming AB in an amount proportional to ab; a is reduced to a' and r is converted to ab, and the final proportions of minerals are those of the starting composition X.

A liquid Y with a composition between AB and B but richer in AB than R would first crystallize A. For such a composition, however, the final equilibrium assemblage of crystals of bulk composition Y must be made up of only AB and B in the proportions a'b' and b, respectively. Although crystals of A are formed initially, they are entirely consumed by reaction to form AB when the liquid reaches R. After they disappear, a quantity of liquid still remains and continues crystallizing AB until the eutectic is reached and B is formed.

If such a liquid were to crystallize under natural conditions, equilibrium might not be achieved, and reaction of the liquid with early-formed crystals might not go to completion. If, for example, the crystals of A form on the floor or walls of a magma chamber and are buried there as more crystals accumulate, the remaining liquid may cool without access to the earlier crystals. In this case, it will cool and pause at R only long enough to consume those crystals to which it is exposed before going on to precipitate AB down to the eutectic where it will crystallize B along with AB. The end result is an assemblage of three crystalline phases, one of which would not have crystallized under equilibrium conditions. More important, the course of evolution of the liquid would be greatly extended.

A comparison of such a system with one without incongruent melting shows several important differences. In Figure 3-11, in which a eutectic lies between each pair of crystalline phases, the course of any liquid is fixed; it must reach its appropriate eutectic. The high melting temperature of AB forms a thermal barrier resembling a divide at a watershed, and just as a drop of rain falling on the west side of the Continental Divide must ultimately reach the Pacific Ocean, a liquid only slightly richer in A than pure AB must reach the eutectic E_1 where A crystallizes. Two liquids that are initially very close in composition but on opposite sides of the thermal divide must diverge in composition as they crystallize. A similar relation holds for melting. An assemblage of crystals of AB containing even a trace amount of A will begin to melt at E_1 and form a liquid that is very different from E_2, the first liquid to form from an assemblage of AB with a small amount of B. This simple relation illustrates an important aspect of igneous differentiation, namely that liquids of very different compositions can result from crystallization or melting of very similar starting compositions if the system has two eutectics separated by a thermal divide.

This is not to say that a thermal divide is essential to the evolution of widely differentiated liquids. A similar difference can result from crystallization in systems in which an intermediate compound melts incongruently and does not constitute a thermal divide. In systems of this latter type (Fig. 3-12), we saw that when equilibrium is not maintained and reaction is incomplete liquids can evolve through much greater ranges of composition than they would with complete reaction, and the final assemblage may differ from that expected from equilibrium crystallization and reaction.

Systems with Solid Solution

Few igneous minerals are simple compounds; most are composed of two or more end-members that substitute for each other through a continuous range of compositions known as a solid solution series. Plagioclase and olivine are important examples.

Because minerals of a solid-solution series are not simply mechanical mixtures of their end-members but, as their name implies, are more like liquid solutions in which the components are combined in an intermediate composition, their free energy varies in a nonlinear way with composition. The entropy of intermediate compositions is greater so that the free energy of intermediate compositions is less than that of the end-members (**Fig. 3-13**).

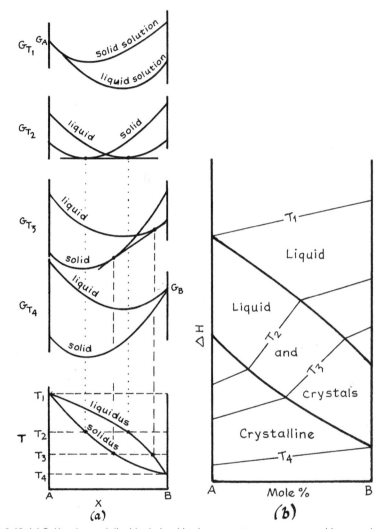

Figure 3-13 (a) G–X and crystal–liquid relationships in a two-component system with a complete range of solid solution between the two end-members. (After S. A. Morse, 1980, *Basalts and Phase Diagrams,* p. 91.) (b) An H–X diagram for the same system shows temperature as a function of enthalpy and composition. Because the thermal properties of the end-members differ, the temperature contours are not horizontal, and the amount of cooling and crystallization or melting for a given input or loss of heat differs according to the proportions of the two components. (Adapted from H. S. Yoder, Jr., 1976, Generation of Basaltic Magma. *Nat. Acad. Sci.,* p. 265.)

Crystallization of minerals of this kind differs from that of crystals of fixed composition. The higher-melting component in an intermediate liquid is below its pure melting temperature when crystallization begins, but the low-melting end-member is not. The higher-melting component is therefore favored by the solid phase, whereas the low-melting one tends to be concentrated in the liquid. For this reason two separate curves, a *liquidus* and a *solidus*, are required to define the compositions of crystals and liquids that coexist at a given temperature. The curves for this "solid-solution loop" can be calculated in a manner similar to that of the simple system with a binary eutectic, but the equations are more complex.

$$\ln \frac{x'}{x} = \frac{\Delta H_A}{R}\left(\frac{1}{T_A} - \frac{1}{T}\right) \tag{3-27a}$$

$$\ln \frac{1-x'}{1-x} = \frac{\Delta H_B}{R}\left(\frac{1}{T_B} - \frac{1}{T}\right) \tag{3-27b}$$

where x' and x are the molar fractions of the high-temperature end-member in the liquid and crystals respectively, and ΔH_A and ΔH_B are the heats of crystallization of the high- and low-temperature end-members, respectively. T_A and T_B are the melting temperatures of the pure end-members.

Consider an intermediate liquid, such as X in **Figure 3-14**, cooling from

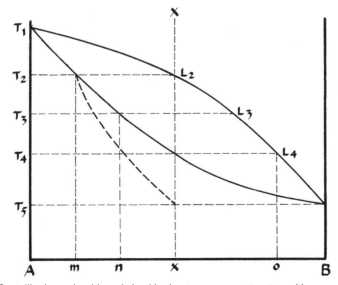

Figure 3-14 Crystallization and melting relationships in a two-component system with a complete range of solid solution. See text for discussion.

an initial temperature above the liquidus. On reaching temperature T_2, it precipitates crystals of composition m, which, because they differ from the liquid, cause the liquid to change composition as the temperature falls and more crystals grow from the liquid. At a lower temperature, T_3, the composition of both the liquid and the crystals in equilibrium with it have changed and are now richer in B. The earlier-formed crystals are poorer in B than the equilibrium composition, and if conditions favor equilibration, they will be changed to the new composition by reacting with the liquid. Under ideal conditions, this equilibration process keeps pace with cooling, so that at all stages the crystals have a uniform composition and, when the temperature reaches T_4, the last drop of liquid is consumed, and all the crystals have the composition of the original liquid, X.

Under conditions that prevent complete reaction, early-formed crystals do not equilibrate with the evolving liquid and are not made over to new compositions. Less of the low-temperature form, B, is extracted from the liquid to re-equilibrate early crystals, and instead, the residual liquid is more strongly enriched in B. When this happens, liquid still remains when the temperature reaches T_4, and crystallization may continue, with the final liquid reaching the composition of B. The resulting crystals are zoned from cores of composition m to rims of pure B, and their average composition follows the course shown by the curved broken line.

In nature, the course of crystallization is normally between these two extremes. Olivine tends to equilibrate well and has little zoning because reaction requires only an exchange of Mg and Fe ions. Plagioclase, however, can only equilibrate by a paired exchange of Ca for Na and Al for Si. Both of these latter ions are strongly bound in large framework structures and diffuse so slowly that equilibration is possible only in very slowly cooled magmas. For this reason, zoning is much more common in plagioclase than in other solid-solution minerals.

For homogeneous, unzoned, solid-solution crystals, the course of melting with rising temperature is essentially the reverse of equilibrium crystallization, provided, of course, that equilibrium is maintained between the crystals and liquid. The first liquid formed from crystals of composition X in Figure 3-14 has composition L_4, and with rising temperature, the crystals and liquid change composition along the solidus and liquidus curves. The last crystal, now made over to composition m, disappears when the liquid reaches composition L_2. With incomplete equilibration, however, the effect of melting is the opposite of that during crystallization; it reduces the range of temperature between the completely crystalline and liquid states. The crystals, if they fail to equilibrate, are richer in the low-melting component and disappear completely at a lower temperature than they would if they were able to adjust their composition to the rising temperature.

Zoning can have an important effect on both crystallization and melting. We have seen that crystallization without reaction produces zoned crystals and drives the residual liquid to compositions much richer in the low-temperature component, B, than one evolving under equilibrium conditions with complete reaction. In the same way, the first liquid produced by melting of zoned crystals will be richer in B and appear at a lower temperature than one resulting from melting of unzoned crystals of the same average composition.

The effects of contamination can be seen by considering an intermediate liquid, such as L_3. If crystals are added with a composition, such as m, which would be precipitated at an earlier stage of crystallization, they tend to react with the liquid, depleting it slightly in B in order to attain the equilibrium composition, n, for the liquid at T_3. This reaction tends to drive the liquid to the left of the liquidus composition, L_3, and causes it to precipitate additional crystals of n, in order to maintain the composition appropriate for that temperature. If, on the other hand, the added crystals have a composition, such as o, which would only be precipitated later in the course of cooling, those crystals are unstable in L_3 and tend to dissolve and make the liquid slightly richer in B. As a result, some of the earlier-formed crystals of n are dissolved, and the liquid maintains its composition appropriate for temperature T_3. The first process is exothermic; it shifts the bulk composition slightly toward the high-temperature component A and reduces the range of crystallization on subsequent cooling. The second is endothermic; the added crystals shift the liquid toward the low-temperature end and tend to extend the range of possible liquids produced during cooling.

Systems with Limited Solid Solution

In many minerals that form a solid solution series, the crystal structure of the end-members may differ to the extent that the amount of substitution of one component in the other is limited. This is the case, for example, in the monoclinic and orthorhombic pyroxenes and in the monoclinic and triclinic feldspars. The phase diagram for this type of two-component system with limited solid solution has the form shown in **Figure 3-15a**. It has a temperature minimum at M between two solid solution loops. The point M differs from the eutectic of a binary system without solid solution in that liquids reach such a composition only under conditions of strong fractionation (i.e., when crystals and liquid are separated and prevented from equilibrating). In a binary system without solid solution, they must always reach

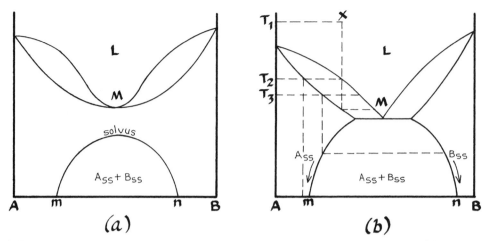

Figure 3-15 A two-component system with limited solid solution between the end members. In (a), the solvus defining the compositions of coexisting crystals is completely below both solidus curves, whereas in (b), it intersects them. Note the different forms of the temperature minimum in (a) and (b). This difference can be understood if the phase rule is used to test the conditions at the temperature minimum. In (a), where the minimum liquid coexists with only one crystalline phase, $F = 2 - 2 + 2 = 2$, and the minimum must be a divariant surface extending back in the third dimension, pressure. In (b), the liquid co-exists with two crystalline phases, A_{SS} and B_{SS}, and $F = 2 - 3 + 2 = 1$ so that in three dimensions the minimum is a univariant line.

the eutectic, but with solid solution, they do so only under certain conditions, as when no reaction takes place between the liquid and early-formed crystals. In a similar way, the first liquid produced by melting has a composition corresponding to the temperature minimum only if the crystals have this exact composition in their outermost layers. Otherwise, the first liquid is higher on the liquidus curve at a composition appropriate for that of the crystals.

Solid solution between two species of differing crystal structure decreases with falling temperature so that a crystal that is homogeneous at high temperature may, on cooling, separate into two structurally and compositionally different forms. This may take place well below the solidus, as in Figure 3-15a, or in the range of crystallization, as in Figure 3-15b. The boundary curve defining the limits of solid solution is known as a *solvus* and encloses a region in which the bulk composition can only be made up of two coexisting solid phases.

A liquid of composition X, cooling to temperature T_2, initially crystallizes a composition, which, if no reaction takes place, would cool without intersecting the solvus and have a homogeneous composition. As the remaining liquid migrates down the liquidus, it eventually precipitates crystals with compositions richer in B than m, and these, when they intersect the solvus on cooling, exsolve increasing amounts of B-rich solid solution. In the case where the solidus curves do not intersect the solvus (Fig. 3-15a), crystal fractionation can drive the liquid to the temperature minimum at M, where a single phase of intermediate composition will crystallize and split into two forms, A_{SS} and B_{SS}, only when it cools to below the solvus. In the case where the solidus is low enough to intersect the solvus (Fig. 3-15b), the two separate forms will crystallize simultaneously from a liquid at the eutectic, and on cooling, each will exsolve the other as they become progressively purer with falling temperature. This behavior is an important feature of crystallization of pyroxenes and alkali feldspars.

Systems with Immiscibility

In some compositional ranges, the limits of complete mixing may extend above the liquidus into the liquid region. Two liquids may form physically distinct phases that resemble oil and water in that they will not mix to form intermediate compositions. This phenomenon, long thought to have only limited significance in geology, has recently been shown to extend to geologically reasonable compositions and conditions and is an important factor in the evolution of certain types of magmas.

The cause of immiscibility can be visualized in terms of the form of the free-energy relations of certain nonideal liquids (**Fig. 3-16a**). Instead of having a form that is concave upward throughout the entire range of compositions, as it is in ideal liquids, the free-energy of a nonideal liquid may be higher in the intermediate range of compositions than that of liquids somewhat richer in the end-member components. Thus, the intermediate liquid will be less stable than two coexisting liquids and will split into two conjugate compositions.

A possible binary relationship is shown in Figure 3-16b. A liquid X cooling from a temperature, such as T_1, begins to split when it reaches the two-liquid field at T_2, and with further cooling, the two coexisting liquids diverge in composition. At temperature T_3, crystals of A begin to grow from liquid L_1, and the proportion of liquid L_1 diminishes while that of L_2 increases. Only when L_1 disappears completely can liquid L_2 crystallize A and continue its descent to lower temperatures.

At T_3, the crystals of A formed from liquid L_1 are also in equilibrium with liquid L_2 because the liquids themselves are in equilibrium. (If two phases are in equilibrium and one of the two is also in equilibrium with a third, so too is the other.) This relationship provides a rigorous test for immiscibility: two magmas cannot have an immiscible relationship if both are not in equilibrium with the same minerals.

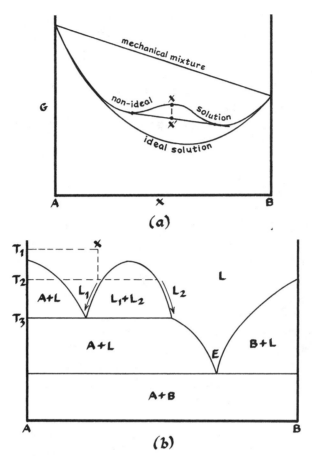

Figure 3-16 (a) G–X relationships for a nonideal liquid that is immiscible in its middle range of composition. The free energy of a mixture, X′, of two liquids, L_1 and L_2, is lower than that of a single liquid X. (b) A possible phase diagram for a two-component system in which a field of immiscibility intersects the liquidus of one of the crystalline phases, A.

■ Three-Component Systems

The basic principles we observed in two-component systems still hold when another component is added, but their graphical representation in phase diagrams becomes more complex. Addition of another component requires that we sacrifice one of the variables when we plot the compositions on a two-dimensional diagram.

Geometrical Relationships

If the three components, A, B, and C, are plotted in a triangle, such as **Figure 3-17a,** the compositional range of the liquidus of each mineral, which was a

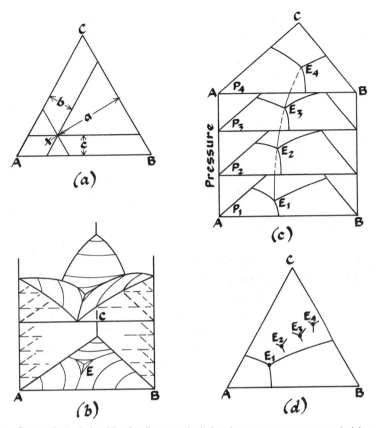

Figure 3-17 Geometrical relationships for diagrams depicting three-component systems. In (a), a composition X has the proportions of A, B, and C shown by the lengths of the lines a, b, and c, respectively. (b) shows the three-dimensional form of a prism, in which temperature is the vertical axis. The liquidus surfaces descend toward the interior of the triangle and can be contoured like the sides of valleys. The base of the figure shows these contours projected onto a two-dimensional surface. The effect of pressure can be shown in a series of diagrams, each for a different pressure, as in (c). If the boundaries of the fields are projected onto a single plane, their intersections at successive pressures can be shown as in (d).

curve in the binary system, now becomes a surface. Temperature can be shown by contours on the liquidus surface so that the interior of the triangle resembles a topographic relief map on which the field boundaries correspond to stream courses down valleys (Fig. 3-17b).

Points such as E are commonly referred to as ternary eutectics or quadruple points because they correspond to compositions in which four phases—three crystalline and one liquid—can coexist. (The vapor phase is usually ignored in "condensed systems.") We see from the phase rule that, with three components and four phases, there can be only one degree of freedom, but because all three variables, temperature, pressure, and composition, cannot be shown in a single two-dimensional diagram, it is necessary to illustrate this relation in a series of diagrams (Fig. 3-17c).

Systems with a Eutectic

The simplest case is that in which a eutectic falls between the three components (**Fig. 3-18a**). A composition X in the primary field of A crystallizes that mineral when it reaches its liquidus temperature, and as A is removed, the course of the liquid crosses the field away from A. It descends across temperature contours until it intersects the field boundary at d, at which point it begins to crystallize an additional mineral, B. As this phase is now removed along with A, the liquid takes a new direction down the thermal valley toward the ternary eutectic, E, where C joins the crystallizing phases. The curve followed by the crystallizing liquid is more properly called a *cotectic* because it defines the course of liquids co-precipitating two phases together. Its shape is determined by the proportions of crystals A and B that are being separated at any particular temperature. If this proportion is constant, it is a straight line. If not, it curves away from the phase being removed in greater quantities. Thus, the proportions of crystals being removed at any stage can be determined by projecting a tangent to the cotectic curve from the composition of the liquid, e, back to f, and by applying the lever rule to determine the proportions of crystallizing phases. In this

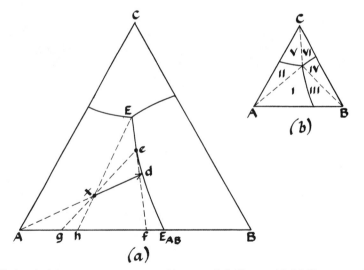

(a)

(b)

Figure 3-18 A typical three-component system with a eutectic is illustrated in (a). The course of crystallization of liquid X will be A, A + B, and A + B + C. A total of six different orders of crystallization can be defined for liquid compositions in the fields I through VI in (b).

example, the ratio of A to B is as fB is to fA. The composition of the total assemblage of crystals precipitated between X and e is obtained by projecting a straight line from the liquid e back through the original bulk composition X to the point g. The proportion of the remaining liquid to crystals at this stage will be as gx is to ex, and ex is composed of A and B in the proportions gB to gA. When the liquid reaches its final composition at the eutectic, E, and C begins to crystallize, the proportion of liquid will be hX and the crystals EX. As C is removed along with A and B, the bulk composition of the crystals moves from h toward X and will reach X at the instant the last drop of liquid disappears. The composition of the liquid, of course, does not change after reaching E.

By inspection, we can see various possible paths of crystallization for compositions within the triangle ABC. All compositions in the primary field of A (I and II of Fig. 3-18b) will crystallize that mineral first, but in field I the next mineral will be B and then C, whereas in II, C will appear before B.

In melting, the process is simply the reverse of the one we have just followed for crystallization, provided the liquid and crystals are not separated as melting advances. Under conditions of crystal–liquid segregation, however, a very different sequence of liquids is obtained. We saw that crystallization produces a continuous series of liquids from the initial composition X to d and finally to E, and if no phases react with the liquid, the course would not be altered if crystals were completely removed as soon as they formed. This is not true of melting. A crystalline assemblage of composition X would begin to melt with an initial liquid E at the ternary eutectic, but if that liquid is removed as quickly as it forms, an abrupt discontinuity will occur when the last crystal of C is consumed. The only crystals remaining are now A and B, and the temperature must rise until it reaches that of the binary eutectic between those two minerals before a new liquid of that composition is formed. If the same process continues and the liquid is removed as soon as it forms, B will eventually disappear, and the next liquid will have the composition and melting temperature of pure A. This process is known as *fractional melting*; its significance will become apparent when we come to consider processes of magma generation in a later chapter.

Systems with an Intermediate Compound

Ternary systems with an intermediate compound that melts congruently (**Fig. 3-19a**) can be treated as two separate ternary systems separated by a thermal divide. They have many of the features of the corresponding binary systems. The join between C and AB divides the system into two parts, each with its end point at a eutectic, and compositions crystallizing on one side of that division have no way of crossing to the other.

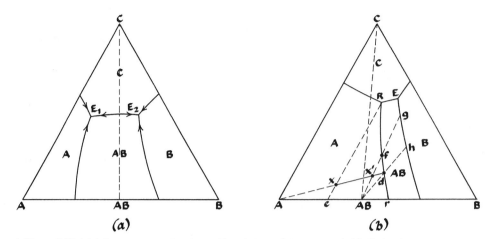

Figure 3-19 (a) A three-component system with an intermediate compound, AB, that melts congruently, has two eutectics and can be treated as two independent, three-component systems. (b) A three-component system with an intermediate compound, AB, that melts incongruently can be recognized by the fact that the boundary between the fields of two crystalline phases (A and AB) does not fall within the triangle of which those phases are corners (A–AB–C) but outside it, in AB–B–C. Instead of two eutectics, as in (a), the system has one eutectic and a reaction point, R.

As in the binary system, a mineral that melts incongruently may cause a liquid to crystallize a phase that is not an equilibrium component of the original composition at low temperatures. In Figure 3-19b, the mineral AB melts incongruently so that instead of a eutectic within the triangle A–AB–C a reaction curve, or *peritectic*, rR, and a reaction point, R, fall within the triangle AB–B–C.

It is a simple matter to distinguish a reaction point from a eutectic, even if a diagram has no temperature contours. Note that in Figure 3-19a, temperatures always decline along the cotectic curves in the directions indicated by arrows away from the line joining the two phases that are crystallizing. In the same way, temperature must fall along the entire curve from R toward E because the two phases crystallizing on that curve, AB and C, are joined by a line that lies to the left of R and does not intersect the curve. If the line AB–C intersects the curve, as it does in Figure 3-19a, the temperature falls in both directions away from that intersection toward both E1 and E2.

Even though crystallization drives the course of a liquid X outside the subtriangle A–AB–C, the final crystalline product must consist of A, AB, and C if equilibrium prevails. Similarly, even though a liquid such as X′ initially crystallizes A, the fact that it is not in the triangle having A as a corner indicates that the final equilibrium assemblage of crystals cannot include that mineral. The way this comes about will be clear if we trace the course of crystallization of these two liquids.

First consider liquid X. The first mineral to appear is A, and removal of A drives the composition of the liquid along the course XX'd. When the liquid reaches d, the proportion of liquid to crystalline A, by the lever rule, is AX to Xd. As it follows a course along the boundary with the field of AB toward R, the liquid precipitates AB, but its composition cannot stay on the boundary curve unless A is dissolved. Removal of AB or any combination of AB and A would drive the liquid away from those compositions and off the curve. As in the binary system, crystals of A are dissolved at the same time that AB is being crystallized, and the vector that determines the course of the liquid has one component away from AB and another toward A so that the resulting vector of the liquid composition coincides with a tangent to the curve at d. A boundary of this kind is known as a *reaction curve*, or *peritectic*.

As the liquid descends toward R, the composition of the total crystalline assemblage, which is obtained by projecting from the liquid composition back through the total original composition X to an intersection with the join A–AB, migrates toward AB. When the liquid reaches R, C begins to crystallize, and the composition of the crystals leaves the join A–AB and follows a course along eX until the last drop of liquid R is consumed and the crystal assemblage has a composition X.

If the original liquid composition were X', the course of crystallization would be similar until the liquid reaches f. At that point, the crystalline assemblage has reached AB, and A is entirely consumed. From that point on, the liquid can only crystallize AB and must leave the boundary curve and move across the field of AB until it intersects another field boundary at a point such as g. The liquid then crystallizes AB and B as it descends toward E. At E, C begins to crystallize, and the final crystalline assemblage is composed of AB, B, and C in the proportions of X'.

Both of these examples assume complete reaction and re-equilibration of early-formed A along the peritectic. If the reaction is prevented, the liquids would not follow the peritectic but would immediately cross it at d and take a new course from d to h as they crystallize AB. On reaching h, B begins to crystallize, sending the liquids down to an end-point at the eutectic E.

Systems with Solid Solution

A ternary system in which two of the components form a solid solution series cannot have a ternary eutectic. Instead, a boundary curve divides the system into two fields, as shown in **Figure 3-20**. A liquid with an initial composition X is equivalent to a crystalline assemblage of C and an intermediate solid

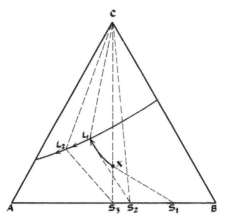

Figure 3-20 A three-component system in which two components, A and B, have a complete range of solid solution. See text for discussion.

solution of AB having the composition S_3. If crystallized under ideal conditions, the final product will have this mineral composition.

Because the composition of the solid-solution crystals in equilibrium with liquid X must be richer in the high-temperature component, B, than the liquid, the first crystals to form have a composition such as S_1. With falling temperature, the liquid follows a curving course toward the field boundary, and at any instant, the course of the liquid is directly away from the composition of AB that is currently being removed. Because the crystals of AB are steadily becoming richer in A and moving from S_1 toward S_3, the course of the liquid is concave away from A. When the liquid reaches the cotectic field boundary, C appears and crystallizes along with AB as the liquid moves toward the temperature minimum between A and C. The sides of the triangle L_1–C–S_2 join the three coexisting phases, and the bulk composition of the system must fall within its boundaries.

With complete reaction, early-formed crystals of AB, such as S_1, are made over to a composition in equilibrium with the changing liquid. At the instant the last drop of liquid, L_2, disappears, it will be precipitating C together with AB of composition S_3. The bulk composition falls on the join between C and S_3 and is composed of those components in the proportions S_3X and CX, respectively. Every liquid starting in the field AB has a unique path that cannot be determined from the ternary diagram alone. If there is no reaction between the liquid and early-formed crystals, A is removed at a slower rate, and the liquid has a straighter path across the field. With extreme fractionation, it can descend to the temperature minimum between A and C.

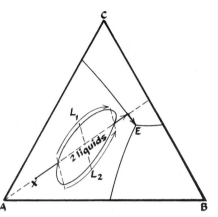

Figure 3-21 A three-component system with a field of liquid immiscibility intersecting the liquidus surface of one of the crystalline phases. A section through the triangle from the corner A and through X to the opposite side of the triangle would have the same form as the two-component system illustrated in Figure 3-16b.

Systems with a Field of Immiscibility

The course of crystallization of a three-component system that includes a field of immiscibility (**Fig. 3-21**) is similar to that of the corresponding binary system. A liquid with an initial composition, X, first crystallizes A and moves away from that component until it intersects the two-liquid field. It then splits into two liquids, L_1 and L_2, the compositions of which follow the boundary of the two-liquid field. At any point, the compositions of the coexisting liquids must, of course, fall on the same isotherm, and their proportions can be determined by the lever rule. As the system cools and crystallizes A, the composition of one of the liquids, L_2, returns to intersect a linear projection from the original composition, X, away from A and on the low temperature side of the two-liquid field. At that stage, all the liquid L_1 will have been converted to crystalline A and liquid L_2. Thereafter, the course of crystallization is identical to a normal three-component system.

It is interesting to note that, unless the course of crystallization is interrupted and the identity of the two liquids is in some way preserved, the final products of crystallization will show no evidence of the system having passed through a two-liquid field.

■ Systems of Four or More Components

Systems with more than three components follow the same principles as simpler systems, but they are more difficult to depict in diagrams. Although a quaternary system can be illustrated in three dimensions (**Fig. 3-22**), the course

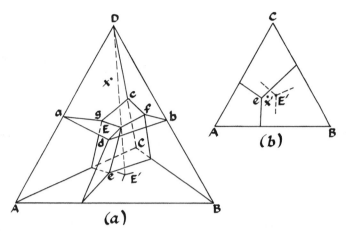

Figure 3-22 (a) Four-component systems can be depicted by a three-dimensional tetrahedron having the components, A, B, C, and D, as corners. The base of the tetrahedron illustrated in (b) shows the three-component A–B–C with the four component eutectic and field boundaries projected from D onto the same surface (dashed lines). The composition X, which would appear to precipitate B first if the component D is ignored, would in fact be in the region in which A crystallizes first in the four component system.

of a liquid cannot be traced precisely in the interior of the tetrahedron. For this reason, it is customary to project points within the interior of the three-dimensional volume to a plane, which may either be a face of the tetrahedron or an interior section. The plane can then be treated as a ternary system in which each of the three components is also accompanied by the fourth, with which the liquid is assumed to be saturated at all times.

For example, if a magma with an initial composition X in the system A–B–C–D is in the volume D–a–b–c–E, D is the first phase to crystallize, and the magma is saturated with that phase throughout its subsequent cooling history. One can therefore project the interior boundaries away from the corner D on to a plane such as the base of the tetrahedron. The three-dimensional spaces are now the two-dimensional fields A + D, B + D, and C + D. Liquid X projects to point X′ and follows a course of crystallization: D, D + A, D + A + B, and finally, D + A + B + C.

Projections of this kind can lead to serious errors. The interior boundaries of the quaternary system projected from D do not coincide with those of the ternary system A–B–C in which D is not present. This is a reflection of the fact that the behavior of a liquid saturated with D differs from that of one lacking that component. Consequently, the order of crystallization of A, B, and C from a liquid X may be different from what one would deduce by ignoring component D and plotting compositions in the three-component system. In the example shown in Figure 3-22b, it appears that A would crystallize after B, whereas in fact, it precedes B. In later sections, we shall see examples of how this problem can lead to confusion when dealing with natural rocks.

Selected References

Bowen, N. L., 1928, *The Evolution of Igneous Rocks*. Mineola, NY: Dover, 334 p. A classic work in which Bowen first laid out the basic concepts by which igneous rock are interpreted in terms of crystal fractionation. This has been by all measure the most influential book in igneous petrology, and although somewhat outdated, it still remains one of the most elegant treatments of crystal–liquid relationships in natural silicates.

Darken, L. S., and R. W. Gurry, 1953, *Physical Chemistry of Metals*. New York: McGraw-Hill, 535 p. Although written for metallurgists, this well-written text has long been considered a standard reference for petrologists.

Morse, S. A., 1980, *Basalts and Phase Diagrams*. Berlin: Springer-Verlag, 493 p. A very thorough and clearly presented treatment of phase diagrams with examples of their use in the interpretation of differentiated basaltic magmas.

Philpotts, A. R., 1990, *Principles of Igneous and Metamorphic Petrology*, Upper Saddle River, NJ: Prentice Hall, 498 p. Contains an excellent treatment of basic thermodynamics and phase equilibria.

Yoder, H. S., Jr., ed., 1979, *The Evolution of Igneous Rocks, Fiftieth Anniversary Perspectives*. Princeton, NJ: Princeton University Press, 588 p. An updating of Bowen's 1928 book. Chapters 3 through 7, in particular, deal with many of the principles discussed in this chapter.

4

Igneous Minerals and Their Textures

The igneous rocks making up the vast bulk of the earth's crust consist of a remarkably small number of common minerals. About half a dozen minerals accommodate the principal elements, O, Si, Al, Ca, Mg, Fe, Ti, Na, and K, and account for all but a percent or two of the entire range of volcanic and plutonic rocks. Most other igneous minerals occur only as accessories or in rare rocks of unusual composition. The purpose of this chapter is to outline the occurrences, phase relationships, and textural features of these common minerals and relate them to the concepts set out in the preceding chapter.

■ Common Igneous Minerals

Silica Minerals

Although silica is the most abundant oxide component of almost all igneous rocks and is an essential constituent of most rock-forming minerals, it seldom forms a separate phase in rocks having less than about 55 to 60 weight percent SiO_2. Of the various polymorphs—quartz, tridymite, cristobalite, coesite, and stishovite—quartz is by far the most common. It occurs as anhedral grains or in intergrowths with alkali feldspars in felsic plutonic rocks and as phenocrysts in siliceous volcanic rocks, especially ignimbrites. The high temperature form, tridymite, is confined to volcanic and hypabyssal rocks. It almost always inverts to quartz on cooling so that it has the optical properties of quartz even though it may retain the bladed morphology of tridymite. Cristobalite is found mainly in fractures and vesicles where it has been deposited hydrothermally, usually as a metastable phase. Coesite (monoclinic) and stishovite (tetragonal) are high-pressure forms of silica with densities of 2.93 and 4.35 g cm^{-3}, respectively. They are formed only at very high pressures in the deep mantle or in rocks that have been subjected to intense shock by meteoritic impact. Coesite has been found as inclusions in diamonds brought to the surface from depths of 100 km or more.

Quickly cooled silicate liquids and glasses can contain large amounts of SiO_2 without nucleating and precipitating a silica mineral, and because of this, the amount of quartz in intermediate and felsic volcanic rocks is always less than that in chemically equivalent plutonic rocks. Silica has a high solubility in water-rich solutions, particularly at high temperatures and pressures. It is readily leached from volcanic glass by hydrothermal solutions or circulating meteoric water and may be redeposited in cavities as amorphous opal and chalcedony. It is also concentrated in water-rich fluids formed in the late stages of crystallization of plutonic rocks and is deposited as euhedral crystals, commonly of great size, when these fluids move into gas pockets and open fractures in and around intrusions.

Quartz and tridymite are virtually the only common igneous minerals that do not form solid solution series. Their phase relationships are, therefore, relatively simple. The stability relationships of the polymorphs have already been shown in Figure 3-6. Other systems in which silica minerals are possible phases are shown in several of the following sections.

Feldspars

Feldspars are by far the most abundant minerals in the earth's crust and are major components of almost all crystalline igneous rocks. Because of their range of solid solution and the diverse conditions under which they form in nature, they have many forms and compositions. Their basic properties are given in **Table 4-1**.

Plagioclase is far more abundant than potassium feldspar, especially in volcanic rocks. Most of the potassium in common volcanic rocks is incorporated as a minor element in plagioclase or is concentrated in interstitial glass. In the subalkaline series, it rarely forms a separate feldspar in any but the most felsic rocks, such as dacites and rhyolites, and in most of these, the alkali feldspar is a high-temperature form in the intermediate compositional range of anorthoclase. In more alkaline series, however, sanidine, orthoclase, and anorthoclase may be important constituents of quite mafic rocks, especially in highly potassic series, such as the shoshonitic and lamprophyric rocks discussed in Chapter 11. Potassium feldspar is more common in plutonic rocks in which the conditions of protracted cooling and higher water contents favor better equilibration and crystallization of two separate feldspars. The high-temperature potassium feldspar, sanidine, is found only in very potassic volcanic rocks and in reheated xenoliths that originally contained orthoclase or microcline. The low-temperature form, microcline, is characteristic of plutonic rocks that have been extensively recrystallized under metamorphic conditions.

The end-members of the feldspar system (**Fig. 4-1**) differ in crystal structure and in their ranges of solid solution. The sodium and calcium feldspars

Table 4-1   Compositions and properties of the common feldspars

Feldspar	Formula	Crystal System	Optic Angle	Density
Albite	$NaAlSi_3O_8$	Triclinic	45°(–)	2.60
Anorthite	$CaAl_2Si_2O_8$	Triclinic	78° (+)	2.76
Orthoclase	$KAlSi_3O_8$	Monoclinic	69°–72° (–)	2.59
Sanidine	$(K,Na)AlSi_3O_8$	Monoclinic	0°–12° (–)	2.59
Anorthoclase	$(Na,K)AlSi_3O_8$	Triclinic	18°–54° (–)	2.59

Plagioclase Compositions

Typical Natural Plagioclases

	Pure Albite	An 20	An 40	An 60	An 80	Pure Anorthite
SiO_2	68.74	63.35	58.03	53.38	47.67	43.20
Al_2O_3	19.44	22.89	25.81	29.71	33.46	36.65
Fe_2O_3	–	0.09	0.68	0.19	trace	–
CaO	–	4.09	8.01	11.86	16.23	20.16
Na_2O	11.82	8.90	6.47	4.44	2.19	–
K_2O	–	0.65	0.46	0.18	0.07	–

Alkali Feldspar Compositions

Typical Natural Feldspars

	Pure Orthoclase	Orthoclase	Sanidine	Anorthoclase
SiO_2	64.76	64.28	64.03	64.33
Al_2O_3	18.32	19.40	19.92	20.94
Fe_2O_3	–	0.34	0.62	0.78
CaO	–	0.48	0.45	2.01
Na_2O	–	2.74	4.57	7.22
K_2O	16.72	11.80	10.05	4.71

(Ab and An) form a complete series of plagioclase compositions (**Fig. 4-2**) and crystallize in the manner of the solid-solution series discussed in Chapter 3. The sodic (Ab) and potassic (Or) feldspars have a limited range of solid solution that declines with falling temperature and increasing water pressure, and An and Or are separated by a wide compositional gap at all temperatures. The relationships are further complicated by the fact that the potassium feldspar has high-, intermediate-, and low-temperature forms (sanidine, orthoclase, and microcline) and melts incongruently to form leucite ($KAlSi_2O_6$) and liquid. For simplicity, the potassium feldspar is labeled orthoclase on most phase diagrams, and its incongruent melting is ignored because it disappears with addition of small amounts of other com-

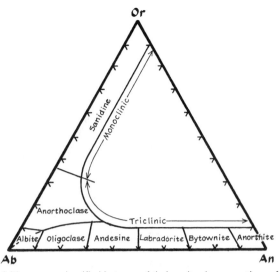

Figure 4-1 Volcanic feldspars are classified in terms of their molecular proportions of the three end-members, anorthite (An), albite (Ab), and potassium feldspar (Or). No feldspars fall in the intermediate range between plagioclase and alkali feldspars. The range of plutonic feldspars is shown in Fig. 4-3 and 4-6.

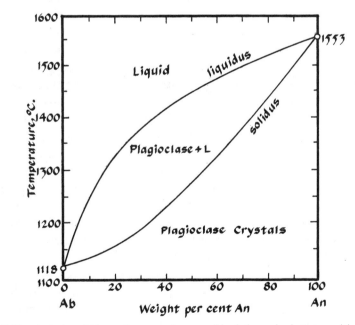

Figure 4-2 The plagioclase feldspars form a continuous solid solution series between calcium-rich anorthite and sodium-rich albite.

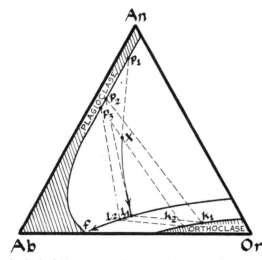

Figure 4-3 Crystallization in the feldspar system is governed by a cotectic separating the field of plagioclase from that of potassium feldspar. Typical tie-lines connecting compositions of coexisting feldspars and liquids are shown by broken lines. The course of crystallization of common liquids is discussed in the text.

ponents and plays little or no role in crystallization of any but a small group of very potassic magmas.

Most mafic magmas have initial compositions on the plagioclase side of the cotectic curve that descends across the feldspar triangle (**Fig. 4-3**). With progressive crystallization, such a liquid follows a curved path away from plagioclase that becomes more albitic with falling temperature. On reaching the cotectic, the liquid begins to crystallize an alkali feldspar; the composition of which also changes as the liquid descends toward the temperature minimum on the Ab–Or join at f. At any point along its course, the liquid is in equilibrium with two feldspars, as indicated in the diagram by the successive triangles, but the exact compositions of coexisting phases cannot be determined without more information than is given in Figure 4-3. With complete reaction, the last drop of liquid would disappear when the join between the two feldspars coexisting in equilibrium with the liquid overtakes the initial composition X. Normally, however, the reaction is incomplete, and the liquid continues to descend toward f.

Because plagioclase requires large amounts of undercooling to nucleate and, after crystallizing, is very slow to equilibrate by reaction, it characteristically has a wide range of compositions, even in individual crystals. **Figure 4-4** illustrates a common form of oscillatory zoning that is especially characteristic of plagioclase.

Figure 4-4 Oscillatory zoning in plagioclase feldspar. Repeated overgrowths of layers of different compositions resemble the annual growth rings in trees. Each sequence becomes more albitic outward then reverts back to a more anorthitic composition to begin the next layer. In this specimen, narrow zones of sharp zoning at W, X, Y, and Z alternate with more weakly zoned layers. The area B is an inclusion of basaltic glass. Width of the view is about 0.25 mm. (Courtesy of A. T. Anderson.)

Oscillatory zoning of the kind shown in Figure 4-4 has been attributed to repeated loss of volatiles from the magma because of recurrent volcanic eruptions. Another possible (and more likely) explanation is that the crystals were carried from one level and pressure to another in a column of convecting magma. Because of the pressure dependence of the liquidus temperature, crystals will tend to grow in a descending magma and remain stable or be resorbed in an ascending one.

In most respects, plagioclases in igneous rocks show relationships broadly consistent with those of the simplified system of Figure 4-3. The

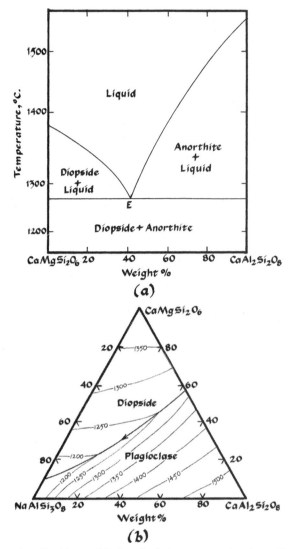

Figure 4-5 (a) The system diopside-anorthite is a simple two-component system with a binary eutectic. (Because small amounts of alumina enter diopside, the system is not strictly binary.) When albite is added to the system (b), crystallization follows the principles outlined in Chapter 3 for three-component in which two of the components form a solid-solution series.

anorthite content of early-crystallizing phenocrysts is greater than that of the groundmass and declines as cooling and crystallization advance. One cannot, however, make a direct quantitative correlation between the natural and simple systems. The temperatures of crystallization are lower in magmas

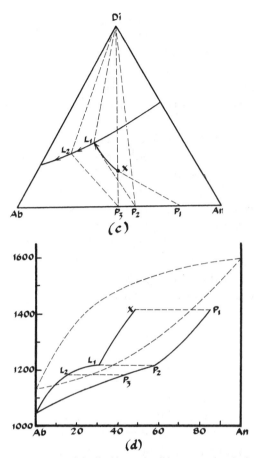

Figure 4-5 (Cont.) The temperatures of the liquidus and solidus curves for plagioclase (c) now vary according to the amount of the third component, Di, in the liquid. During the early course of plagioclase crystallization, a liquid X approaches the cotectic and becomes richer in Di. The liquidus and solidus of plagioclase are depressed to progressively lower temperatures, but when Di begins to crystallize, depleting that component in the liquid, the liquidus and solidus curves rise to temper-atures approaching those of the pure two-component system. (After P. J. Wyllie, 1963, *Min. Soc. Am. Spec. Paper* 1:204–212.)

because of the presence of other components in the liquid, and the forms of the liquidus and solidus curves differ when other phases are coprecipitated.

The effect of adding another mineral is illustrated by the addition of pyroxene to the plagioclase system (**Fig. 4-5**). The liquidus and solidus temperatures of plagioclase are lowered by amounts that are a function of the proportion of diopside in the liquid, and for that reason, the amount of plagioclase that crystallizes in a given temperature interval also changes. When plagioclase crystallizes alone (from X to L_1 and P_1 to P_2), the slopes of the liquidus and solidus are steepened as diopside is residually enriched in the liquid. They then flatten when the two phases crystallize simultaneously (from L_1 to L_2 and P_2 to P_3) and the proportion of diopside in the liquid diminishes. Note the large compositional change resulting from a small change of temperature when the liquid is on the cotectic, and compare this with the change over a similar interval in the simple system for albite-anorthite alone.

The alkali feldspars (**Fig. 4-6**) form a continuous solid solution series at high temperatures and if quickly cooled may preserve a homogeneous intermediate composition. Thus, one or the other of the two high-temperature alkali feldspars, anorthoclase or sanidine, may be found in alkali-rich volcanic rocks, whereas slowly cooled plutonic rocks of similar chemical composition do not contain these high-temperature forms. Instead, they have two feldspars, orthoclase and albitic plagioclase, either as intergrown or distinctly separate crystals. The behavior of the alkali feldspars under these latter conditions is illustrated in Figure 4-6b; details of the textural relationships and the effects of water pressure on the feldspars are discussed in Chapter 10.

Feldspathoids

Highly alkaline rocks may have too little silica to form feldspar as the only alumino-silicate phase. Instead, they have one or more of the alkaline feldspathoids, nepheline or leucite, or a related mineral, such as analcite, sodalite, noselite, or zeolites. These minerals (**Table 4-2**) can be thought of as silica-deficient equivalents of feldspar and, as such, cannot coexist in equilibrium with a silica mineral. Leucite is unstable at elevated pressures and is all but unknown in plutonic rocks, but it is an abundant constituent of leucite basanites, wyomingites, and a number of other types of uncommon potassic peralkaline rocks that are found in small but widespread occurrences, mainly in continental regions. Nepheline forms under a wider range of conditions and is found in plutonic rocks ranging from mafic and ultramafic volcanic rocks to nepheline syenites. It is an equally important constituent of very sodic volcanic rocks ranging from mafic basanites to felsic phonolites. Both of the feldspathoids tend to be unstable at low temperatures. Leucite breaks down on cooling to an intergrowth of orthoclase and nepheline, and nepheline alters readily to analcite, sodalite, and zeolites. The genetic relationships of the feldspathoids are discussed in Chapter 11.

Pyroxenes

Pyroxenes are second only to feldspar in abundance in crustal rocks. They have several different compositional and structural forms (**Fig. 4-7 and Table 4-3**).

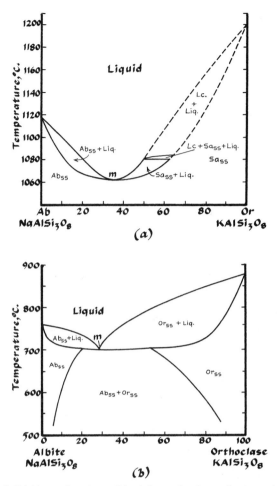

Figure 4-6 (a) The alkali feldspars form two solid-solution series descending to a minimum temperature at m. In volcanic or shallow intrusive magmas, homogeneous feldspars can form with compositions anywhere between pure albite and pure sanidine. The field of leucite (Lc) disappears in most natural liquids of subalkaline compositions. (After D. R. Waldbaum and J. B. Thompson, 1969, *Amer. Min.*, 54:1249–1298.) (b) Elevated pressures of water depress the temperatures of both the liquidus and solidus curve, and at 5 kilobars, they are low enough to intersect the solvus so that two separate feldspars, one sodic and the other potassic, are separated by a field in which intermediate compositions are not stable. Crystallization of the alkali feldspars is dealt with in greater detail in Chapter 10. (After S. A. Morse, 1970, *Jour. Petrology* 11:221–251.)

Ca-rich pyroxenes are common constituents of almost all igneous rocks, but the Ca-poor pyroxenes can form only in subalkaline and certain types of ultramafic rocks. Pyroxenes with intermediate calcium contents, usually referred to as subcalcic augites, are found in many if not most volcanic rocks, especially as a component of the groundmass, but only rarely in plutonic rocks.

Table 4-2	Compositions and properties of feldspathoids and analcite

Mineral	Formula	Crystal System	Optic Sign	Density
Nepheline	$Na_3K(AlSiO_4)$	Hexagonal	(−)	2.61
Leucite	$KAlSi_2O_6$	Pseudoisometric	isotropic	2.47
Sodalite	$Na_4Al_3Si_3O_{12}Cl$	Cubic	isotropic	2.30
Analcite	$NaAlSi_2O_6 \cdot H_2O$	Cubic	isotropic	2.27

Chemical Compositions

Pure Minerals

	Nepheline	Leucite	Sodalite	Analcite
SiO_2	39.00	53.12	34.49	48.82
Al_2O_3	33.08	22.53	32.92	20.71
Na_2O	18.98	–	20.01	15.84
K_2O	8.94	24.35	–	–
H_2O	–	–	–	14.63
NaCl	–	–	12.58	–

Typical Natural Minerals

	Nepheline	Leucite	Sodalite	Analcite
SiO_2	44.65	56.39	36.72	54.23
Al_2O_3	32.03	23.10	31.63	23.67
Fe_2O_3	0.59	–	0.55	–
CaO	0.71	0.27	0.28	–
Na_2O	17.25	2.17	24.02	13.81
K_2O	3.66	18.05	0.46	trace
H_2O	0.96	–	0.28	8.67
Cl	–	–	5.56	–

Both the Ca-poor and Ca-rich pyroxenes form solid-solution series between Mg and Fe end-members and have limited solid solution with each other. The Ca-rich series, diopside-hedenbergite, is a continuous solid-solution series similar to those of plagioclase or olivine. Very iron-rich calcium pyroxenes crystallize initially with the triclinic form of Fe-wollastonite (or ferrobustamite) and invert on cooling to the monoclinic structure of hedenbergite. Few magmas reach this degree of iron enrichment, however, and natural examples of the inversion are rare.

An aluminous (tschermakitic) component may enter both monoclinic and orthorhombic pyroxenes, as $CaAl_2SiO_6$ and $(Mg,Fe)Al_2SiO_6$, respectively, but it is most important in Ca-rich monoclinic pyroxenes. The same

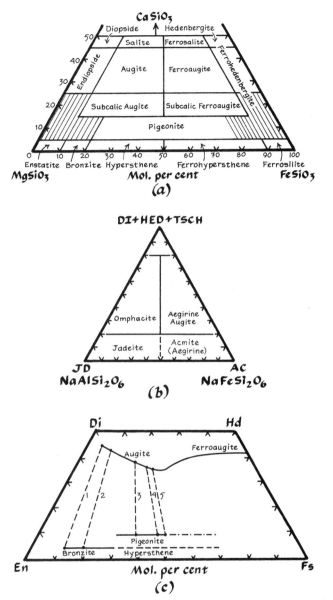

Figure 4-7 (a) The principal pyroxenes are defined in terms of the their molecular proportions in the pyroxene quadrilateral. No compositions fall within the shaded areas, and the iron end-member of the Ca-poor pyroxenes, ferrosilite, is not found in igneous rocks. (b) Sodic pyroxenes are classified in terms of the end-members, jadeite, acmite, and the Ca-rich clinopyroxenes (diopside, hedenbergite, and Ca-tschermak molecule). (c) The trends of compositions of common igneous pyroxenes, and the numbered lines join typical coexisting Ca-rich and Ca-poor compositions in a series of differentiated rocks.

Table 4-3   Compositions and properties of pyroxenes

Mineral	Formula	Crystal System	Optic Sign	Density
Enstatite	$Mg_2Si_2O_6$	Orthorhombic	(+) 56°	3.21
Hypersthene	$(Mg,Fe)_2Si_2O_6$	Orthorhombic	(–) 52°–90°	3.40–3.80
Pigeonite	$(Mg,Fe,Ca)(Mg,Fe)Si_2O_6$	Monoclinic	(+) 0°–25°	3.30–3.46
Ferrosilite	$Fe_2Si_2O_6$	Orthorhombic	(+) 55°	3.96
Diopside	$CaMgSi_2O_6$	Monoclinic	(+) 59°	3.30
Augite	$Ca(Mg,Fe)Si_2O_6$	Monoclinic	(+) 25°–60°	3.23–3.52
Hedenbergite	$CaFeSi_2O_6$	Monoclinic	(+) 52°–62°	3.56
Jadeite	$NaAlSi_2O_6$	Monclinic	(+) 67°–70°	3.42
Acmite (aegirine)	$NaFeSi_2O_6$	Monclinic	(–) 60°–70°	3.58

Chemical Compositions

Pure End Members

	Enstatite	Ferrosilite	Diopside	Hedenbergite	Jadeite	Acmite
SiO_2	59.84	45.54	55.49	48.44	57.19	50.28
Al_2O_3	–	–	–	24.26	–	
Fe_2O_3	–	–	–	–	–	33.41
FeO	–	54.46	–	28.96	–	–
MgO	40.16	–	18.62	–	–	
CaO	–	–	25.89	22.60	–	
Na_2O	–	–	–	–	18.55	16.31

Typical Natural Minerals

	Augite	Ti-augite	Ferroaugite	Hypersthene	Pigeonite	Acmite
SiO_2	49.68	46.20	49.73	53.18	51.53	52.48
TiO_2	0.56	3.21	0.77	0.21	0.58	0.57
Al_2O_3	0.78	5.38	1.39	3.08	1.41	0.96
Fe_2O_3	3.29	3.22	1.50	0.25	0.12	31.74
FeO	18.15	5.41	19.20	18.05	23.17	0.93
MgO	16.19	12.90	9.40	23.26	16.10	0.15
CaO	9.90	23.16	17.75	2.09	7.05	0.28
Na_2O	0.65	0.47	0.24	–	0.26	12.05
K_2O	0.15	0.02	–	0.23	0.35	

is true of Ti and to a lesser degree Cr. The sodic, ferric-iron component, $NaFeSi_2O_6$ (called aegerine or acmite), is abundant in the clinopyroxenes of many peralkaline rocks, whereas jadeite ($NaAlSi_2O_6$) is an important constituent of the clinopyroxenes of mafic or ultramafic rocks formed at high pressures.

Two effects may cause alumina to enter pyroxene—the composition of the liquid and the pressure at which the mineral crystallizes. When pyroxene crystallizes from liquids of low silica contents, Al substitutes for Si in the tetrahedral site (with fourfold coordination) and results in an increase of the c dimension of the unit cell. This substitution of Al for Si is often coupled with substitution of Ti for Mg and Fe to maintain a charge balance. The Al that enters pyroxenes at high pressures does so in a different way. It is substituted simultaneously for both Si and Mg or Fe, and by entering the octahedral site in six-fold coordination, it reduces the b and c crystallographic dimensions. It may also enter as the jadeite molecule, $NaAlSi_2O_6$, in which case it leads to an even greater reduction of volume, as shown by the volumes of three end-member molecules:

Diopside	$CaMgSi_2O_6$	438.0 Å^3
Ca-tschermakite	$CaAl_2SiO_6$	421.7 Å^3
Jadeite	$NaAlSi_2O_6$	405.0 Å^3

Thus, substitution of either of the two Al-bearing pyroxene components, as it reduces the volume of diopside, would be expected in pyroxenes formed at high pressures.

Crystallization of the Ca-poor pyroxenes (**Fig. 4-8**) resembles that of other solid-solution minerals but is complicated by the two polymorphic Ca-poor forms found in natural rocks—a high-temperature, monoclinic form, clinoenstatite or pigeonite, and a low-temperature orthorhombic form, enstatite or hypersthene. The temperature of inversion between these two forms declines with increasing iron content at a greater rate than the liquidus and solidus so that the two sets of curves intersect in the intermediate compositional range.

A liquid of Mg-rich composition that initially crystallizes in the orthorhombic form becomes richer in iron as it cools. After crossing the inversion curve, it crystallizes initially in the monoclinic form, pigeonite, which on cooling tends to change to hypersthene, the stable orthorhombic form at low temperatures. In slowly cooled rocks, pigeonite almost always inverts to hypersthene, but most volcanic rocks cool too quickly for this inversion to take place, and pigeonite is preserved in a metastable state.

As one would expect from their differing crystalline structures, the range of solution between the monoclinic and orthorhombic forms is limited (**Fig. 4-9**); a broad solvus separates their coexisting subsolidus compositions, and on cooling, each of the end members tends to become purer as it exsolves the other. Exsolution of the two forms of pyroxenes has a distinctive crystallographic pattern (**Fig. 4-10**). When the stable form of the Ca-poor pyroxene is orthorhombic,

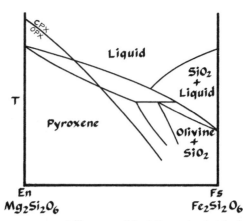

Figure 4-8 The Ca-poor pyroxenes crystallize as a solid solution series, but the inversion curve separating the stability field of monoclinic pigeonite (cpx) from the orthorhombic form (opx) crosses the liquidus and solidus in the middle range of compositions. The iron end-member, ferrosilite, is replaced by fayalitic olivine and a silica mineral, such as quartz or tridymite. The diagram is greatly simplified.

the lamellae of exsolved pyroxene form along (100) planes of the host crystal; if the stable Ca-poor pyroxene is monoclinic, the lamellae form along (001) planes. These relationships hold regardless of which pyroxenes is the host. If exsolution begins when the stable pyroxenes are both monoclinic and continues through an interval in which the Ca-poor pyroxene inverts to the orthorhombic structure, the orientation of lamellae will first be along (001) and then later on (001), and the final crystal will have two intersecting sets.

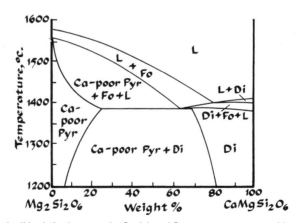

Figure 4-9 Limited solid solution between the Ca-rich and Ca-poor pyroxenes resembles the relations in the alkali feldspars, but the solvus intersects the solidus at all pressures, and relationships are complicated by the initial crystallization of olivine from Ca-poor liquids. Fields of various forms of Ca-poor pyroxene are omitted.

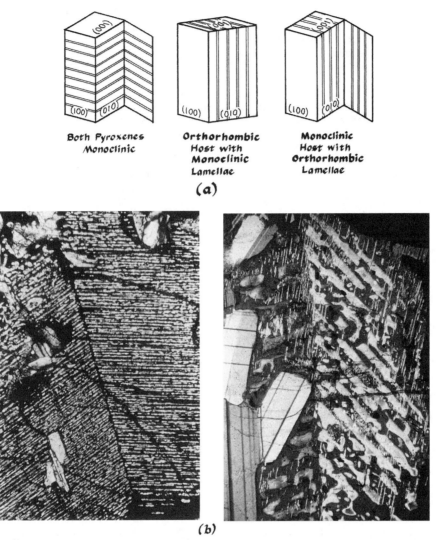

Both Pyroxenes
Monoclinic

Orthorhombic
Host with
Monoclinic
Lamellae

Monoclinic
Host with
Orthorhombic
Lamellae

(a)

(b)

Figure 4-10 (a) Crystallographic relations of exsolution lamellae in pyroxenes. (b) Examples of exsolution in inverted pigeonite. The monoclinic crystals of pigeonite were twinned, as can be seen from the inclined exsolution along the original (001) planes of their host. In the crystal on the left, all exsolution occurred while the host was still monoclinic, but the crystal on the right continued to exsolve after the inversion so that a second set of lamellae has formed parallel to the (100) plane of the orthorhombic host.

Enstatite melts incongruently to form olivine plus liquid in the manner explained in the discussion of binary systems with intermediate compounds in Chapter 3. For this reason, a liquid with the composition of enstatite does not crystallize that mineral directly. Instead, olivine crystallizes first

Figure 4-11 (a) The liquidus surface of the pyroxene system has several phases other than pyroxene. Olivine is the first phase to crystallize from liquids with compositions close to enstatite, and a silica mineral crystallizes from liquids close to ferrosilite. The first phase to crystallize from a liquid of hedenbergitic composition has the structure of wollastonite (or ferrobustamite) and inverts on cooling to a pyroxene structure. (b) The positions of the three sections illustrated schematically in (c), (d), and (e). Triangles connect coexisting pyroxenes to liquids descending along the cotectic shown in (a). The initial liquid, X, would precipitate olivine until it reaches a boundary with pyroxene. (c) Liquid L_1 precipitates a diopsidic Ca-rich pyroxene, which on cooling exsolves the Ca-poor form. Because the inversion temperature is above the solidus of the Ca-poor pyroxene (Fig. 4-8), the latter crystallizes initially in the orthorhombic form. Even though the bulk composition is in the field of primary crystallization of olivine, liquid L_1 is not so that both liquidus phases are pyroxenes. (d) Liquid L_2 precipitates augite together with pigeonite because the inversion curve for pigeonite to hypersthene (Fig. 4-8) now lies below the solidus. Both pyroxenes exsolve their counterpart, but on cooling, a subsolidus inversion takes place so that only the orthorhombic Ca-poor pyroxene, hypersthene, is stable at lower temperatures. (e) Liquid L_3 has migrated to a more Ca-rich composition, and only a single pyroxene crystallizes, initially with the structure of Fe-wollastonite (or ferrobustamite), which, on cooling, inverts to hedenbergite. Fayalitic olivine and a silica mineral are in equilibrium with the Ca-rich pyroxene.

and later reacts with the evolving liquid to form pyroxene. **Figure 4-11a** shows the liquidus fields within the pyroxene quadrilateral and illustrates the range of liquid compositions that initially crystallize olivine. Also shown are the fields of iron-wollastonite (or ferrobustamite), the high-temperature, triclinic form of hedenbergite, and pigeonite, the high-temperature monoclinic form of Ca-poor pyroxene. Figure 4-11b shows typical tie-lines joining cotectic liquids with the different coexisting forms that crystallize at successive stages of evolution as the liquid descends to lower temperatures and higher iron contents.

Pigeonites normally have about 5 to 9 percent of the calcium component ($CaSiO_3$) in solid solution. This is more than that of the orthorhombic pyroxenes under the same conditions and reflects the greater range of solid-solution between forms sharing a monoclinic structure. Because of this difference, the inversion curve that was a simple line in the Ca-free system (Fig. 4-8) becomes a set of curves, one for each of the coexisting pyroxenes, that form a curved wedge extending into the diagrams for more iron-rich pyroxenes (Figs. 4-11d and 4-11e).

The crystallization behavior through a range of pyroxene compositions is best illustrated by a series of hypothetical examples. The sketches of crystals shown in Figure 4-11 represent the three stages of differentiation in Figures 11c, 11d, and 11e. At an early stage (Fig. 4-11c), diopsidic augite crystals have small exsolution lamellae parallel to (100), while a coexisting enstatite has embayed cores of olivine and exsolved diopside along (100) in its outer

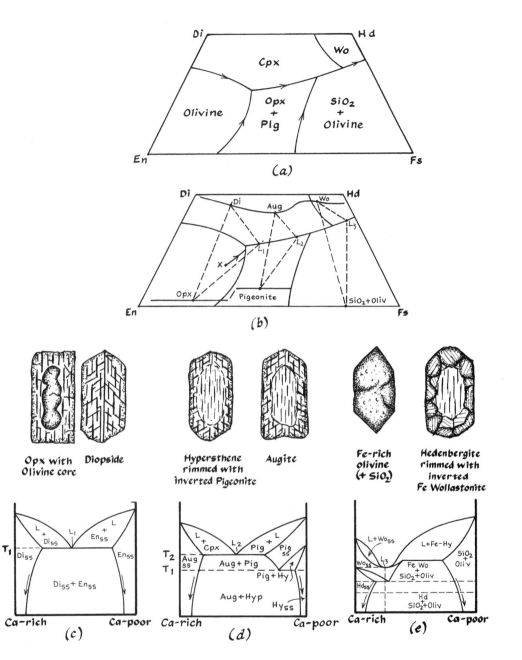

(a)

(b)

Opx with
Olivine core   Diopside

Hypersthene
rimmed with
inverted Pigeonite   Augite

Fe-rich
olivine
(+ SiO₂)   Hedenbergite
rimmed with
inverted
Fe Wollastonite

(c)   (d)   (e)

rims. Several deductions could be made from these relationships. It is clear that crystallization took place below the monoclinic–orthorhombic inversion temperature because all exsolution is parallel to (100). Second, the initial liquid was probably on the Ca-poor side of the low-temperature trough so that olivine crystallized first, reacted with the liquid to form enstatite, and was later joined by diopside. This we know from the embayed cores of olivine, which would be less likely to form if the liquid had first crystallized diopside and approached the trough from the Ca-rich side.

A later stage of crystallization is illustrated in Figure 4-11d. The liquid has now migrated down the thermal valley to a more Fe-rich composition. Each pyroxene has two sets of lamellae, one parallel to (100) and the other along (001). This relation shows that crystallization began above the inversion temperature where both forms were originally monoclinic, but after cooling and exsolving one set of lamellae on (001), the pigeonitic pyroxene was replaced by hypersthene and subsequent exsolution was on (100).

The late stages of iron enrichment are illustrated in Figure 4-11e. The magma precipitates only a single pyroxene, ferroaugite, containing two sets of exsolution lamellae, and an iron-rich olivine and quartz crystallize in place of the Ca-poor pyroxene. On approaching the Hd corner of the quadrilateral, the thermal trough has migrated toward the Di-Hd join until it has "fallen off" the solvus, and from that point on, only a single pyroxene can precipitate directly from the liquid.

Olivine

In subalkaline rocks, olivine is confined almost entirely to compositions more mafic than andesite or diorite. Although normally absent from the groundmass, it may be an abundant phenocryst, especially in basalts. When it is, it commonly shows evidence of reaction with the groundmass. Iron-rich olivine is a stable phase in highly evolved felsic volcanic and hypabyssal rocks, such as rhyolites, granophyres, and ferrogabbros, but it is rare in felsic plutonic rocks.

In alkaline rocks, olivine can occur either as phenocrysts or as part of the groundmass, regardless of the degree of differentiation. Indeed, stable groundmass olivine is a useful petrographic criterion by which the alkaline rocks can be distinguished from sub-alkaline types (Chapter 7).

Olivine forms a solid-solution series between forsterite and fayalite (**Table 4-4 and Fig. 4-12**) similar to that of plagioclase, but because it has a simpler structure and equilibrates more readily, olivine is less likely to be zoned. It also differs from plagioclase in the reaction that occurs between Mg-rich crystals and liquids to form pyroxene (**Fig. 4-13**). The reaction relationships will be recognized as an example of a system with an intermediate

|Table 4-4   Compositions and properties of olivines

Mineral	Formula	Optic Sign	Density
Forsterite	Mg_2SiO_4	(+) 82°	3.22
Fayalite	Fe_2SiO_4	(−) 46°	4.39
Montecellite	$Ca(Mg,Fe)SiO_4$	(−) 72°–82°	3.08–3.27

Chemical Compositions of Pure and Natural Olivines

	Pure Forsterite	Fo 96	Fo 86	Fo 47	Fo 3	Pure Fayalite	Montecellite (in peridotite)
SiO_2	42.71	41.07	39.87	34.04	30.15	29.49	36.67
TiO_2	–	0.05	0.03	0.43	0.20	–	0.13
Al_2O_3	–	0.56	–	0.91	0.07	–	0.75
Fe_2O_3	–	0.65	0.86	1.46	0.43	70.51	0.10
FeO	–	3.78	13.20	40.37	65.02	70.51	7.57
MnO	–	0.23	0.22	0.68	1.01	–	0.17
MgO	57.29	54.06	45.38	20.32	1.05	–	21.11
CaO	–	–	0.25	0.81	2.18	–	32.56

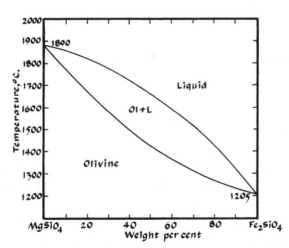

Figure 4-12 The common igneous olivines form a continuous range of solid solution between magnesium and iron end-members, forsterite and fayalite. Their crystallization and melting relationships are similar to those of plagioclase. Effects of oxidation or reduction are ignored, and all iron is assumed to be in the ferrous state. (After N. L. Bowen and J. F. Schairer, 1935, *Amer. Jour. Sci.* 29:151–217.)

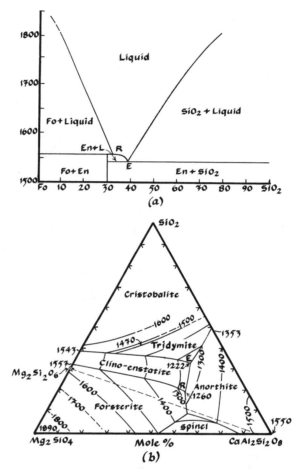

Figure 4-13 (a) The system forsterite-silica is an important example of a binary system with and intermediate compound, enstatite, that melts incongruently. (After N. L. Bowen and O. Andersen, 1914, *Amer. Jour. Sci.*, 37:487–500.) (b) When anorthite is added to the forsterite-silica system, the intermediate compound, En, occupies a field extending into the interior of the three-component system. The boundary between the pyroxene and olivine field is a peritectic descending to an invariant reaction point, R. (After O. Andersen, 1915, *Amer. Jour. Sci.* 39:407–454.)

compound, in this case enstatite, that melts incongruently. Crystallization and melting follow the principles outlined for such systems in Chapter 3 (Figs. 3-12 and 3-19b).

These reaction relations do not extend to the full range of olivine compositions. With greater iron contents, the crystallization behavior changes in several ways, depending on the degree of enrichment and oxidation state of the iron. Crystallization of olivine from reduced Fe-bearing liquids is illustrated by the system forsterite-fayalite-silica (**Fig. 4-14**). Several possible

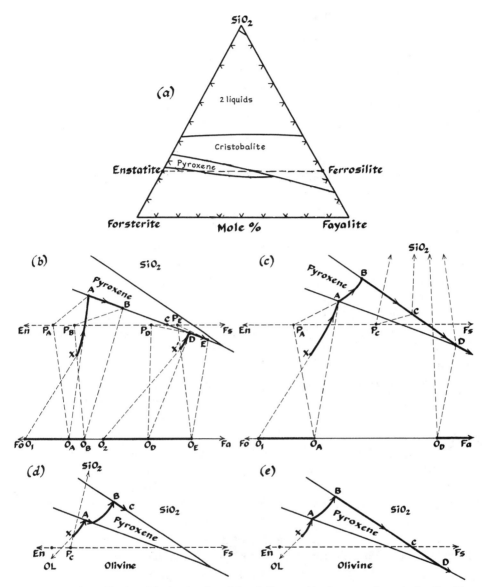

Figure 4-14 Crystallization of the iron-bearing olivines is illustrated by the system forsterite-fayalite-silica. Depending on the composition of the starting liquid and the degree of fractionation, several courses of crystallization are possible. The most important of these are discussed in the text.

courses are shown in expanded schematic diagrams of the central part of the system.

Consider first liquids with compositions in the field between the olivine and pyroxene joins, Fo-Fa and En-Fs. With complete reaction and equilibration, such liquids crystallize to assemblages of olivine and pyroxene with the same bulk composition as the starting liquid, but depending on their initial composition, they may do this in different ways. The liquid X in the left part of Fig. 4-14b first crystallizes olivine of composition O_1 and follows a curved course across the olivine field as the composition of the crystals becomes more fayalitic. On reaching the peritectic boundary of pyroxene at A, the olivine has a composition O_A (indicated by a tangent from the liquid at A). Pyroxene of a composition, such as P_A, then begins to crystallize, and at the same time the liquid reacts with early-formed olivine and converts part of it to pyroxene. The composition of both the pyroxene and remaining olivine become more iron rich. When the liquid reaches a composition, such as B, at which the compositions of the pyroxene and olivine have reached P_B and O_B, respectively, a line joining these phases passes through the bulk composition X. At that point, the last liquid crystallizes, and the crystalline assemblage consists of X-P_B parts olivine and X-O_B parts pyroxene.

A liquid X' in the right-hand part of the same diagram is somewhat richer in iron. It follows a similar course of crystallization, but when the olivine-pyroxene boundary crosses below the En-Fs join, it becomes a cotectic. The reaction relationship disappears, and olivine and pyroxene crystallize together. The reason for this change can be seen by examining the effects of crystallization or reaction on the composition of the liquid. The early pyroxene P_D lies below the boundary, and if reaction did not take place, it would cause the liquid to leave the boundary; however, when the pyroxene composition crosses point C to a composition, such as P_E, this relationship changes, and reaction is no longer required to keep the liquid on the boundary.

Without reaction, the course of both of these liquids, X and X', will be extended to more iron-rich compositions (Fig. 4-14c). The liquid X, for example, would again follow a curved path across the olivine field, but on reaching the pyroxene boundary, it would not follow it. Instead, it would embark on a curved path across the pyroxene field crystallizing increasingly more iron-rich pyroxene until it reaches the boundary of the silica field. It then descends along the pyroxene-silica cotectic, crystallizing a silica mineral, first together with pyroxene and then with olivine. (Note that during this sequence, the silica content of the liquid first increases and then decreases as crystallization proceeds.)

Liquids with compositions above the En-Fs join, such as X in Fig. 4-14d initially precipitate olivine, but with complete reaction, they should crystallize to a final assemblage consisting of only pyroxene and SiO_2. The liquid, after crossing part of the olivine field, reaches the peritectic pyroxene boundary and follows it as pyroxene crystallizes and olivine reacts with the liquid. The small amount of olivine is soon consumed, however, and the liquid must leave the peritectic, cross the pyroxene field, and reach the cotectic boundary with SiO_2. It then crystallizes pyroxene together with quartz or tridymite until the composition of the pyroxene reaches P_C, at which point the last drop of liquid has crystallized. Without reaction (Fig. 4-14e), the course is similar to that outlined in Figure 4-14c.

The iron-rich pyroxene, ferrosilite (Fs), is unstable at low pressures and has never been found in igneous rocks. The stability field of pyroxene gradually narrows with increasing proportions of iron and gives way to a cotectic boundary between quartz and iron-rich olivine. Thus, a liquid increasingly enriched in iron eventually ceases to crystallize pyroxene and instead precipitates a second generation of olivine together with a silica mineral. This relationship explains why olivine can crystallize, not only from basic magmas, but also from rhyolites and granophyres in equilibrium with quartz or tridymite.

Iron-Titanium Oxides

Iron and titanium oxide minerals are minor but almost ubiquitous components of almost all volcanic and plutonic rocks. They may occur as primary phases or as products of alteration of other minerals, such as olivine, amphibole, and biotite. Ten percent or so of the total iron content of most natural magmas is in the ferric state, Fe^{3+}, and does not enter appreciably into the crystal structures of common iron-bearing silicates. Instead, it combines with ferrous iron to form magnetite (Fe_3O_4). TiO_2, although a minor component of most igneous rocks, may exceed 3 or 4 weight percent in some basalts and gabbros, particularly those of alkaline composition. At concentrations below about 1 weight percent or so, most of this titanium is taken up in solid solution in titaniferous magnetite, but at greater concentrations, it forms a separate phase, ilmenite.

The principal oxide phases and their compositional ranges are shown in **Figure 4-15**. Magnetites may contain up to 80 percent of the isostructural titanium component, ulvospinel (Fe_2TiO_4), as well as small amounts of magnesium- and aluminum-rich spinel. Magnesium-rich spinels are referred to as magnesioferrites or, more loosely still, spinel. In a similar fashion, ilmenite can contain up to 15 percent or so of hematite (Fe_2O_3).

The amount of solid solution of the titanium component in magnetite and of hematite in ilmenite is a function of temperature and oxidation state (**Fig. 4-16a**). Provided that they have not exsolved and recrystallized their respective components at lower temperatures, the compositions of these minerals record the temperature and partial pressure of oxygen at which the magma crystallized, and they are widely used as a measure of these parameters. Re-equilibration at lower temperatures is rapid, however, and temperatures estimated in this way must be interpreted with caution.

With increasing degrees of oxidation, the ratio of ferric to ferrous iron increases; larger proportions of the iron in magmas enter magnetite, and less goes into the ferrous iron-bearing silicates, such as olivine and pyroxene. The effect is well illustrated by the system olivine-silica at two different oxidation states (**Fig. 4-17**). The reduced system without ferric iron (Fig. 4-17a) is the same as that shown earlier as Figure 4-14. In the oxidized system, fayalite is replaced by magnetite and silica:

$$3Fe_2SiO_4 + O_2 \leftrightarrow 2Fe_3O_4 + 3SiO_2 \tag{4-1}$$
$$\text{fayalite} \quad \text{oxygen} \quad \text{magnetite} \quad \text{silica}$$

Note that the large field of magnetite cuts off the fields of both olivine and pyroxene and restricts the range of differentiated liquids to much lower levels of iron enrichment. In natural magmas, the oxidation state is normally between these two extremes.

A reaction such as Equation 4-1 has an equilibrium constant that is a function of the activities of the various reactants and products. Being a gas, the activity of oxygen is referred to as its *fugacity*, which is simply the partial pressure of the gas multiplied by an activity coefficient that corrects for the deviation of its properties from those of a perfect gas. Its function is

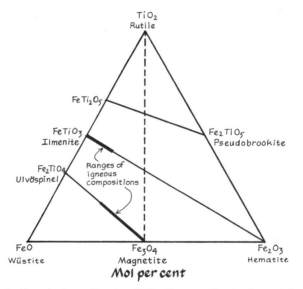

Figure 4-15 The principal iron-titanium oxide minerals. The lines crossing the diagram join minerals that form solid solution series. Intermediate compositions are shown in molecular proportions.

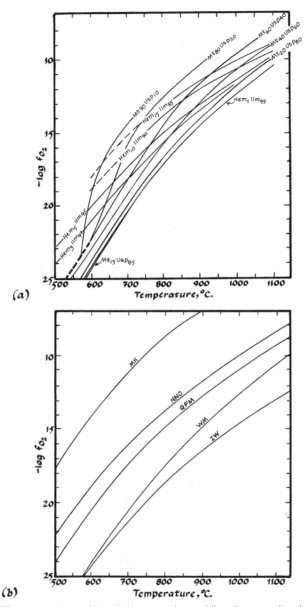

Figure 4-16 (a) The compositions of co-existing magnetites and ilmenites are a function of temperature and the partial pressure or fugacity of oxygen (fO_2 in atmospheres). The intersecting curves for the compositions of the two minerals form a grid from which it is possible to estimate the temperature and oxygen fugacity at which a pair of co-existing oxide minerals last equilibrated. (After A. F. Buddington and D. H. Lindsley, 1964, *Jour. Petrology*, 5:310–357.) (b) Certain oxidation-reduction reactions, often referred to as "buffers," are taken as standard references for the relationship between temperature and oxygen. Those shown here are the most common in the range of magmatic conditions on the Earth or Moon. Abbreviations are as follows: MH, magnetite-hematite; NNO, nickel-nickel oxide; QFM, quartz-fayalite-magnetite; WM, wustite-magnetite; and IW, metallic iron-wustite.

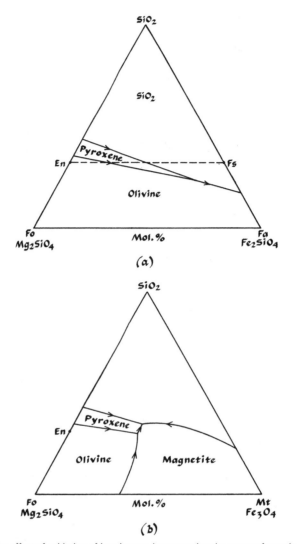

Figure 4-17 The effect of oxidation of iron is seen by comparing the system forsterite-fayalite-silica at two oxidation states, one (a) in which all iron is in the ferrous state (as in Fig. 4-14a) and another (b) at atmospheric conditions where much of the iron is in the ferric state, and magnetite replaces fayalite as a liquidus phase.

basically the same as that of chemical potential discussed in the preceding chapter.

A number of reactions involving oxidation of iron, such as the one described in Equation 4-1, have equilibrium constants that place them within the realm of oxidation conditions in natural magmas. They provide a convenient reference to which the oxidation state of an iron-bearing system can be related. They are referred to as *buffer reactions* because so long as all of the components of the reaction are present, the oxygen fugacity is constrained to a well-defined curve that varies in a fixed way with temperature (Fig. 4-16b). In the case of the reaction in Equation 4-1, for example, an increase in the partial pressure of oxygen in the system drives the reaction toward the right; fayalite goes to magnetite and quartz with an absorption of oxygen. A decrease operates in the opposite direction with a liberation of oxygen. Thus, by the Principle of Le Chatelier, the effect of the change is absorbed or "buffered." From the form of the curve in Figure 4-16b, it should be possible to deduce whether the reaction (4-1) is endothermic or exothermic.

Amphiboles and Micas

The hydrous minerals, amphibole and mica (**Table 4-5**), occur in almost all felsic plutonic rocks and in many mafic alkaline rocks as well. They are also found in many dacites and rhyolites, but because they are unstable at high temperatures and atmospheric pressure, they are less common in volcanic than in plutonic rocks of equivalent composition.

The common hornblendes of subalkaline volcanic rocks and most diorites and tonalites are members of the pargasite-hastingsite series. In volcanic rocks they may be the deep reddish brown form, oxyhornblende. The Ca-poor amphibole, cummingtonite, is common in rhyolites and dacites and may coexist in equilibrium with common hornblende. In more alkaline rocks, such as trachytes, phonolites, and equivalent subvolcanic rocks, the sodic amphibole, riebeckite, is more characteristic. The Al-poor amphiboles are, as a rule, most common in felsic rocks. Many alkaline gabbros contain the titaniferous amphibole, kaersutite, but barkevikite and arfvedsonite are more characteristic of alkaline plutonic rocks containing less Ti.

Because of their wide compositional range, the effect of fractionation of amphiboles on the composition of a differentiating magma can be predicted only if the composition of the specific amphibole is known. Crystallization of some amphiboles tends to enrich the remaining liquid in silica, whereas others may have the opposite effect. The composition of the pargasite in Table 4-5, for example, is equivalent to an assemblage of normative minerals including nearly 12-percent nepheline, whereas the cummingtonite in the same table has the equivalent of about 7-percent quartz. Because of this same effect, the first liquids produced by partial melting of rocks containing pargasitic amphiboles are more silica rich than a melt produced from a rock of the same chemical composition but with no stable hydrous mineral.

Table 4-5   Compositions and properties of common igneous amphiboles. Because of the great variation of optical properties, only the general chemical features of amphiboles are given here. For more details on the identification of individual minerals, reference should be made to standard mineralogical tables.

Mineral	Formula	Color	Density
Oxy-hornblende	$Ca_2(Na,K0)_{.5-1}(Mg,Fe^{2+})_{3-4}(Fe^{3+},Al)_{2-1}Si_6Al_2O_{22}(O,OH,F)_2$	dark brown	3.2–3.3
Hornblende	$(Ca,Na,K)_{2-3}(Mg,Fe^{2+},Fe^{3+},Al)_5(Si_6Al)_2O_{22}(OH,F)_2$	green or brown	3.0–3.4
Pargasite	$Na,Ca_2Mg_4(Al,Fe^{3+})(Si_6Al_2O_{22}(OH,F)_2$	pale brown	3.05
Cummingtonite	$(Mg,Fe^{2+})_7(Si_8O_{22})(OH)_2$	pale green	3.36
Ferrohastingsite	$NaCa_2Fe_4^{2+} (Al,Fe^{3+})(Si_6Al_2O_{22})(OH,F)_2$	dark green	3.50
Kaersutite	$Ca_2(Na,K)(Mg,Fe^{2+},Fe^{3+})_4Ti (Si_6Al_2O_{22})(O,OH,F)_2$	dark red-brown	3.2–3.3
Barkevikite	$Ca_2(Na,K)(Fe^{2+},Mg,Fe^{3+},Mn)_5Si_{6.5}Al_{1.5}O_{22}(OH,F)_2$	dark brown	3.35–3.44
Riebeckite	$Na_2Fe^{2+}_3,Fe^{2+}_2(Si_8O_{22})(OH,F)_2$	dark blue	3.0–3.4
Arfvedsonite	$Na_{2.5}Ca_5(Fe^{2+},Mg,Fe^{3+}Al)_5Si_{7.5}O_{0.5}O_{22}(OH,F)_2$	green	3.50

Typical Chemical Compositions of Natural Amphiboles

	Hornblende (gabbro)	Hornblende (tonalite)	Pargasite (peridotite)	Cummingtonite (diorite)	Ferrohastingsite (granite)	Kaersutite (Alkali gabbro)	Riebeckite (granite)
SiO_2	48.71	44.99	43.61	50.78	39.56	40.67	51.01
TiO_2	0.32	1.46	1.15	0.40	1.46	6.22	0.96
Al_2O_3	9.48	11.21	15.06	1.77	12.18	10.45	0.80
Fe_2O_3	2.33	3.33	1.59	1.88	4.10	3.86	16.41
FeO	9.12	13.17	5.14	29.64	23.18	7.56	17.62
MnO	0.23	0.31	0.08	0.14	0.09	0.15	0.48
MgO	14.43	10.41	16.53	11.83	4.43	11.54	0.22
CaO	11.93	12.11	11.80	1.33	9.98	15.55	0.19
Na_2O	1.16	0.97	2.78	0.00	1.81	1.83	7.89
K_2O	0.15	0.76	0.11	0.00	1.38	1.14	1.80
H_2O	1.83	1.52	1.74	2.01	1.26	1.28	0.91
F	0.23	–	–	–	1.20	–	1.70

Most analyses given here are taken from W. A. Deer, R. A. Howie, and J. Zussman, 1963, *Rock-Forming Minerals*, Vol. 2, London: Geological Society.

The micas of volcanic rocks tend to be magnesian biotites. Muscovite is found mainly in pegmatites and plutonic rocks formed at moderately high pressures. Phlogopitic mica is especially characteristic of ultramafic and highly potassic alkaline rocks. The typical micas of alkali gabbros are a deep red titaniferous variety.

Micas, like the pargasitic amphiboles, are silica-poor minerals **(Table 4-6)**. The composition of biotite, for example, is equivalent to the anhydrous assemblage olivine + leucite + kaliophilite (the strongly silica-deficient equivalent of orthoclase—$KAlSiO_4$). Hence, fractionation of mica tends to enrich the remaining liquid in silica, whereas melting or assimilation of the same mineral has the opposite effect.

Because water is an essential component of their crystal structures, the temperatures at which amphiboles and micas are stable increase with an increasing vapor pressure of water. A typical example is the curve for crystallization of hornblende from basaltic liquids **(Fig. 4-18)**. Note the contrast between the negative slopes for the upper stability limits of anhydrous minerals and the positive slope for that of hornblende at increasing water pressure.

| Table 4-6 | Compositions and properties of igneous micas |

Mineral	Formula		Color	Density
Muscovite	$K_2Al_4Si_6Al_2O_{20}(OH,F)_4$		colorless	2.8–2.9
Phlogopite	$K_2(Mg,Fe)_6Si_6Al_2O_{20}(OH,F)_4$		pale brown	2.8–2.9
Biotite	$K_2(Mg,Fe^{2+})_{6-4}(Fe^{3+},Al,Ti)_{0-2}Si_{6-5}Al_{2-3}O_{20}(OH,F)_{4-2}$		dark brown	2.7–3.4

Chemical Compositions

	Muscovite	Phlogopite	Biotite	Ti-Biotite
SiO_2	46.77	40.16	36.67	36.52
TiO_2	0.21	0.45	3.39	8.77
Al_2O_3	34.75	12.65	17.10	13.29
Fe_2O_3	0.71	5.52	4.58	0.58
FeO	0.77	4.39	16.36	15.91
MnO	–	0.03	0.04	0.20
MgO	0.92	22.68	9.50	11.28
CaO	0.13	0.24	0.38	0.68
Na_2O	0.47	–	0.20	0.73
K_2O	10.61	9.62	9.17	9.00
H_2O	4.48	3.00	1.98	3.02
F	0.16	–	1.37	–

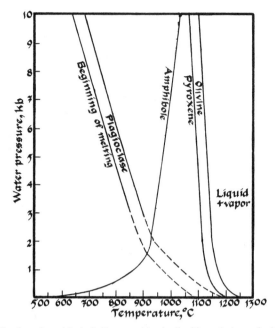

Figure 4-18 Crystallization of amphibole is illustrated by the liquidus relations of a basaltic magma as a function of water pressure. With increasing water pressure, the stability of amphibole increases to higher temperatures, whereas the anhydrous minerals crystallize or melt at lower temperatures than they do under dry conditions. (After H. S. Yoder and C. E. Tilley, 1962, *Jour. Petrology* 3:342–532.)

This difference illustrates the contrasting role played by water in crystallization of the anhydrous and hydrous minerals.

The reason for this difference can be understood in terms of the familiar Clapeyron equation,

$$\frac{dT}{dP} = \frac{T\Delta V}{\Delta H} \tag{3-21}$$

and the volume changes on melting (or reaction) in the presence of a separate water-vapor phase. As we have seen, an anhydrous mineral in the presence of water vapor melts with a decrease of volume at constant pressure because the water dissolves in the melt with a large negative ΔV. However, when a hydrous phase breaks down to an anhydrous assemblage plus water vapor, the volume change is positive because the water has a much greater molar volume as a gas than it has in solution. All dehydration reactions must require an input of heat, for we know by experience that one heats a substance to drive out its water, and the ΔH of a dehydration reaction must be positive, just as it is in melting an anhydrous phase. It can be seen, therefore, that if P in the Clapeyron equation is pressure of water vapor the slope dT/dP for an anhydrous phase will be negative (because ΔV is negative), whereas that of a hydrous mineral will be positive for the opposite reason.

It can be taken as a rule that if water does not enter the crystalline structure, it acts as an unsaturated constituent of the liquid and lowers the melting temperature, as would any other added component. If it is an essential component of a crystalline phase, however, the increased pressure of water acts to stabilize the mineral and increases the temperature range in which it can coexist with liquids.

The stability relationships of biotites are affected not only by water pressure but by the oxidation state as well. With increasing partial pressure (or fugacity) of oxygen, the temperature at which the iron-rich micas break down decreases. The effect is illustrated by the reaction

$$3K(Mg,Fe)_3AlSi_3O_{10}(OH)_2 + 3SiO_2 + 2O_2 \leftrightarrow 3Mg_2SiO_4 + Fe_3O_4 + 3KAlSi_3O_8 + 3H_2O$$

biotite   silica   oxygen   olivine   magnetite sanidine   water

The same equation shows that the activity of silica must also affect the stability of biotite so that the mineral would be stable at higher temperatures in silica-poor magmas. This is in accord with the more common occurrence of biotite in alkali gabbros than in more silica-rich rocks.

■ Crystallization of Igneous Minerals

The great compositional and textural variety of igneous rocks attests to the diverse conditions under which magmas cool and crystallize. The rates of nucleation and growth of crystals differ from one mineral to another and are sensitive to the rate of cooling and the physical and chemical properties of the system. As a result, two rocks that crystallized from the same magma may have very different textures and outward appearances depending on the manner in which the crystals nucleated and grew.

Nucleation and Growth of Crystals

We noted in the preceding chapter that cooling liquids seldom begin to precipitate crystals at the precise temperature at which they become saturated with a particular mineral. The rate of nucleation increases with the degree of undercooling (**Fig. 4-19a**). The rate of growth on existing crystals is also a function of the amount of undercooling, but it reaches a maximum beyond which more rapid cooling results in no further growth. The remaining liquid becomes a glass. Even after nuclei begin to form they must reach a certain

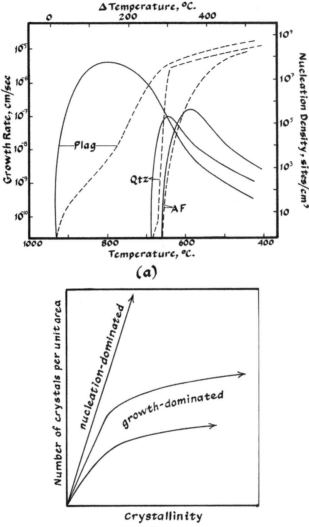

Figure 4-19 (a) Rates of nucleation (broken lines) and growth (solid lines) of plagioclase, quartz, and alkali feldspar from a granodioritic liquid with 6.5 wt percent H_2O as functions of the amount of undercooling, ΔT. (After S. E. Swanson, 1977, *Amer. Min.* 62:966–978.) (b) The distribution of crystals sizes will depend on the relative rates of nucleation and growth on existing crystals. With strong undercooling, there will be an increasing number of crystals per unit volume as crystallization advances, while with less undercooling, the number of crystals approaches a constant value as growth continues on existing nuclei. (After K. Cashman and B. Marsh, 1988, Crystal size distribution (CSD) in rocks and the kinetics and dynamics of crystallization. *Contrib. Mineral. Petrol.* 99:292–305, and Cashman, personal communication.)

critical size if they are to survive. Because it has a larger proportion of surface atoms with unbalanced charges, a very small crystal has a high surface energy and is only marginally stable. Until its dimensions are of the order of 10^{-5} to 10^{-6} cm, it is at a disadvantage with respect to larger crystals and tends to go back into solution. In very slowly cooled bodies, this leads to a process known as *Ostwald ripening* in which the crystals of the individual minerals tend to take on a coarse, nearly uniform grain size.

The relative importance of nucleation and growth on existing crystals is reflected in the distribution of grain sizes (Fig. 4-19b). Crystallization that is primarily by continued formation of new nuclei produced serial sizes, but with less undercooling, more of the growth is on existing crystals and the sizes of crystals tend to be more uniform. A lava flow, because it is relatively thin and loses heat rapidly, continues to nucleate large numbers of crystals until crystallization is totally arrested. This is in contrast to the magma in the lava lake we examined earlier (Fig. 3-1). In that case, the lava cooled more slowly, and after an initial burst of nucleation, crystals grew almost entirely on a nearly constant number of early-formed nuclei.

Growth on the faces of a stable nucleus is governed largely by the rate at which the essential elements diffuse through the adjacent liquid and is limited by the rate at which the component with the slowest diffusivity is supplied to its surface. For a given cooling rate, the crystals that reach the largest size are those formed from components with the highest diffusion coefficient and largest ionic volume. Diffusion is less important when there is relative motion of the liquid and crystal, and fresh liquid is brought to the crystal face. Because of the different surface energies of the various faces, some may grow faster than others. Sharp edges tend to grow more rapidly in a quickly cooled liquid because they can draw on a large volume of the surrounding liquid.

With very slow cooling, crystal growth may continue even after the temperature has fallen below the solidus. Interstitial glass will devitrify by forming myriads of tiny crystallites, and in the case of large plutonic bodies in which temperatures can remain for centuries within two or three hundred degrees of the solidus, large crystals grow at the expense of small ones, while minerals with limited solid solution steadily approach end-member compositions by exsolving their secondary components. This is particularly true of the iron oxides, magnetite and ilmenite, and other solid solution series, such as alkali feldspars and pyroxenes, which show extensive exsolution in slowly cooled rocks.

Theoretically, a very slowly cooled assemblage of crystals would eventually recrystallize to a rock with only a single large crystal of each phase because this would be the lowest energy state. Under natural conditions, however, diffusion is so slow, and the differences of surface energy of crystals are so small that after they reach half a centimeter or so in size this process comes to an end. There are other factors, however, that can contribute to very large crystals. Rocks with exceptionally large crystals, broadly referred to as *pegmatites*, owe their course-grained textures to crystallization in the presence of a fluid phase, usually one rich in water. The fluid facilitates the transfer of components to the growing crystals, either by flowing over them or by providing a medium in which diffusion is greatly accelerated over what it normally is in volatile-poor silicate melts. Other components, such as fluorine or chlorine, may have a similar effect.

■ Textures of Igneous Rocks

Many of the factors governing the nucleation and growth of crystals described in the preceding section are reflected in the textures of volcanic and plutonic rocks. The principal effects fall into several general categories.

a. *High rates of cooling and supersaturation* (**Fig. 4-20**) tend to cause rapid growth of slender crystals that are greatly elongated in one crystallographic direction. The crystals may radiate from a common center or form curved and branching clusters of tiny microlites. Because the edges and corners of crystals can draw on larger volumes of adjacent liquid than can planar faces, they grow more rapidly and produce crystals with hollow cores and "swallow-tail" longitudinal sections.

b. *Differing rates of nucleation and growth* (**Fig. 4-21**) can cause one mineral to form large numbers of small crystals while another forms larger crystals growing on a smaller number of nuclei. If one of the minerals crystallizes earlier or faster than the others, it may form large *phenocrysts* that give the rock a *porphyritic* texture. If most crystals nucleate and grow at the same rates, however, they tend to develop *intergranular* textures, in which grains of pyroxene, olivine, and oxide minerals lie between laths of plagioclase. If, as often happens in slowly cooled basalts, pyroxene forms few nuclei but larger individual grains, it may partly encase the laths of plagioclase and form what is referred to as *ophitic* textures. In extreme cases, one of the minerals may grow from a small number of nuclei and completely enclose many other crystals of smaller size to form what is called *poikilitic* texture.

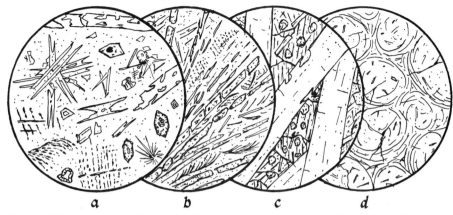

Figure 4-20 Textures produced by rapid cooling and high supersaturation. (a) Crystals in rapidly quenched basaltic lavas are commonly very elongated and may have hollow cores. Small microlites form curved and branching clusters. (b) *Spinifex* textures in ultramafic lavas are characterized by long, slender needles of olivine or pyroxene. In this case, elongation may result from the very low viscosity and high diffusivity of the liquid. (c) Similar textures are produced by elongated plagioclase or olivine crystals in gabbros, which may have had high water contents that facilitated diffusion. The texture illustrated here is that of the "perpendicular feldspar rock" of the Skaergaard intrusion, but similar textures are found in some olivine-rich gabbros, such as the *harrisitic* rocks of Rhum, Scotland. (d) Quenched siliceous liquids rarely nucleate large crystals but may have small microlites. The *perlitic* texture common to many rhyolites erupted in a wet environment probably forms as a result of hydration, expansion, and cracking of glass after it cooled.

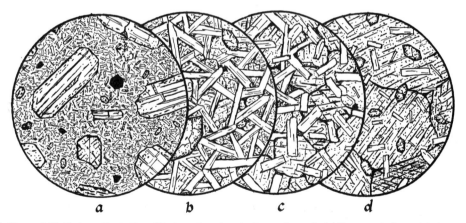

Figure 4-21 Textures illustrating differing rates of nucleation and growth. (a) Phenocrysts formed during an early period of crystallization, usually before eruption, give rocks a *porphyritic* texture. (b) If the principal groundmass minerals grow from many nuclei and at similar rates, the mafic minerals tend to occupy the spaces between plagioclase laths and give the rock an *intergranular* texture. (c) If pyroxene grows from a smaller number of nuclei and partly enclose the plagioclase crystals, as it does in many dolerites, the texture is said to be *ophitic*. (d) *Poikilitic* textures represent an extreme case in which large crystals grow through extensive parts of the rock and completely enclose smaller grains.

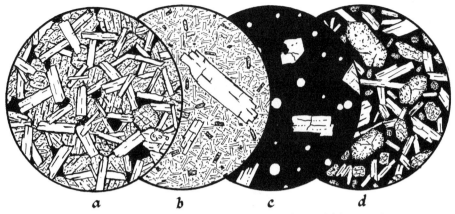

Figure 4-22 Textures resulting from differing proportions of residual glass. (a) *Intersertal* textures are common in basaltic lavas in which small pools of glass remain between crystals of plagioclase, pyroxene, and olivine. (b) The groundmasses of many andesites and dacites are composed of clear glass with many randomly oriented small crystals. Such textures are called *hyalopilitic*. (c) *Vitrophyric* textures in which large phenocrysts are scattered through a glassy groundmass are common in pyroclastic rocks. In this example, the glass has small gas cavities, which make the rock *vesicular*. With larger proportions of crystals, vitrophyric rocks grade into those with *hyaloophitic* textures (d), in which glass partly or completely encloses the crystals.

c. *The amount of residual glass* (**Fig. 4-22**) may vary between wide extremes. If small pools of glass occupy the interstices between crystals, the texture is said to be *intersertal*, but if the glass is continuous through the groundmass, it produces a *hyalophitic* texture, in which glass partly surrounds individual crystals. *Hyalopilitic* textures are common in siliceous lavas, such as dacites and trachytes. They are distinguished by a glassy matrix crowded with many small crystals, but this term would not be applied to a rock with only glass in its groundmass; instead, a rock with large phenocrysts suspended in a matrix that is wholly glassy, or very nearly so, is said to be *vitrophyric*.

d. *Flow structures* in highly viscous magmas that were moving at the time they solidified may align the crystals in wavy or planar layers (**Fig. 4-23**). Many andesitic and dacitic lavas have a *pilotaxitic* groundmass crowded with small, crudely aligned crystals. The alignment of feldspars may be so pronounced that they give the rock a felted appearance that characterizes *trachytic* textures. If the rock has much glass, however, flow-banding may develop along planes of shear, and if vesicles have formed, they tend to be stretched parallel to the direction of flow.

e. *The relative development of crystal faces and mutual interference of grains* (**Fig. 4-24**) results in a variety of textures ranging between those

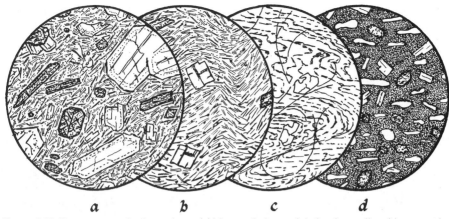

Figure 4-23 Flow structures in viscous lavas. (a) Many andesites and dacites have *pilotaxitic* textures in which the groundmass consists of crudely aligned small crystals that swirl around larger phenocrysts. (b) In *trachytic* rocks, the alignment of small crystals is stronger and they give the entire rock a felted appearance. (c) *Flow-banding* in rhyolite results from concentration of tiny crystals or vesicles along planes of shear and deformation. (d) The vesicles of basaltic lavas may be stretched if the lava is too viscous for them to maintain a normal spherical shape.

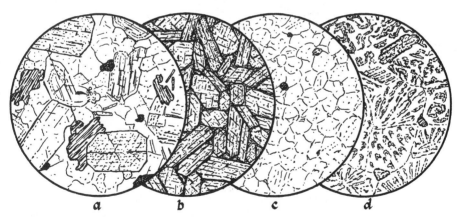

Figure 4-24 Textures resulting from differences of crystal form. (a) Most granitic rocks have *hypidiomorphic granular* textures in which some crystals have euheral forms, some poorly developed faces, and others no crystal form at all. (b) If most of the crystals are euhedral, the rock is said to be *panidiomorphic*, but (c) if crystal faces are rare the texture is *allotriomorphic*, as it is, for example, in many aplites. (d) Rapid eutectic crystallization of siliceous liquids produces *granophyric* textures characterized by intergrowths of quartz and alkali feldspars. The patterns of intergrowth are crudely controlled by crystallographic features of the host crystal, and if this produces a coarse texture with strong alignment of the enclosed crystals, the texture is said to be *graphic*.

that are *panidiomorphic* and have a large proportion of euhedral grains, through *hypidiomorphic* rocks, in which some minerals form well-developed crystals while others do not, to *allotriomorphic* textures in which few if any of the crystals have euhedral outlines. Rocks with textures that have adjusted during prolonged slow cooling tend to develop nearly equidimensional anhedral grains with boundaries that intersect at angles approaching 120 degrees. Such rocks contain few minerals that preserve their crystal faces, but some phases, particularly apatite, zircon, sphene, and other accessory minerals, have well-developed forms, just as they do in metamorphic rocks. A particular feature of intergrowths developed from rapid eutectic crystallization is *granophyric* or *myrmekitic* texture, in which one mineral forms wormy or cuneiform blebs within another.

    f. *The degree of crystal–liquid fractionation* (**Fig. 4-25**) during slow crystallization has an important influence on the modal proportions, compositions, and textural relationships of individual minerals. These effects are best seen in *cumulate* rocks formed from crystals that *accumulate* near the cooling margins of a crystallizing magma. The

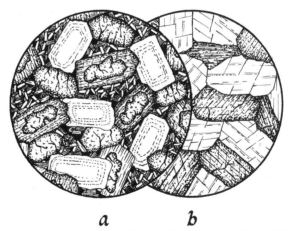

a   *b*

Figure 4-25 The effects of separation or re-equilibration of the intercrystalline liquid between early-crystallizing minerals can be thought of in terms of two hypothetical extremes. With no separation (a), the liquid between the liquidus minerals is "trapped" and, as it cools, precipitates minerals with compositions different from the initial ones. The resulting rock is referred to as an *orthocumulate*. If, as in (b), continued exchange between the interstitial liquid and a much larger reservoir of magma makes it possible for the primary precipitates to continue to grow without changing composition, components excluded from these crystals are expelled and are not precipitated by later crystallization as they would have been if trapped. The final assemblage consists of only the primary precipitates, unchanged from their liquidus compositions, and is known as an *adcumulate*.

original definition of these rocks assumed that they resulted from gravitational segregation, but similar textural relations are observed in rocks that appear to have crystallized in place.

The main types of cumulates are distinguished by the amount of growth of early-formed crystals, or *primary precipitates*, before the appearance of later crystals that grew from the *interstitial* or *"trapped" liquid*. If minerals nucleate and grow from the trapped liquid, the rock is said to be an *orthocumulate*. If the first-formed crystals continue to grow with a constant composition through efficient equilibration with the main reservoir of liquid, the residual components are effectively excluded and the final assemblage, known as an *adcumulate*, contains few if any crystals precipitated from a trapped liquid. In this latter case, in which diffusion keeps up with crystallization and the concentration gradient between the interstitial liquid and the main reservoir of magma is weak, the interstitial liquid may never nucleate new phases. The result is a rock consisting of unzoned crystals of one or more minerals. Intermediate cases may result from a combination of adcumulus enlargement, zoning of early crystals, and nucleation and growth of later ones through the interstices; this type of crystallization produces *mesocumulates* that may differ widely in the proportions and grain sizes of their crystals.

Many cumulate rocks, especially those of gabbroic and syenitic compositions, have pronounced layering, normally parallel to the floor and walls of the intrusion. The layering may result from differences of the proportions of dark and light minerals or from differences of grain size. Various kinds of layering have been attributed to different proposed mechanisms. Examples are discussed in Chapter 6.

Tabular or prismatic minerals having a preferred orientation may give a rock *igneous lamination* (**Fig 4-26a**). The orientation is usually parallel to the cooling surface of the floor, wall, or roof, but under certain conditions of cooling and strong supersaturation, crystals may grow with a marked elongation perpendicular to the solidification front. The name *crescumulate* has been coined for textures of this kind (Fig. 4-26b). Long slender crystals of olivine, plagioclase, or pyroxene may grow like blades of grass on the floor or walls of the intrusion. Crescumulate textures are most common in mafic plutonic rocks, but they have also been found in the margins of granitic plutons.

Growth of slender crystals normal to the cooling surface has been attributed to a phenomenon known as *constitutional supercooling*. If the rate of cooling is rapid compared to that of diffusion, a condition can result in which liquid in a zone adjacent to the crystals becomes supersat-

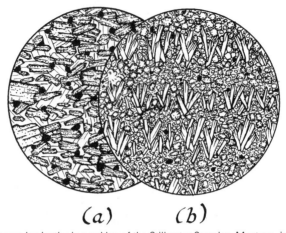

(a)   (b)

Figure 4-26 (a) *Igneous lamination* in a gabbro of the Stillwater Complex, Montana, is the result of pyroxene having a preferred orientation in a plane of layering. Elongated crystals rarely have a linear orientation within the same plane. (b) Crystals may also have a preferred orientation perpendicular to the plane of the layering. In this case, the elongated, subparallel crystals of plagioclase give the rock a *crescumulate* texture. Layers of crystals oriented in this way are referred to as "Willow Lake type layering."

urated because the profile of temperature and composition falls below the liquidus curve. The mechanism is explained more fully in the legend of **Figure 4-27.**

A Word of Caution

One final point before going on: in dealing with these mineral assemblages and textures, it is well to remember that what we see or measure in an igneous rock is the final product of a long and complex process of cooling and crystallization under conditions that can only be guessed by indirect means. No magma has ever been observed while crystallizing under conditions of slow cooling and equilibration. Laboratory experiments are too brief, and even the deep lava lakes described earlier crystallize far too rapidly to provide a true analog of processes taking place in the crust or mantle. One cannot always be certain, therefore, which of the textural features of plutonic rocks can be attributed to crystallization from a magmatic liquid and which reflect the later cooling history of an essentially solid rock. Recent work on plutonic intrusions has revealed that many rocks that were once interpreted as unaltered products of magmatic crystallization have, in fact, recrystallized and no longer retain their original compositions or textures.

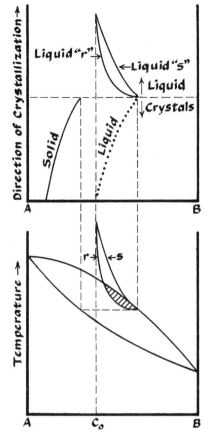

Figure 4-27 If a front of crystallization advances into a liquid with an initial composition C_0 and crystal are richer in the high-temperature component A than the coexisting liquid, the components will have a distribution similar to that shown schematically in the upper diagram. The liquid immediately adjacent to the crystals is enriched in B relative to the solid and the main magma reservoir, and a compositional gradient extends away from the crystals into the hotter liquid above. If the front advances rapidly, the gradient is steep, as in curve "r," but with slower crystallization diffusion reduces the gradient to a more gentle form like "s." In the lower diagram, the temperature in the liquid is shown in relationship to liquidus and solidus curves. Under conditions of rapid advance, the compositional gradient may be steep enough to cause the concentration of B to fall below the saturation level defined by the liquidus curve, and an unstable zone of supersaturation (shown by diagonal ruling) can cause the liquid to be "constitutionally supercooled." Any crystal that extends into this zone provides a site for rapid growth in the direction normal to the main crystal–liquid boundary. With slower crystallization (curve "s"), the concentration of B does not fall below the saturation level, and crystal growth is restricted to a narrow zone parallel to the front.

Selected References

Cashman, K. V., 1990, Textural constraints on the kinetics of crystallization of igneous rocks, *Modern methods of igneous petrology: Understanding magmatic processes*, J. Nicholls and J. K. Russell, eds., *Rev. in Mineral.* 24:259–314. An excellent review of the origins and interpretations of igneous textures with many well-documented examples.

Hargraves, R. B., ed., 1980, *Physics of Magmatic Processes*. Princeton, NJ: Princeton University Press, 585 p. Contains especially useful chapters on the theory and experimental studies of crystal growth.

MacKenzie, W. S., C. H. Donaldson, and C. Guilford, 1982, *Atlas of Igneous Rocks and Their Textures*. New South Wales, Australia, Halstead Press, 148 p. A magnificently illustrated reference on the textures and mineral assemblages of igneous rocks. Although it is purely descriptive and offers no interpretations of textural relations or paragenetic associations, the book is very useful for identifying the petrographic features of all of the common types of volcanic and plutonic rocks.

5 Magmatic Differentiation: Mechanisms and Effects

When Charles Darwin visited the Galapagos Islands in 1835, he observed lavas flows in which many of the crystals had been segregated by gravity. He commented that removal of these crystals could bring about a change in the composition of the remaining magma and that such a process could account for the diversity of igneous rocks. This simple observation was the first recorded statment of the principle of magmatic differntiation by crystal fractionation. We now recognize a number of mechanisms by which magmas differentiate, but they all have the two essential features that Darwin recognized: a chemical difference between two discrete phases and a physical process by which they are segregated. The first determines the *trend* of magmatic differentiation, the second the *degree* to which that trend can evolve. We now use the term *fractionation* for the process by which different phases are mechanically separated, but the word is not *synonymous* with *differentiation* because it is only one element of a dual process.

■ Mechanisms of Differentiation

It will be recalled that, by definition, a phase has specific physical properties by which it can be mechanically separated. All processes of mechanical segregation depend on contrasts of these physical properties—density, viscosity, diffusivity, or geometric form—and derive their driving force from the thermal or mechanical energy of the system. They can be divided into several types according to the nature of the phases being fractionated.

Crystal–Liquid Fractionation

The compositions of igneous minerals are much simpler than those of the natural liquids from which they form. They consist of a relatively small number of essential constituents, most of which have proportionately greater concentrations in the crystal than in the liquid. Because of these differences, elements that have a high concentration in a crystal are depleted from the liquid in which it grows, whereas those that are excluded become concentrated in the diminishing volume of residual liquid. The former are referred to as *included* or *compatible* elements, whereas the latter are said to be *excluded* or *incompatible* elements. If the crystals and liquid are separated, the remaining liquid can continue to crystallize as a rock with a composition quite different from that of the original liquid. This, in essence, is the basic principle of crystal fractionation: separation of crystals from a magma depletes those components having greater concentrations in the crystals than the liquid and enriches the elements that are left in the reduced volume.

Consider again the Hawaiian lava lake mentioned earlier and the compositional changes that developed in the liquid as it crystallized (**Fig. 5-1**). As crystals grew in its cooling margins, they extracted certain components from the magma while leaving others to accumulate in the remaining liquid, and in this

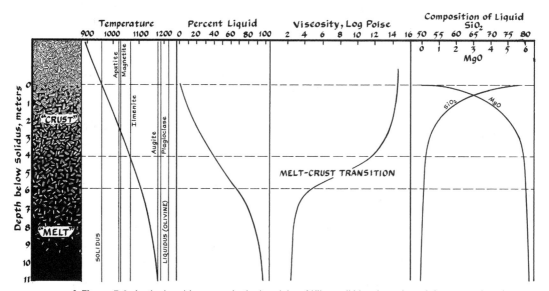

Figure 5-1 As the basaltic magma in the lava lake of Kilauea Iki lost heat through its crust, a broad zone of solidification extended downward into the liquid interior. It advanced more rapidly than small crystals could sink, and thus, there was little or no crystal fractionation by crystal settling. The viscosity of the magma increased most sharply when the proportion of liquid declined below about 40 percent, and the crust took on the properties of solid rock. At the same time, the composition of the liquid began to change at an accelerating rate. Most of the increase of SiO₂ and decrease of MgO, for example, occurred in the last 10 to 20 percent of crystallization. At this stage, the differentiated liquid was too far from the solidification front for diffusion to be an effective mode of mass transfer between interstitial liquid and the main reservoir below. As a result, the trapped liquid crystallized in place, and the bulk composition of the final rock remained close to that of the original magma. (Based on a summary of the crystallization behavior of three Hawaiian lava lakes by T. L. Wright, D. L. Peck, and H. R. Shaw, 1976, *Amer. Geoph. Union Geophysical Monograph* 19:375–392, and analytical data obtained from Kilauea Iki by W. C. Luth, personal communication.) (Viscosities are from measurements by T. Murase.)

way the liquid close to the growing crystals took on a composition different from that of the hotter, still-liquid interior. These differences resulted in compositional gradients in the liquid, and elements removed by the crystals diffused down these gradients to feed continued growth, whereas rejected ones diffused in the opposite direction. The slower the diffusion of these components and the more rapid the growth of the crystal, the greater the compositional contrast between the liquid near the crystal–liquid interface and the main reservoir of liquid some distance away. Unless the crystals and altered liquid are mechanically separated, the bulk composition of the assemblage remains unchanged.

This is what was happening in the lava lake. As in most small bodies of magma erupted at the surface or intruded at shallow levels of the crust, crystallization was so rapid and diffusion so slow that the effects of differentiation were confined to the late liquid trapped in interstices between early-formed crystals. The crystals became zoned, and the remaining liquid was enriched, first in iron and titanium and then in silica, alkalies, and phosphorus. When these residual components reached critical concentrations, new minerals were formed. Magnetite and apatite, for example, crystallized when the remaining liquid became strongly enriched in iron and phosphorus. Had the liquid been separated (or fractionated) at this stage, it would have crystallized to form a rock very different from its parent. Because it was not, there was little or no change in the bulk composition.

Because of the limitations imposed by slow diffusion, differentiation is greatly enhanced if the growing crystals are brought into contact with fresh, undepleted liquid. If they sink, float, or otherwise move with respect to the liquid as they grow, they can extract the elements essential to their growth from a larger volume of magma than would be possible by diffusion alone. The same effect can result from flow of liquid over static crystals growing on the walls of a magma chamber. Regardless of which is moving, crystals or liquid, the efficiency of fractionation is greatly enhanced by any relative motion of the two phases.

Take, for example, a trace element such as chromium, which, in basaltic magmas, seldom has a concentration greater than about 100 parts per million. This metal is mined from large gabbroic intrusions in which layers of almost pure chromite have thicknesses of several tens of centimeters. This amount of chromium would have to be extracted from a volume of liquid at least 10,000 times greater than that of the layer in which it now resides. It could not have done this by any known process of diffusion. Because the diffusivity of heat (about 4×10^{-3} cm^2 sec^{-1}) is several orders of magnitude greater than chemical diffusivity (10^{-6} to 10^{-8} cm^2 sec^{-1}), the time required to concentrate chromium by chemical diffusion alone would be far greater than the time required for the intrusion to cool and crystallize. It seems clear, therefore, that crystals can only concentrate a component to such a remarkable degree by moving through a large volume of magma or, alternatively, by a similarly large amount of magma flowing over the zone of crystallization. In either case, the crystallizing phase must be exposed to enough undepleted liquid to extract the element in the relatively short time that crystals grow.

Crystal Settling

An obvious way of achieving such a concentration would be for crystals nucleated under the roof of the intrusion to sink through the entire thickness of melt and scavenge elements as they descend. Such a mechanism involving relative motion of crystals and liquid is strongly dependent on the viscosity of the magma and the relative densities of the phases. Settling rates, U, of these crystals are governed by the simplified form of Stokes' Law from Equation 2-6.

$$U = \frac{2\Delta\rho g r^2}{9\eta} \qquad (5\text{-}1)$$

in which $\Delta\rho$ is the density difference between the liquid and crystals, r is the radius of the crystal, and η is the viscosity of the liquid. Settling of newly formed crystals must be very slow because of their small sizes, but as they continue to grow, their velocities increase with the square of the radius. At the same time, cooling results in an increase in the viscosity of the liquid and in the volumetric proportion of crystals it contains. Depending on their rates of sinking relative to the rate of advance of the front of solidification and increasing viscosity, some crystals may escape, whereas others may be trapped. The crystals most likely to be segregated by gravity are those having a large density contrast with the liquid and a large size at an early stage of cooling. In the case of the lava lake, the only crystals to settle were phenocrysts of olivine that were suspended in the magma when it erupted; those nucleated under the thickening crust were held in place.

If most crystals that nucleate and grow under the roof of a cooling intrusion are locked into the advancing zone of solidification, some other mechanism of fractionation is required to separate the crystals and evolving liquid. Recent experiments (**Fig. 5-2**) show that growing crystals, particularly plagioclase, tend to form an interconnected network from which the liquid can easily be separated by compaction or convective exchange with the main reservoir. If a magma crystallizing under the roof becomes enriched in iron and titanium and hence denser (as is the case with most basic tholeiitic magmas), the liquid may drain from the network of crystals and be replaced by a less dense, more primitive one. The reverse can occur on the floor if the interstitial liquid becomes less dense than the overlying magma or if it is

Figure 5-2 In an experiment designed to test the ability of liquids to separate from a mesh of crystals, a sample of basalt, 1 cm^3 in size, was heated until it was approximately 70-percent liquid. As seen here, the melt drained from the crystals with no other force but gravity, and the framework of interlocking crystals maintained its original form. (From A. R. Philpotts, et al., 1998, *Nature* 395:343–346.)

expelled by compaction of the crystals under their own weight, just as water is squeezed from a thickening pile of compacting sediments.

The chemical variations produced by such a process involving separation of an interstitial liquid from a mush of crystals growing in situ differ from those caused by crystal settling. The compositions of both the liquid and crystals are not uniform but vary with the extent of crystallization at any given level. The liquid returned to the main reservoir is therefore more differentiated than one produced by crystal settling.

Flow Segregation

Segregation can also occur when a magma containing suspended crystals flows along the walls of a dike or a convecting pluton. As noted in Chapter 2, the velocity gradient of a viscous fluid is steepest at its contact and declines to a nearly uniform velocity near the center. The shearing produced by a gradient of this kind results in a weak force, usually referred to as *dispersive shear pressure*, that tends to drive crystals out of the zone of maximum shear next to the walls and into the interior of the flow. The result is to redistribute suspended crystals in such a way that their concentration is an inverse function of the velocity gradient. Effects of this kind can be seen in porphyritic dikes and, less commonly, in the basal zones of sills and lava flows.

Before this mechanism was recognized, dikes with concentrations of large crystals in their centers were thought to be formed by slower cooling and crystallization of the liquid in the interior of the dike, but if this were the case, the bulk composition of the porphyritic interior would be the same as that of the fine-grained margins, and this is rarely the case.

Similar concentrations of crystals in the interior of dikes can result from intrusion of a compositionally graded magma. The first liquid to enter a fissure and solidify against its walls may come from the upper, crystal-free part of a deeper reservoir, whereas the crystal-rich interior could be from a lower level of the same reservoir. We shall see examples of this in later chapters.

In Situ Crystallization and Compaction

Crystals growing slowly in situ maintain a narrow range of composition because the components of the liquid have time to diffuse down concentration gradients between the crystal faces and the main reservoir of magma. If the liquid becomes gravitationally unstable, it can also be replenished by convective exchange. When growing slowly on a floor, crystals can compact under their own weight so that most of the interstitial liquid is mechanically expelled and returned to the overlying reservoir. In the early stages, the rate of compaction is governed primarily by the rate at which the viscous liquid can flow upward through the interconnected spaces between crystals. At first, the crystals are able to align themselves mechanically, but after they form a self-supporting network, further compaction is achieved by deformation or recrystallization. Crystals are dissolved where stresses are concentrated at points of contact, and the material is redeposited in the remaining space.

Because the expelled liquid comes from a wide range of depths and temperatures, its composition may be more evolved than it would be if the crystals were growing in a dispersed suspension (**Fig. 5-3**). If the temperature gradient is very gentle, however, the interstitial liquid will be expelled long before the temperature reaches the solidus, and the crystals and expelled liquid will have a more restricted range of composition.

Partial Melting

The processes of melting are in many ways the reverse of solidification. The compositions of liquids vary as melting advances, much as they do with crystallization, and they differ depending on whether the crystals are zoned or uniform and whether the liquid is produced over a range of temperatures

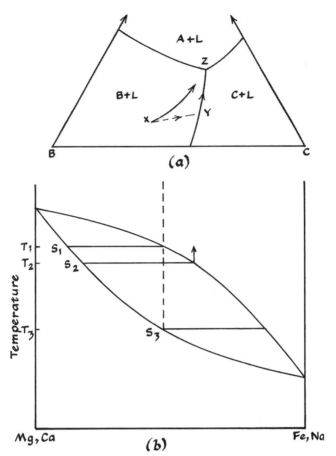

Figure 5-3 (a) As a zone of cooling liquid crystallizes in situ, it evolves through the full range between the liquidus and solidus. As it does so, it crystallizes minerals with which the main magma is not yet saturated, and if this liquid is expelled, it will change the trend of the main magma. Instead of following the normal path for fractional crystallization, X–Y–Z, it will take a course such as that shown by the curved solid line. (Adapted from C. H. Langmuir, 1989, *Nature* 340:199–205.) (b) If the crystallizing magma is analogous to a solid-solution system and is compacting as it slowly cools in a very gentle temperature gradient, the interstitial expelled liquid will be expelled long before the temperature reaches the solidus at T_3. If an intermediate liquid, such as one at T_2, is expelled, it may react with the overlying crystals as it rises, but it cannot deposit rims on them because it is above its liquidus temperature. As a result, the range of compositions of crystals at any given level is greatly restricted.

(**Fig. 5-4a**). They also differ according to the frequency with which the liquid is removed. In a multicomponent system, such as the one shown schematically in Figure 5-4b, the first liquid has the composition of the eutectic; if melting advances without removal of the liquid, the composition leaves the eutectic when one or the crystalline phases is exhausted, and subsequent liquids will evolve, first up the cotectic and then when the next mineral is consumed, into the field of the last remaining crystalline phase. The course

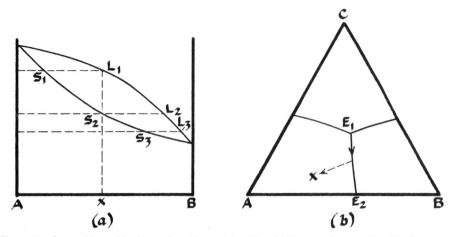

Figure 5-4 Compositions of liquids produced by partial melting. (a) If a rock of composition X is formed by equilibrium crystallization, all crystals would have composition S_2, and the first liquid to be produced by melting would be L_2. If the rock contained crystals zoned from S_1 to S_3, the first melt would have the composition L_3. (b) Melting of a rock of composition X would begin at eutectic E_1, and assuming the liquid is not removed, the composition would leave the eutectic when C is consumed and follow the cotectic until B is completely melted. Thereafter, the liquid crosses the field of A until it reaches the starting composition, X. If the liquid is removed as rapidly as it forms, it will start at E_1, but the next composition to form after C disappears will be at E_2. After B is consumed the next liquid will have the composition of pure A.

of melting is very different if the liquid is withdrawn as rapidly as it is formed. The first liquid is still that of the eutectic composition, but when the first crystalline phase is exhausted, the remaining crystals constitute a system of only two solid phases, and the next liquid will be that of the binary eutectic. No intermediate compositions will be produced; instead, there will be a large step in both composition and temperature. Again, if the liquid is removed as melting proceeds at the binary eutectic, melting will stop temporarily when the next crystalline phase is exhausted and will not resume until the temperature reaches the melting temperature of the last remaining crystalline component. Thus, fractional melting, as this latter process is called, would be expected to yield discontinuous rather than smoothly changing compositions and would have steps of time and temperature separating successive stages of melting.

Zone Refining

A process known to metallurgists as *zone refining* is used as an industrial process to purify certain types of alloys. A bar of impure metal is passed through a heating element so that a fraction of the metal is melted, and as the bar slowly moves, this zone of melt passes from one end to the other carrying with it elements that are preferentially concentrated in the liquid. With repeated passes, these elements are swept to one end of the bar. One can visualize how a similar mechanism could lead to compositional changes in magmas.

Suppose that a lens-like zone of melt is rising through the mantle in such a way that crystallization at its base is accompanied by melting of its roof. This could happen if volatile components were more concentrated at the top or even if the magma has a nearly constant composition provided the pressure difference between its roof and floor were enough to make the liquidus temperature at the base of the column a degree or so higher than that at the top. Heat liberated by crystallization at the floor can be transferred by conduction or convection to the top of the melt zone where it provides the heat necessary to melt the roof. In this way, a zone of melting would pass upward with little net loss of heat. As it does so, elements that are preferentially concentrated in the low-melting fraction would tend to become progressively enriched in the pocket of rising melt.

The concentration of a component in the liquid, C_L, is governed by this equation:

$$\frac{C_L}{C_0} = \frac{1}{K} - \left(\frac{1}{K} - 1\right)e^{-Kn} \tag{5-2}$$

in which C_0 is the initial bulk concentration, K the bulk distribution coefficient (i.e., the ratio of the concentration in the solid to that in the liquid), and n the ratio of the mass of solid processed to the mass of liquid. As n becomes large, C_L/C_0 approaches a constant value of $1/K$. This is a limiting value at which the concentration of the component ceases to increase because the amount entering the liquid from melting at the top is the same as that removed by the crystals at the base.

In natural conditions, the amount of liquid in the remelted zone will increase as the zone advances and collects excluded components, such as H_2O, which have the effect of lowering the melting temperature and increasing the proportion of liquid. This tends to increase the effectiveness of the process because the liquid does not become saturated as quickly as it would be at constant volume.

The efficiency of zone refining as a mechanism of differentiation is strongly dependent on the relative rates of diffusion and motion of the zone of liquid. Unless the zone of melting moves very slowly or unless the process is repeated many times, the degree of segregation and differentiation is likely to be small. The process would be most effective at depths where the rocks are already at high temperatures and little heat is consumed in raising them to their melting temperature.

■ Liquid–Liquid Fractionation

Most differentiation in magmatic intrusions takes place in a narrow zone, or *boundary layer*, at or near the margins where the magma interacts thermally and chemically with its surroundings. Because the liquids in this zone are characterized by gradients of temperature, crystallinity, and composition, their mechanical instability is conducive to several possible processes of differentiation.

Boundary Layer Fractionation

A potentially important mechanism is one in which a layer adjacent to the wall of an intrusion becomes lighter or heavier than the main mass of magma and is segregated by gravity-driven flow to the roof or floor (Fig. 2-15). Such a process could result when crystallization, melting, or absorption of volatiles at the wall changes the composition of the adjacent liquid and causes its density to differ from that of the interior. The simplest case is that in which the boundary layer absorbs water from its surroundings so that the density of the liquid is lowered by an amount that exceeds the effect of falling temperature and thermal contraction. A buoyant, water-rich layer may then rise along the wall and collect under the roof. The same effect could result from any other process, such as crystallization or melting, that changes the relative concentration of heavy components and causes the boundary layer to be gravitationally segregated from the main body of magma in the interior.

A process of this kind may contribute to differentiation of shallow intrusions of magma in which dense components tend to be preferentially concentrated in the crystalline phases, thus making the residual liquid lighter. Andesitic magma, for example, is less dense than basalt, even though its temperature is lower, and rhyolitic magma, because it is poor in iron and other dense components, is lightest of all. Crystallization against the wall of such an intrusion could produce a thin layer of compositionally different liquid that may flow along the contact to collect at the roof of the intrusion and eventually produce a compositionally zoned column of magma.

Immiscibility

Three types of magmatic systems are known to have compositional ranges in which two immiscible liquids may separate under geologically reasonable conditions. Sulfide liquids may separate from mafic silicate magmas, even at low concentrations of sulfur; highly alkaline magmas rich in CO_2 can split into two immiscible fractions, one rich in alkalies and silica and another rich in CO_3, and very iron-rich tholeiitic magmas may form two separate liquids, a felsic one rich in SiO_2 and a mafic one rich in iron. We shall encounter examples of each of these in later chapters.

The densities of two coexisting immiscible liquids may be sufficiently different for them to separate gravitationally, just as oil does when it rises and floats on water. Sulfide liquids, rich in iron and other heavy metals, may separate as small droplets and settle through basic magmas to collect on the floor. As they sink, they scavenge the chalcophile metals, such as iron, copper, nickel, and cobalt, and the platinum group for which they have a strong affinity, and form rich concentrations of these elements in economic ore deposits.

The felsic liquids of tholeiitic systems have much lower densities than the iron-rich basic magmas from which they separate and could have a strong tendency to rise and collect in the upper levels of a magmatic reservoir. Much depends, however, on the stage of crystallization at which immiscibility develops. Often the magma has crystallized to a viscous mush of entangled crystals so that the felsic liquid, when it finally forms, does not separate as a continuous mass but remains as small, dispersed droplets locked in the interstices between early-crystallizing minerals.

Immiscibility may be an attractive explanation for closely associated liquids of contrasting compositions, but it is not the only one. Contamination, mixing, and a variety of crystallization phenomena can produce textural features suggestive of immiscible globules of one liquid in another. On closer scrutiny, many clot-like inhomogeneities that were originally thought to be immiscible have been found to have compatible and completely miscible compositions. A simple but rigorous test can be applied to these associations. Because two immiscible liquids are in equilibrium with each other, the crystals in one liquid must also be in equilibrium with the other as well. Therefore, if the two proposed liquid compositions have precipitated different crystalline phases, they could not have been an immiscible pair and probably have another origin.

This rule does not apply to elements that are not essential components of the crystallizing minerals. Certain minor or trace elements that are partitioned in different concentrations between two coexisting liquids can be very diagnostic. In the case of tholeiitic magmas, for example, most of the elements concentrated in felsic immiscible liquids are the same that are residually enriched by normal crystal fractionation. One notable exception is phosphorus. During normal crystallization, this element accumulates in felsic liquids and increases in abundance until apatite begins to crystallize. It is also concentrated in the first liquids produced by partial melting. In immiscible systems, however, it is strongly partitioned into the iron-rich, silica-poor liquid and is relatively depleted from the coexisting felsic member. In this way, the distribution of phosphorus provides a useful way of distinguishing immiscible siliceous liquids from those evolved by crystal fractionation or partial melting.

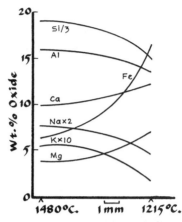

Figure 5-5 The distribution of the components of a basaltic liquid changes in response to a strong thermal gradient in the manner shown in this experimental example. The elements enriched at the hot end of the sample (SiO_2, Al_2O_3, and alkalies) are the same that are concentrated in low-temperature liquids that evolve by crystal fractionation. The sample was held in the temperature gradient at 10 kilobars pressure for 137 hours. (After D. Walker and S. E. DeLong, 1982, *Contr. Mineral. Petrol.* 79:231–240.)

The Soret Effect

At least passing mention should be made of another process of liquid fractionation that could be important in magmas subjected to very strong temperature gradients. Because of the differing effects of temperature on the chemical potential of the various components of magmas, a weak difference of free energy causes some elements to be more concentrated at the hot end of a temperature gradient, whereas others tend to migrate toward low temperatures. Laboratory experiments indicate that the components moving toward high temperatures include SiO_2, Al_2O_3, Na_2O, and K_2O, whereas those drawn toward the low-temperature end of the gradient would be FeO, MgO, TiO_2, and CaO (**Fig. 5-5**). Thus, the distribution is the opposite of what would be expected in gradients produced by fractional crystallization or melting. Although no clear evidence has yet been found to show that this effect is strong enough to be important on a large scale, the unusual distribution of elements it entails may account for a number of anomalous effects that are difficult to explain by more conventional mechanisms.

■ Assimilation

When one considers the great variety of crustal rocks and the large compositional differences between these rocks and the magmas rising through them, it would seem that assimilation would be a common phenomenon capable of producing a wide range of igneous compositions. Evidence that crustal material is caught up in magmas is not hard to find. Many igneous rocks contain *xenoliths* and *xenocrysts* (rocks and crystals of foreign origin) that come from fragments of the crust in various stages of assimilation. In order to understand how these contaminants affect the compositions of magmas, we must first examine the basic principles underlying the process of assimilation.

Consider first a simple hypothetical system, such as the one illustrated in **Figure 5-6**. Suppose that a magma of composition L is precipitating crystals of A and is suddenly contaminated with a rock consisting of crystalline B. Obviously, the liquid cannot be in equilibrium with both crystalline phases at the temperature T_1. This is apparent from the phase diagram, which shows that liquid L cannot be saturated with crystals of B at any temperature above the eutectic. We can also see from the phase rule that a system with two components (A and B) and three phases (liquid, A, and B) should have only one degree of freedom

$$F = C - P + 2 = 2 - 3 + 2 = 1$$

whereas the liquidus surface of the diagram (if we include the third dimension, pressure) has two. This tells us that one of the phases is unstable.

As long as the liquid is not yet saturated with B, it will react with the added crystals and dissolve them. The heat required to do this can be derived from crystallization of A, and the total effect will be to lower the temperature of the magma without leaving the liquidus of A. If the amount of B that

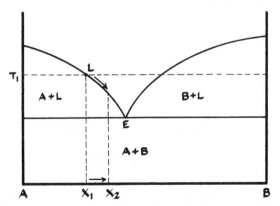

Figure 5-6 If a liquid L at temperature T_1 is saturated with crystals of A and crystals of B are added, the added crystals will dissolve because the liquid is not saturated with that phase. The heat required to dissolve the added B is gained by cooling and crystallization of more A, and the liquid evolves down its normal line of descent, as it would had no crystals of B been added. The bulk composition is shifted toward B, however, and the end result is increased proportions of B in the rock.

is added is small enough to be absorbed without completely crystallizing the liquid, the eventual result will be a liquid, somewhat more evolved but with a composition no different from what it would have reached in due course had assimilation never taken place. The only difference is that the bulk composition will be shifted in the direction of B by an amount proportional to the added component and can be calculated by the lever rule.

A slightly different situation is illustrated in **Figure 5-7**. Suppose that a liquid, L_1, is crystallizing an intermediate compound, AB, and with falling temperature approaches eutectic E_1. If it is contaminated with a small amount of crystalline B, the liquid, as it is not saturated with B, will react with the added crystals and gain the heat to do so by crystallizing additional AB. The crystals of B combine with A from the liquid to precipitate more AB, the only phase with which the liquid is saturated. The effect is to drive the composition of the liquid farther from AB so that, with falling temperature, it becomes more *unlike* the composition of the added component, B. Of course, if the amount of added crystalline B is large enough to make the bulk composition richer in B than AB, and the temperature remains nearly constant, all L_1 would be consumed in the reaction, and another liquid, L_2, would form in equilibrium with the crystals of AB.

The important point to note here is that one cannot assume that the major-element composition of an evolving magma will reflect the effect of

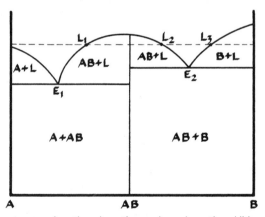

Figure 5-7 In some systems, such as the schematic one shown here, the addition of a contaminating phase may cause the liquid to be depleted in the components that have been added. If liquid L_1 is precipitating AB and crystals of B are added, the latter will react with the liquid, and A will be taken from the liquid to form additional AB. At the same time, the heat absorbed by the added crystals cools the liquid and causes it to evolve further down its line of descent, becoming poorer in B, the component that was added. Under conditions far from equilibrium, a liquid of composition L_3 may form around the crystals of B, whereas the liquid adjacent to nearby crystals of AB would have the composition L_2, and the liquid between these two crystalline phases could range between L_2 and L_3. Elsewhere in the same magma, the liquid in equilibrium with AB would still have the composition L_1.

an added component in any simple way. Its composition may actually become more depleted in the same components that were assimilated. As long as the liquid is in equilibrium with crystals it remains tied to the liquidus surface and does not diverge from the line of descent it would have followed had contamination never taken place.

This limitation applies only to equilibrium conditions, and if the magma is poorly mixed, local variations may develop in the liquid. In the case just discussed, for example, a liquid of composition L_3 may form adjacent to the added crystals of B and produce a range of liquid compositions grading toward another liquid, L_2, adjacent to the earlier crystals of AB. Thus, the local compositions of the liquid may differ from that of the main mass at some greater distance from the contaminant, and if the assemblage crystallizes before these disequilibrium assemblages disappear, the result will be a series of compositions.

The effects of different contaminants on the end products of a differentiated magma can best be seen by examining what happens to a simple solid-solution series under two conditions, one in which an added crystal has a composition that could have been produced earlier in the history of the liquid and another in which the phase would only crystallize from a more evolved liquid. Consider an example (**Fig. 5-8**) in which a cooling liquid, L_1, is currently precipitating crystals of S_1. If a small crystal of S_2 is added, we see that its composition is one that would have been precipitated at a higher

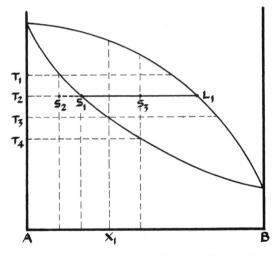

Figure 5-8 Assimilation of crystals, such as S_2, that would be in equilibrium with an earlier liquid of a solid-solution series tend to shift the bulk composition in the direction of the high-temperature component, A, and reduce the range of differentiated liquids that would evolve on cooling. Addition of crystals, such as S_3, that would only appear later has the opposite effect.

temperature, T_1, at an earlier stage of crystallization. To equilibrate with the present liquid, it cannot dissolve (because the liquid is already saturated with this composition) but must react and add to its composition the low-temperature component, B, in order to attain a composition in equilibrium with the liquid at the temperature T_2. In doing so, it draws B from the liquid and tends to make the liquid poorer in that component. In order to maintain the composition L_1, the liquid must precipitate more S_1 and give up an equivalent amount of heat. The bulk composition of the system, originally at X_1, is shifted slightly toward A so that the end result must be a smaller proportion of the low-temperature component, B, in the final product of crystallization. Hence, if the crystals continue to equilibrate as the temperature falls, the range of crystallization is reduced accordingly.

Addition of a crystal of composition S_3, richer in B than those in equilibrium with L_1, will have the opposite effect. It will be dissolved, and the heat of solution will be gained from crystallization of an additional amount of S_1. The added crystal shifts the bulk composition toward the low-temperature end of the series. The result will be that crystallization will produce a larger proportion of B-rich crystals, and the final temperature of equilibrium crystallization is lower. The amounts of crystalline material that dissolve and crystallize are determined by their relative heats of fusion and crystallization. **Figure 5-9** illustrates some of the textural relations that can result.

From these relations, we can draw the general conclusion that assimilation of material that would have crystallized at an earlier high-temperature stage of crystallization tends to reduce the proportion of differentiated products, whereas the addition of crystals with which the liquid has not yet become saturated will have the opposite effect; it will extend the range of evolution of the liquid and produce a larger proportion of low-temperature end-members. In terms of natural conditions, this means that a mafic magma that assimilates felsic crustal rocks will tend to differentiate with larger proportions of more evolved products than would the same magma if it had assimilated more mafic material or nothing at all. In both cases, however, the course of the liquid, in terms of its major-element composition and mineral phases, will tend to follow the same line of descent that would be possible without assimilation. Only the proportions of the products are changed.

Mixing of Magmas

Just as they assimilate rocks, magmas can also assimilate other liquids by mixing with them. Mixing is in some respects the opposite of fractionation. The latter process entails separation of an originally homogeneous magma

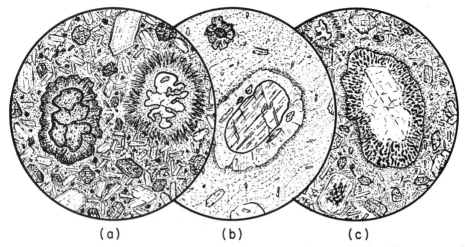

Figure 5-9 Textures of contaminated volcanic rocks. (a) Xenocrysts of quartz are jacketed by pyroxene, and nearby crystals of olivine have reaction rims of the same mineral. Specimen is from a basaltic lava erupted through granitic basement rocks in southeastern Guatemala. (b) Rounded and resorbed xenocrysts of plagioclase have reacted with a rhyolitic liquid and precipitated a rim of potassium feldspar. A small grain of olivine is rimmed with pyroxene. The specimen is from a group of mixed rhyolitic and basaltic lavas near the Gardiner River, Yellowstone National Park. (c) An alkali basalt from the vicinity of Gharian, Libya, contains xenocrysts of potassium feldspar and quartz derived from granitic rocks through which the magma rose. The feldspar is rimmed with a vermicular zone of glass.

to form rocks or liquids of differing compositions, whereas mixing combines two initially different liquids to produce a less differentiated product. It has been well documented in contrasting magmas mixed during eruptions from the same or closely connected vents (**Fig. 5-10**).

Mixing is most efficient when magmas of similar physical properties are turbulently intermingled. The density differences of strongly contrasting liquids inhibit mixing and tend to cause them to become gravitationally stratified with light liquids overlying denser ones. Any new magma intruded into such a body must seek a level at which it is gravitationally stable, and unless convection causes wholesale overturn and mixing, the new and old magmas may have very little interaction. Because chemical diffusion is so slow, thorough mixing requires strong, turbulent stirring so that the different liquids are physically associated on an intimate scale. Two mechanisms have been proposed as possible ways that this might come about.

If, as we noted earlier (Fig. 2-15), crystallization produces a differentiated liquid that is less dense than the original magma from which it is derived, a buoyant boundary layer may rise along the walls of the intrusion and accumulate under the roof. If the flow of this rising layer is turbulent, it may entrain magma of the interior and by mixing with the undifferentiated

Figure 5-10 A mixture of basalt and rhyolite in a bomb ejected from a cinder cone on the flank of Newberry Volcano, Oregon. The mixing must have taken place during rapid eruption from two magmatic bodies in close proximity at a shallow depth beneath the volcano.

liquid produce a range of intermediate compositions. This mechanism does not involve mixing of two independently derived magmas, but rather back-mixing of a derivative liquid with its parent.

If hot, dense, mafic magma lies beneath a cooler, lighter one and begins to crystallize and differentiate (**Fig. 5-11**), the resulting compositional change may lower its density until it is lighter than the overlying liquid and is able to rise and mix with it turbulently. Such a process has been proposed, for example, for mixing of a primitive tholeiitic magma that is initially rich in magnesium and iron and becomes lighter by precipitating olivine. This could lower the density of the magma enough to reverse the density relations. A variety of textural features of inhomogeneous lavas, such as crystals out of equilibrium with their ground mass or with other crystals in the same rock, point to some such process of mechanical mixing, but the physical plausibility of the process is still a matter of debate.

The effect of mixing of magmas, if it leads to homogenization, is to produce intermediate compositions and temperatures that will be determined by the proportions, compositions, and heat contents of the two original liquids. A series of mixed liquids will have a linear variation of their components between two end-members of the series. The range of intermediate compositions may not be unlimited, however, if either of the liquids contains abundant phenocrysts. Referring again to Figure 5-7, we see that an intermediate liquid cannot result from mixing of a liquid of composition L_2 with another, L_3, that contains crystals of B, unless the crystals are dissolved. If the mass

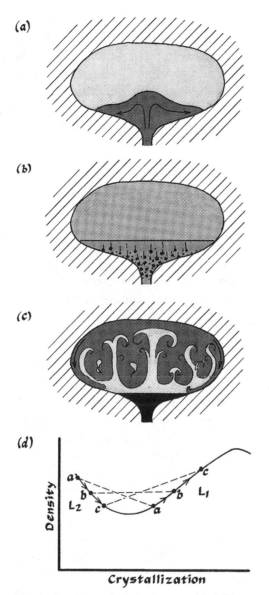

Figure 5-11 Because of the density minimum in the basic range of tholeiitic magmas (Fig. 2-12), it is possible that a new injection of hot basic magma L_2 could be denser than a slightly more evolved magma, L_1, that had already begun to differentiate in a shallow chamber. On intrusion, the denser new magma would pond at the base of the reservoir (a), but on cooling and crystallizing, the new magma could fractionate olivine and become lighter, whereas the original magma could fractionate olivine and plagioclase and evolve toward *a* more iron-rich and denser composition (b). If this process continues, the density relationships might be reversed (c), in which case the two liquids could overturn; if mixed turbulently, the resulting magma could have an intermediate composition and density. The schematic diagram (d) at the bottom shows the density relationships with tie-lines connecting the two magmas at each of the stages above. (Based on a model by R. S. J. Sparks, P. Meyer, and H. Sigurdson, 1980, *Earth & Planet. Sci. Ltrs.* 46:419–430.)

of crystals is large compared with the mass of L_2, the second liquid may be consumed by reaction with the crystals. Even if the two liquids are both crystallizing the same mineral but are separated by a thermal divide, as L_1 and L_2 would be, they cannot combine to form an intermediate composition but would crystallize the mineral with which they are saturated until one or the other of the two liquids is consumed. In either of these cases, if the mixed magmas cool before equilibration is complete, a variety of compositional and textural can result.

■ Liquid–Vapor Fractionation

The differences between the physical properties of gases and silicate liquids are much greater than those between solids and liquids. This, together with the great mobility of volatile components, can make fractionation in a separate vapor phase a potentially effective way of transferring components. A number of processes involving gases have been proposed, but much less is known about them than about the more conventional mechanisms discussed in the preceding sections.

Absorption from Wall-Rocks

Most of the water in shallow magmas is probably absorbed from crustal rocks when intrusions quickly heat the rocks through which they rise. With a sudden increase of temperature, the water in pores and fractures is vaporized and may even be superheated if it is heated faster than it can diffuse away from the contact zone. The greatly increased vapor pressures can also drive the water into the margins of the intrusion. Under these conditions of elevated temperature and pressure, water vapor can dissolve large amounts of silica, alkalies, and other components of wall-rocks, which it carries into the magma.

The effect is to decrease the density and viscosity of the contaminated magma at the margins of the intrusion. As a result, a thin, buoyant boundary layer may rise along the wall to accumulate at the top of the column. With time, this can produce a zone that is much richer, not just in water but in silica and other components derived from the wall rocks.

Gaseous Transfer

Because of the differing effects of pressure and temperature on the chemical potential of each individual component of a magma, these components cannot be uniformly distributed in any magmatic body that differs in temperature or pressure from one level to another. To attain a uniform chemical potential, large light ions with weak bonding energies tend to diffuse toward zones of low pressure and temperature near the top of a column of magma. In general, however, rates of diffusion are too slow for important compositional differences to develop by diffusion over distances of more than a meter or so. Heat diffuses so much faster than chemical components that a magma is more likely to crystallize before diffusion of volatile components can have large compositional effects. Some other mechanism of transport is required, such as segregation of a separate vapor phase that rises through the column of magma as bubbles of gas.

Figure 5-12 illustrates the equilibrium distribution of water in an isothermal column of magma of large vertical dimension. It shows how the concentration of a component, such as water, will tend to vary with depth. The solubility of water in a silicate melt is a direct function of the pressure of water; that is, as the pressure of water vapor increases, more water is driven into the melt and absorbed. However, the solubility of water also varies inversely with the total pressure on the melt. When the latter exceeds the

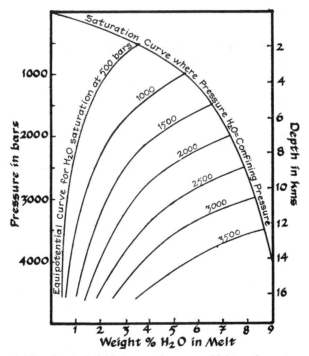

Figure 5-12 The solubility of water in felsic magmas increases with increasing water pressure, but at constant water pressure, increasing load pressure on the magma decreases the solubility in the manner illustrated by the equipotential curves for different water pressures. Because water pressure increases more slowly with depth than load pressure in the magma, a vertical column of magma will tend to have a greater concentration of water at the top. (After G. C. Kennedy, 1955, *Geol. Soc. Amer. Spec. Paper* 62: 489–504.)

partial pressure of water, it tends to "squeeze" the water out of solution. If the load pressure and pressure of water on the melt vary independently, the former normally increases more rapidly with depth. The density of magma is much greater than that of water, and with increasing depth, the pressure on the magma must increase more rapidly than would the pressure on a column of water of the same dimensions. For this reason, the ratio of water pressure to load pressure, and hence the water content, must decrease with depth. The result is an equilibrium distribution having the form shown by the lines of equal chemical potential in Figure 5-12.

It can be seen from these relations that a column of magma of initially uniform water content could be oversaturated at depth and undersaturated at its top. It could therefore nucleate bubbles at some tower level, and these bubbles could rise and be resorbed in upper levels of the column. Any components having a higher solubility in the gas than in the liquid would tend to be carried in the bubbles and transferred upward.

The importance of this effect, as well as that of absorption from wall-rocks, is seen in explosive volcanic eruptions in which the opening phases erupt magma that is richer in volatiles than later stages coming from deeper levels. Certain components, such as silica and alkalies, are concentrated along with volatiles at the top of the column, and their degree of enrichment is often a function of the length of repose between eruptions.

Volatiles in Late, Interstitial Liquids

During shallow crystallization, water and most other volatiles are excluded from the principal crystalline phases and tend to be concentrated in the remaining liquid. As crystallization advances and dissolved volatile components are concentrated in progressively smaller amounts of liquid, their concentrations will eventually reach saturation levels, so that a separate gas phase can evolve. This phenomenon is often referred to as "retrograde boiling" because the vapor phase evolves with cooling rather than with heating.

A separate gas phase is less likely to develop at deep levels where hydrous minerals are stable. Amphibole or mica, because they take in water, can absorb much of the effect of the increased concentration of that component in the liquid. We examine some of these relations in greater detail in Chapter 10.

■ Combined Mechanisms of Differentiation

In nature, many of the processes of differentiation we have been considering can take place concurrently. Rarely do magmas evolve by one mechanism alone during the full course of generation, intrusion, and cooling. The initial composition resulting from melting at a source in the mantle will be modified by cooling, decompression, and reaction with the rocks through which the liquid rises. After it is emplaced in a crustal reservoir, it may assimilate wall rocks as it cools and crystallizes and at periodic intervals part of the magma may be erupted and replaced by a new influx of fresh, undifferentiated magma. The individual contributions of all these different processes may be combined and superimposed on each other to produce magmas of complex, hybrid origins.

Although it is difficult to distinguish the individual effects of these different processes in a complex magma, one can assess the effects of some of the simpler combinations by use of certain diagnostic trace elements and geochemical ratios. References to some of the more commonly used models are given at the end of this chapter, and examples of their applications will be found in later chapters. These techniques provide ways to relate the members of a series of magmas in terms of combinations of as many as three or four different processes.

With so many possible mechanisms to draw on, one can usually find some combination that could account for the compositional relations of various groups of rocks, including some that are in fact unrelated. Because they are so poorly constrained, these schemes inspire little confidence.

■ Tracking Compositional Effects of Differentiation

The compositional changes of differentiating liquids, regardless of how they are produced, depend primarily on the manner in which individual components are partitioned between crystals and liquid during crystallization or melting. An element that is preferentially concentrated in fractionated crystals will be depleted in the remaining liquid, whereas one that is alien to the crystallizing phases will be enriched because it is concentrated in a diminishing amount of liquid. Thus, the degree of enrichment or depletion depends on the compositions and amounts of the phases being subtracted or added.

In a multisaturated magma, that is, one that is precipitating more than a single crystalline phase, the entire crystalline assemblage must be considered in order to determine whether an individual component is enriched or depleted. One cannot assume, for example, that because a particular element, such as aluminum, has a lower concentration in the liquid than in a fractionated mineral like plagioclase that the element will necessarily be depleted by removal of that mineral. The concentration of the element in the liquid may decline if plagioclase is the only mineral being fractionated, but if another mineral with little or no aluminum, such as olivine, is also being removed, the liquid may actually be enriched in that component. Thus, depending on the proportions of the crystallizing minerals, individual components can be either enriched or depleted. The effect can best be seen in variation diagrams representing the changing concentration of components in crystallizing liquids.

Variation Diagrams

It is often helpful to depict the changing abundances or ratios of elements graphically by plotting components against each other or against some measure of differentiation. A common technique is to plot selected components against another component, such as silica, that varies with differentiation. Such a plot, usually called a "Harker diagram" after Alfred Harker, the petrologist who first used it nearly a century ago, is useful for subalkaline igneous series that increase markedly in SiO_2. The variation of SiO_2 may be quite irregular, however. In some series, it actually declines with progressive differentiation, and in such cases other indices must be used.

Figure 5-13 is an example of a two-component Harker diagram. Assume a magma having an initial SiO_2 content of 50 percent and an Al_2O_3 content of 15 weight percent, as indicated by the point M_1. Crystallization of olivine with 45 percent SiO_2 and essentially no Al_2O_3, (point O) causes both the SiO_2 and Al_2O_3 contents of the residual liquid to increase along a trend indicated by a vector from M_1 directly away from O. The distance the liquid composition is displaced along this "control line" is a function of the fraction of the original liquid that crystallizes. Suppose, next, that plagioclase of composition P begins to crystallize when the liquid reaches point M_2. With removal of this new mineral, the course of the liquid must change and follow a new path that is the resultant of two vectors, one away from olivine and the other away from plagioclase. Depending on the proportions of the two crystallizing phases, Al_2O_3 may increase or decrease in the subsequent evolution of the liquid.

In multicomponent systems in which the compositions and proportions of fractionating phases are constantly changing, variations of individual components follow curves rather than straight lines. The slopes of curves change whenever a new phase begins to crystallize or another is no longer precipitated. In Figure 5-13a, for example, the change of slope for TiO_2 marks the beginning of crystallization of ilmenite, and the change of FeO + Fe_2O_3 corresponds to the appearance of magnetite. This pattern of curved lines differs from that produced by mixing two liquids (Fig. 5-13c). In the latter case, the abundances of the components are simple linear functions of the proportions of two end-members.

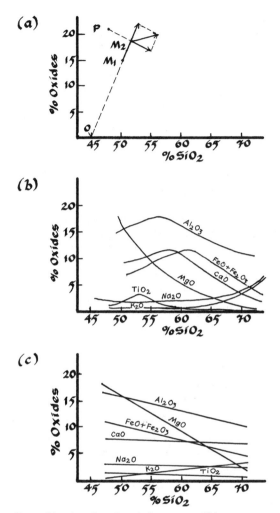

Figure 5-13 (a) The effects of fractionating minerals from a crystallizing magma can be illustrated by a variation diagram showing the changes of concentrations of the oxide components with increasing silica content (see the text for explanation). The variations of individual components of a differentiating magma always plot as curved lines on a Hacker diagram (b) because the compositions and proportions of the crystallizing phases change as cooling progresses. Inflections mark the appearance or disappearance of a phase that affects the abundance of that component. (c) A straight-line variation would result from mechanical mixing of two end-members, one basaltic and one granitic.

One can use diagrams of this kind to calculate the composition and proportions of material that must be subtracted or added to produce one composition from another. Suppose, for example, that two magmas, M_1 and M_2 in **Figure 5-14**, are lavas erupted from a single volcano. Assuming that the magma with the smaller SiO_2 content is the most primitive and could be the parent of the other, then M_2 could be produced either by

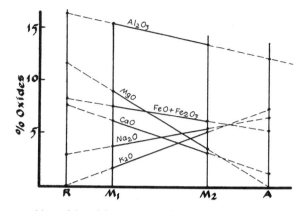

Figure 5-14 The compositions of the minimum amount of material that would have to be removed from (R) or added to (A), an A magma of composition M_1 to produce a second magma, M_2, can be determined graphically by drawing straight lines through the individual components of the two magmas and projecting them to lower and higher silica contents until one of the components intersects the baseline where its abundance goes to zero. The concentrations of the other components can then be read directly from the intersection of the straight lines with a vertical axis at the silica content of R or A. The proportion of A that must be added to M_1 to produce M_2 is M_1M_2, and the proportion of M_1 to which it is added is M_2A. In a similar way, the proportion of composition R that must be removed is M_1M_2, and the amount of remaining liquid, M_2, is proportional to RM_1.

fractionation of a crystalline assemblage having less SiO_2 than M_1 or by addition and assimilation of material, such as crustal rocks, with a composition richer in SiO_2 than M_2. If we ignore physical–chemical constraints and consider only the mass balance of components, an infinite number of possible combinations could be calculated to relate the two liquids by all combinations of these two effects; the only limitation on the proportions of added or subtracted material is that no component can have a concentration less than zero.

Calculations based on this principle have long been used to test various schemes of differentiation. Having determined the oxide proportions of the components in this way, it is a simple matter to calculate the normative composition of the fractionated or added material in order to determine whether the calculated compositions are mineralogically reasonable. Modern computers can perform such calculations quickly and in much more sophisticated ways. It is possible, for example, to calculate the effect of simultaneously fractionating two or more minerals while adding assimilated material in various proportions and then compare the results with the observed compositions of a series of differentiated rocks. Examples of the use of these calculations in interpreting differentiated series will be seen in later chapters.

The computer-based techniques are especially valuable when large amounts of analytical and petrographic data are available for both liquids and crystals, because constraints on the mathematical solutions become increasingly rigorous as the compositions of phases are defined in terms of larger numbers of components. It must be borne in mind, however, that these calculations ignore the principles of phase equilibria and may yield results that are mathematically sound but petrologically absurd. The fact that mass–balance relationships are consistent with a particular scheme of differentiation does not mean that other mechanisms could not have produced the same results.

Plots of chemical data on variation diagrams are sometimes used to test whether a group of rocks is genetically related, the rationale being that if the data fall on smooth curves they are likely to be cogenetic. As in the calculations just mentioned, this technique does not prove that rocks are cogenetic; it only shows that their compositions are not inconsistent with such a relationship.

Smooth curves on variation diagrams can even be misleading, especially when multiple combinations of components are plotted. As a general rule, the more closely the sum of the variables being plotted approaches unity, the stronger will be their negative correlation. If, for example, a number of components, such as A, B, C, D, and E, totals 100 percent, there may be significance in a smooth correlation between A and B, especially if it is a positive one, but there is none whatever between A + B and C + D + E because any plot of these combinations will yield a straight line with a negative slope. For this reason, variation diagrams in which components are plotted against a "differentiation index" that includes SiO_2 and other components that together constitute a large proportion of the total almost always yield good linear trends with negative slopes.

It would seem that trace elements would be immune to such affects because their abundances contribute little to the sum of the components, but they too can be misleading. Here one is dealing with elements that may vary in absolute abundance by factors of a hundred or more from one specimen to another, depending on the distribution of the phases in which the elements occur in the rock. **Figure 5-15** shows how the total variation of a group of trace elements from one sample to the next may be so great that it overshadows the relative variations of individual elements. For this reason, one may find an apparent positive correlation between the absolute concentrations of two elements that in fact have a negative correlation of their relative concentrations. Here the effect is the reverse of what it is in major elements; a negative correlation is more likely to be meaningful than a positive one. The problem is best avoided by carefully examining the mineralogical content of

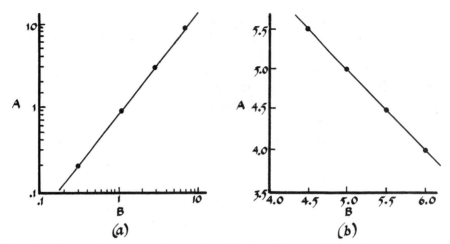

Figure 5-15 Suppose that two trace elements, A and B, are strongly excluded from all major rock-forming minerals but can form an accessory phase that may be present in minute amounts in a wide range of rocks. Suppose also that A and B enter the accessory phase as a solid-solution series so that they have an inverse relation (i.e., as A goes up, B goes down). Assume that in four random rocks the contents of A and B in the accessory mineral are as follows:

	A	B	A/B
Rock 1	4.0	6.0	0.667
Rock 2	4.5	5.5	0.818
Rock 3	5.0	5.0	1.000
Rock 4	5.5	4.5	1.222

If the rocks contain small but differing amounts of the accessory phase, their contents of A and B could be

	Wt % of phase	A in rock	B in rock
Rock 1	0.0005	0.2 ppm	0.3 ppm
Rock 2	0.0020	0.9 ppm	1.1 ppm
Rock 3	0.0060	3.0 ppm	3.0 ppm
Rock 4	0.0160	8.8 ppm	7.2 ppm

If the contents of A and B in the rocks are plotted on a log scale, they show an excellent positive correlation (a); however, if the contents of A and B in the accessory mineral are plotted on a linear scale, they have a negative correlation, as in (b), reflecting the solid-solution substitution of one for the other. Because of the high concentrations of the elements in the accessory phase, a small, random variation of the abundance of that mineral has a greater effect than that of the variation in the mineral alone, and the latter relation is swamped by what may seem to be a genetic relationship of the rocks but in fact could be a meaningless, random fluctuation.

samples and taking care not to use trace-element analyses of rocks containing an accessory phase rich in those elements.

Another solution is to avoid plotting absolute abundances and to use, instead, ratios of cation abundances. These "Molecular Proportion Ratio" or "Pearce Element Ratio" diagrams (**Fig. 5-16**) surmount the closure problem of major-element Harker diagrams by dividing the abundances of two ele-

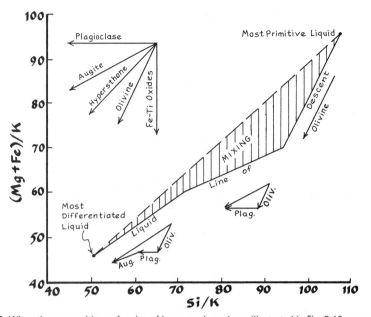

Figure 5-16 When the compositions of series of lavas, such as those illustrated in Fig. 5-13a are plotted in terms of their ratios of (Mg + Fe)/K versus Si/K, they define a line with segments of differing slopes determined by the ratios of these components in the fractionated minerals. As shown in the inset, the effects of removing the minerals olivine, augite, plagioclase, and Fe-oxides are indicated by lines with slopes corresponding to their proportions of (Mg + Fe) to Si (i.e., 0/1 for plagioclase, 1/2 for augite, 1/1 for hypersthene, 2/1 for olivine, and 1/0 for Fe-Ti oxides). The upper segment of the trend of composition corresponds to the slope for olivine, whereas the lower parts reflect the addition of first plagioclase and then augite to the crystallizing assemblage. Unlike SiO_2 in Figure 5-13b, Si/K declines in proportion to the rate of removal of Si from the liquid. The ruled area indicates the range of possible compositions that could result from mixing of any combination of liquids along the liquid line of descent. A composition could fall outside this field if crystals from one liquid were added to another.

ments or groups of elements by a third that does not enter the principal minerals in important amounts and is therefore neither added to nor subtracted from the system. In this way, the ratios of components removed in a mineral of fixed composition remain constant and plot as straight lines. Such a diagram may be useful for identifying the minerals responsible for the variation of a series of cogenetic liquids because the slope of a curve defined by a series of differentiated magmas is an undistorted measure of the bulk ratio of minerals lost or gained. For complex systems with multicomponent minerals, the diagram is less useful than computer-aided calculations, such as those illustrated in later chapters.

The ideal type of variation diagram, if such a thing exists, is one that corresponds to known crystal–liquid relationships and depicts compositional changes in terms of the petrologic processes that produce them. When the proportions of mineral components, either modal or normative, are plotted on an experimentally determined phase diagram, they should fall in the fields of primary crystallization of those minerals that occur as phenocrysts in the rocks. Their distribution should define a trend that is a line of liquid descent resulting from fractionation of the minerals that are crystallizing at a particular stage of differentiation. **Figure 5-17** illustrates the use of such a diagram. Although they can be instructive, most diagrams of this sort are highly simplified; because some of the components must be ignored, the plotted points are in effect projected into the plane of the diagram from all the other components, and the trends they define can be misleading. We noted this effect in Chapter 3 when we considered the problems raised by projections in multicomponent phase diagrams. We will see it again in some of the chapters to come.

Many types of variation diagrams have been devised to depict particular features of related rocks and minerals. Depending on the relationship being explored, any combination of components can be plotted in the way that

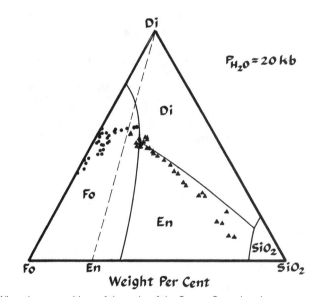

Figure 5-17 When the compositions of the rocks of the Oregon Cascade volcanoes are expressed in terms of the normative components, diopside, olivine, hypersthene, and quartz, and plotted on the phase diagram for the system Di-Fo-SiO$_2$, they correspond to the course liquids would take if basalt (shown by dots) fractionates olivine (Fo) and augite (Di) and more differentiated rocks (▲) fractionate two pyroxenes, augite (Di), and hypersthene (En) and follow the cotectic toward SiO$_2$ as they evolve to dacite, rhyodacite, and finally rhyolite near the eutectic close to the SiO$_2$ corner. In view of the additional components, such as Fe and Al, that are in the natural system, it is not surprising that that the points do not fall exactly along the cotectic. (Based on the system Fo-Di-SiO$_2$ at 20 kilobars determined by I. Kushiro, 1969, *Amer. Jour. Sci.* 267-A:269–294.)

best demonstrates a specific petrologic feature. These diagrams, if properly used, are well suited for visual presentations of data but are by no means a substitute for numerical calculations.

■ Trace Elements and Differentiation

The main questions that confront us in petrology are genetic ones—how magmas are generated and evolve in their particular tectonic settings. By now the student will have perceived that the major-element compositions of rocks give us few unambiguous answers to questions of this kind. In all the processes considered thus far, we have seen that the major-elements in magmas are controlled by the conditions that prevail when the crystals and liquid last equilibrated, and they are insensitive to how the liquid arrived at that state. Trace elements provide ways of surmounting these limitations.

As noted in Chapter 3, the individual components of a differentiating magma are concentrated either in the crystals or in the co-existing liquid, depending on their relative free energies in the two phases. This is true to an even greater degree in the case of certain trace elements. Ni, for example, is strongly partitioned into olivine because it has the same charge (+2) and nearly the same atomic radius (0.69 Å) as Mg (0.66 Å). Its concentration, however, does not follow the same rules that govern elements that are essential constituents of the crystalline structure. The concentrations of Mg and Fe in olivine crystallized from basaltic magmas have a narrow range that is fixed by the Mg/Fe ratio that crystals must have to be in equilibrium with such a liquid. As long as the intensive parameters remain nearly the same, changing the absolute amounts of Mg and Fe in the liquid changes the amount of olivine precipitated but has little effect on its composition. The same is not true of a trace element. The concentration of Ni in olivine will be greater than it is in the liquid by a nearly constant factor, and if the amount of Ni in the liquid is doubled, so too is the amount in the crystals. As a result, olivines with the same Mg/Fe ratios can differ in their Ni contents by large factors, depending on the Ni content of the liquid in which they grew.

The reason for this difference lies in the inherent behavior of very dilute components in liquids. According to Henry's Law, the activity coefficient of such a component has a nearly constant value, and the chemical potential of the component in a particular phase varies directly with its concentration. It follows, therefore, that the concentration, c_S, in the solid will have a constant ratio, K_d, to the concentration in the liquid, c_L.

$$K_d = \frac{c_S}{c_L} \qquad\qquad (5\text{-}3)$$

The value K_d is known as the *distribution coefficient* for a particular element in a given mineral, and depending on whether it is greater or less than 1.0, the element is preferentially concentrated in the mineral or liquid. Ni has a distribution coefficient for olivine in equilibrium with most basaltic liquids of about 10. As long as equilibrium prevails, Ni will be 10 times as plentiful in the crystals as in the liquid.

Typical distribution coefficients for a number of important trace elements are given in Appendix G. These values are only approximate and vary with composition from one magma to another and also with temperature. Moreover, the equilibrium distribution coefficient may differ from the effective one that governs partitioning during rapid crystallization or melting, because the degree of equilibration between crystals and liquid is strongly affected by the rate of diffusion through the liquid.

Trace elements such as Ni that have large distribution coefficients in favor of a particular mineral ($K_d \gg 1$) are strongly depleted from a liquid when that mineral crystallizes. This effect is reduced, of course, if other minerals that exclude the element are crystallizing, for their effect is to increase the concentration of the element in a reduced amount of liquid. Knowing the distribution coefficients of a trace element in the minerals that have been fractionated from an evolving magma, one can calculate the net effect on the abundances of that element that would result from removal of specific amounts and proportions of those minerals.

Elements that are strongly partitioned into only a single mineral may provide a sensitive measure of the amount of that mineral that has crystallized. Conversely, a trace element with a small distribution coefficient ($K_d \ll 1$) is preferentially concentrated in a liquid and reflects the proportion of liquid at a given stage of crystallization or melting because its concentration varies inversely with the amount of liquid diluting its abundance.

These relationships can be seen more clearly by considering the equations governing the distribution of trace elements and how they determine concentrations of elements with different distribution coefficients. (Space does not permit the derivation of these equations but the interested student can find this information in the works of Shaw [1970] and Greenland [1970] listed in the references at the end of this chapter.)

Partial Melting

We noted in an earlier section that partial melting of a multicomponent rock can take place in a number of ways depending on how the melt collects and re-equilibrates with residual minerals before being separated. In fractional melting, the liquid is removed as rapidly as it forms; in batch melting, it is collected and reaches a certain amount before being separated.

Consider first the case of batch melting in which the liquid remains in contact with the residual crystals, equilibrates with them as melting proceeds, and is removed only after a certain amount of melt has collected. In this process, the ratio of the concentration of a trace element in the liquid, C_L, to that in the total original assemblage, C_0, is

$$\frac{C_L}{C_0} = \frac{1}{D + F(1-P)} \tag{5-4}$$

where F is the fraction of melt and D is the bulk distribution coefficient for all the solid phases combined. The latter term is simply the effective distribution between a number of different minerals and a liquid with which they are in equilibrium, It is determined by

$$D = X_\alpha K_\alpha + X_\beta K_\beta + \ldots \tag{5-5}$$

where X_α, and X_β are the initial weight fractions of phases α, β, etc., and K_α and K_β are the corresponding solid–liquid distribution coefficients. The term P represents the combined contributions of the various phases to the melt and is given by

$$P = P_\alpha K_\alpha + P_\beta K_\beta + \ldots \tag{5-6}$$

where P_α and P_β and so on are the fractions of liquid contributed by each individual mineral. If the minerals contribute to the melt in the same proportions as their abundances, P would be the same as D. **Figure 5-18** shows how elements with various distribution coefficients will be concentrated in a liquid (relative to the original solid) as melting proceeds for the case where P = D. In the extreme case of an element that is perfectly excluded (D = 0), the ratio C_L/C_0 from Equation 5-4 becomes a simple inverse function of F:

$$\frac{C_L}{C_0} \approx \frac{1}{F} \tag{5-7}$$

In other words, the amount of melting that produced a given magma can be estimated directly if one knows the concentration of a strongly excluded element in both the source rock and in the melt derived from it.

On the other hand, if the amount of melting, F, is small, Equation 5-4 reduces to

$$\frac{C_L}{C_0} \approx \frac{1}{D} \tag{5-8}$$

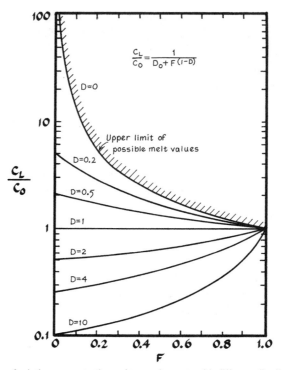

Figure 5-18 Variations of relative concentrations of trace elements with different distribution coefficients during "modal melting" according to Equation 5-4. P is assumed to equal D, and the fraction of melt, F, is assumed to equilibrate with the remaining crystals until it is removed as a "batch."

Thus, the concentration in a source rock from which a small amount of melt has been derived can be estimated from the concentration in the melt and D, even though the source rock is conjectural. The larger the value of D, the greater is the extent of melting, F, for which this relationship applies. For example, if D = 10, then C_L/C_0 has a value of about 0.1 until nearly 30 percent is melted. These relationships will be altered, of course, as soon as the first mineral phase disappears during melting. As soon as an individual mineral is exhausted, it no longer has an effect on the trace-element content of the liquid, and both D and P change accordingly.

In practice, the most valuable trace elements for interpreting the history of a particular liquid will be those with a wide range of distribution coefficients for the individual minerals in the source rock. According to Equation 5-4, an element with a very low value of D for the assemblage as a whole provides an estimate of the extent of melting; however, an element that has a large value of K_d for one specific mineral phase and a small value for all others in the rock provides a way of estimating the amount of that particular mineral in the crystalline residue. This can be seen if D is taken to be approximately $X_\alpha K_\alpha$, in which case Equation 5-8 reduces to

$$X_\alpha = \frac{C_0}{C_L K_\alpha} \qquad (5\text{-}9)$$

By this means, it is possible to estimate the contribution of an individual mineral to the partial melt.

Take, for example, the ratio of potassium to rubidium, which is often used to determine the importance of amphibole in an ultramafic source rock. Olivine and pyroxene contain very little of either of these elements, and thus, their contributions to the bulk distribution coefficient are negligible. Assuming there is no mica in the rock, almost all the K and Rb must reside in the amphibole, which has a distribution coefficient of about 1.0 for K and about 0.3 for Rb. Because amphibole has a larger distribution coefficient for K than for Rb, partial melting of an assemblage including this mineral results in a decrease in the ratio of K to Rb in the liquid over what it was in the original rock. Other factors being equal, a magma produced by partial melting of an amphibole-bearing source rock would have a smaller ratio of K to Rb than would one derived under conditions where that mineral was not stable. This effect is quite different from what would be expected if K and Rb were essential major-element constituents of amphibole. Most amphiboles have large ratios of K to Rb, and melting would produce an equivalent ratio in the melt. As melting progresses, the absolute abundances of these elements in the liquid would approach that in the original rock, but their ratio would remain constant. This is not the case if these elements are not essential constituents. The ratio of K to Rb would then increase with increasing degrees of melting and only reach the K/Rb ratio of the source rock when the mineral is exhausted.

The effects are somewhat different in a process of fractional melting, in which the melt is continuously removed as it is formed. If the minerals melt in the same proportions as they occur in the solid, the distribution of a trace element between liquid and solid is governed by

$$\frac{C_L}{C_0} = \frac{1}{D}(1 - F)^{(1/D) - 1} \qquad (5\text{-}10)$$

Trace-element concentrations can be sensitive to the disappearance of a mineral that is exhausted before the source rock is totally melted. This is especially true if the mineral is very rich in a particular trace element. As long as the mineral is present, it contributes that element to the melt, but once exhausted, the abundance of the element in subsequent liquids drops to a much lower value. This sudden drop might not be detected if the liquid accumulates during protracted melting because the liquid would still contain an amount of the element contributed during earlier stages of melting.

Fractional Crystallization

As in partial melting, trace-element partitioning during crystallization depends to a large extent on the proportions of liquid and crystals that equilibrate during the time the two phases are in close association. Conditions can range between two extremes, one in which crystals remain in the liquid and re-equilibrate continuously as the liquid slowly crystallizes and another in which crystals become separated as rapidly as they form and cannot re-equilibrate. The first case is essentially the reverse of equilibrium melting, but the second is closer to the condition likely to prevail in nature.

Crystallization in which crystals are segregated from the liquid and do not re-equilibrate as crystallization advances is often referred to as *Rayleigh fractionation*, and partitioning of trace elements under this condition is governed by the equation

$$\frac{C_L}{C_0} = F^{(D-1)} \qquad (5\text{-}11)$$

where F is the fraction of liquid remaining. As before, the bulk distribution coefficient, D, is given by

$$D = X_\alpha K_{\alpha\cdot} + X_\beta K_\beta + \ldots \qquad (5\text{-}12)$$

where K_α, K_β, etc., are the distribution coefficients, and X_α, X_β, etc., the weight fractions of individual crystallizing phases, α, β, etc.

Just as in the case of partial melting, when D approaches zero,

$$\frac{C_L}{C_0} = \frac{1}{F} \qquad (5\text{-}13)$$

so that the concentration of an element strongly excluded from all growing crystals is a simple inverse function of the amount of crystallization. It follows that the parameter $(1 - C_0/C_L)$ for an excluded trace element is a good index of differentiation, as it represents the fraction of a parental magma that has crystallized to produce a derivative magma.

The effect of fractional crystallization on elements with different distribution coefficients is shown in **Figure 5-19**. The curve for D = 0 delineates the upper limit of concentration of a totally excluded element in a fractionated liquid as a function of the amount of crystallization. A comparison with the behavior of another element having a large value for D, such as the curve for D = 10, shows how rapidly an element that is strongly partitioned into the

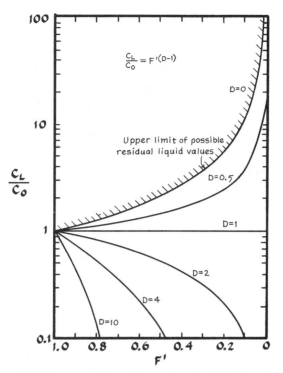

Figure 5-19 Variations of relative concentrations of trace elements with different distribution coefficients during crystallization, according to the Rayleigh-type model of Equation 5-10. The fraction of liquid remaining, F', equilibrates only with the surface of the co-existing crystals and not with the total solid.

crystallizing minerals will be depleted from the liquid as crystallization advances.

Comparing Figures 5-18 and 5-19, we see the contrasting effects of partial melting and crystal fractionation and the differences between the amounts of depletion and enrichment that are possible in the two types of systems. For elements with large distribution coefficients, liquids produced by fractional crystallization are more strongly depleted than those produced by partial melting at comparable proportions of crystals and liquid. Conversely, in the middle range of crystal–liquid proportions, elements with distribution coefficients less than 1.0 are more concentrated in a liquid that has evolved by fractional crystallization than in one produced by partial melting. A small amount of melting produces liquids with large initial concentrations of strongly excluded elements, and with further melting, that element is simply diluted by the growing mass of liquid. In crystal fractionation, however, the concentration of a strongly excluded element increases slowly so long as the amount of remaining liquid is large but rises at an increasing rate as the amount of liquid diminishes. When the fraction of liquid is large, the amount of crystallization required to change its composition is also large.

As an example of how these two processes might affect the compositions of a series of magmas produced by different stages of melting and crystallization, consider the hypothetical case in which melting proceeds in such a way that batches of liquid are extracted periodically and then evolve by crystal fractionation. From the behavior of included and excluded elements just outlined, we can see that the result would be a series of curves with a form somewhat like those shown in **Figure 5-20**. As in the preceding diagrams, these relationships are highly idealized, and in magmas, the patterns are seldom this simple.

Behavior of Specific Trace Elements

Trace elements can be divided into several broad groups according to their behavior in magmatic systems. The first includes large cations with low valences (+1 or +2), such as Rb, Ba, Sr, and Cs. These large-ion lithophile

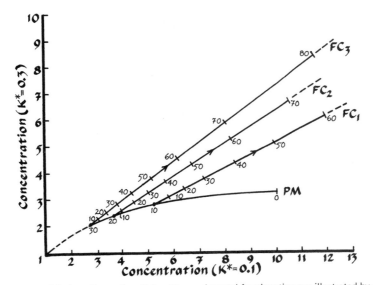

Figure 5-20 The differing effects of partial melting and crystal fractionation are illustrated by a comparison of the variations of two excluded trace elements, one with a distribution coefficient of 0.1 and the other 0.3. The first liquid produced by melting will have the concentrations represented by the point 0 on the line PM, and the elements will become more dilute (at differing rates) as the amount of liquid increases to larger percentages indicated by numbers along the line. A liquid extracted after 10-percent melting and crystallized without reaction will follow the course of FC₁ and melts separated at more advanced stages will follow similar trends, such as FC₂ and FC₃. The amount of crystallization is indicated on each of these lines and shows the accelerating rate of enrichment in later stages of crystallization (B. H. Baker, personal communication).

(LIL) elements have distribution coefficients of less than 1.0 for most minerals; that is, they tend to be enriched in the first products of partial melting or in the residual liquids of crystal fractionation. The effect, however, differs according to the specific minerals involved. Sr, for example, is excluded from most common minerals except plagioclase, and Ba is similarly excluded from most minerals except K-feldspar. Thus, the ratio of Ba to Sr tends to increase with crystallization of plagioclase but may decrease when orthoclase begins to crystallize. In a similar way, Rb is excluded from most minerals but is included in micas and would be a sensitive indicator of fractionation of that phase. Another example would be K and Rb, which, as noted earlier, are sensitive to crystallization or melting of amphibole.

The group of large cations with high charges, sometimes referred to as high-field-strength elements, is more uniformly excluded. They include Zr, Hf, Ta, Nb, Th, and U, all which tend to be rejected by the common igneous minerals and strongly concentrated in residual liquids. They may, however, be concentrated in accessory minerals, such as zircon, apatite, or sphene, especially in highly differentiated rocks, such as granites. For this reason they are useful indicators of the degree of crystal fractionation so long as the liquid is not saturated with one of these minerals. These same elements are also important components of sialic continental rocks and may be enriched in magmas that are strongly contaminated by crustal material. A batch of magma produced by melting of a source rock having a particular ratio of two such elements will preserve that ratio throughout its subsequent differentiation if neither of the elements is extracted more than the other. Hence, the ratios of high-field-strength elements may indicate a cogenetic relationship for rocks having similar ratios and distinguish them from other rocks derived from batches of magma or source rocks with different ratios.

These highly excluded elements tend to be concentrated along grain boundaries between crystals, unless there is an accessory mineral into which they enter, as Zr does in zircon or Cu in sulfides. As noted earlier (Fig. 5-15), the concentration of trace elements in the bulk rock may be very sensitive to the abundance of a minor phase of this kind and may differ so much from one specimen to another that bulk-rock analyses are difficult to interpret.

A third group is made up mainly of transition elements, such as Ni, Co, Cr, and Sc. These elements tend to be included in the mafic minerals, but to different degrees depending on their size. Ni, as we saw, is strongly included in olivine but less so in pyroxene. Cr and Sc, on the other hand, enter olivine only in small amounts but are strongly included in pyroxenes. The ratio of Ni to Cr or Sc, for example, can provide a way of distinguishing the effects of crystallizing olivine and augite. Again, the concentration of these elements in crystalline rocks may be very sensitive to the presence of a minor phase in which they are strongly concentrated. Sulfide minerals have distribution coefficients measured in thousands for certain metals, such as platinum, that enter readily into their structure, and a small variation in the abundance of such a mineral will result in huge differences in the bulk rocks.

Finally, special mention must be made of a group of elements that has proved particularly useful in petrogenetic interpretations, namely the group of elements with atomic numbers between 57 (lanthanum) and 71 (lutetium). These rare-earth elements (REEs) have very similar geochemical properties. Under geological conditions, their valence is +3, except for Eu and Ce, which may be +2, +3, or +4, depending on oxidizing conditions. Because their atomic radii are large, they enter only minerals with sites that can accommodate large atoms. For this reason, their distribution coefficients are very small for minerals such as olivine but larger for others, such as hornblende and apatite, and differ for individual elements according to their atomic radii, which decrease with their atomic number. These properties make the REEs useful tools for identifying minerals that have affected the composition of a rock through crystal fractionation or partial melting.

Because of the analytical difficulties of determining some of these elements, only six or seven are commonly used, but these are enough to establish the pattern for most rocks and minerals. In order to smooth out the effect of their differing natural abundances, their concentrations are usually normalized by dividing their abundances by those in a reference standard, usually chondritic meteorites. When these normalized abundances are plotted on a logarithmic scale against their ionic radius, they form curves with distinctive forms and slopes reflecting the relative enrichment of the individual elements in a particular phase (**Fig. 5-21**).

For example, europium, because of its different valence, goes into plagioclase much more readily than the other REEs, and crystallization of plagioclase can remove so much europium from the liquid that any other minerals that grow in the same liquid inherit a negative europium anomaly (Fig. 5-21). These relationships provide a valuable tool for determining whether large amounts of plagioclase have been fractionated from a particular magma.

A similar technique is used to compare the abundances of diagnostic elements in normalized multicomponent diagrams, often referred to as "spider diagrams" (**Fig. 5-22**). Selected elements are normalized by dividing their abundances in the sample by those in a suitable reference, such as chondritic meteorites, average mantle, midocean ridge basalt, or any other composition to which it is convenient to relate the sample. (A set of commonly used values is provided in Appendix G.) The elements are listed at equal intervals on the X-axis, preferably in the order shown here, and the normalized values

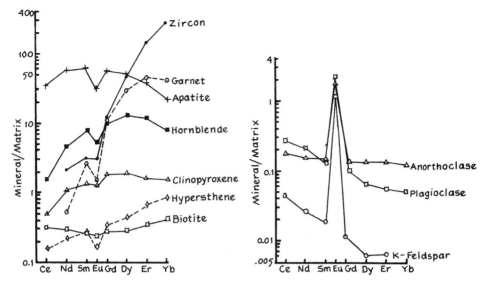

Figure 5-21 The ratios of concentrations of the rare-earth elements in common minerals to their coexisting ground mass composition (which is taken as the liquid from which they formed) is shown in the diagram on the left. The REE on the horizontal axis are plotted according to increasing atomic number. The corresponding patterns for feldspars, shown in the diagram on the right, illustrate the strong positive Eu anomaly that results from the different valence and greater compatibility of that element in the feldspar structure. Minerals crystalized from a liquid that has been depleted of Eu in this way show corresponding negative Eu anomalies, such as those in the diagram on the left. (After G. N. Hanson, 1978, *Earth Plan. Sci. Ltrs.* 38:26–43.)

are plotted on the vertical axis, using either a linear or logarithmic scale. Irregularities of the resulting curves often reflect the distinctive signature of a particular fractionated mineral or mechanism of differentiation. As will be seen from examples in later chapters, the diagrams are most useful for graphic comparisons of related rocks.

■ Kinetic Effects

Many common textural features of igneous rocks, such as zoning and compositional differences between phenocrysts and ground mass minerals, show that equilibrium is seldom achieved in magmas crystallizing under natural conditions. The rates of equilibration between crystals and liquid are severely limited by the rate of diffusion between the interior and surface of crystals and between crystal faces and the main reservoir of liquid some distance away (**Fig. 5-23**). If the rate of crystal growth is rapid, diffusion is unable to supply included elements or remove excluded ones fast enough for crystals to maintain the compositions they would have if they were

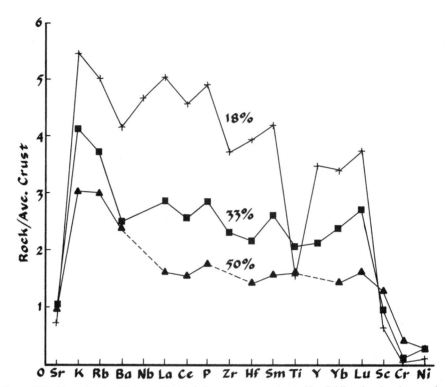

Figure 5-22 Normalized multielement (spider) diagrams are constructed by dividing the abundances of selected elements in a given sample by their abundances in a reference standard, such as those given in Appendix G. The resulting pattern of enrichment or depletion often reveals how the composition of the sample has been affected by crystal fractionation, contamination, or hydrothermal alteration. In this example, three successive liquids segregated from a thick, ponded flow of Columbia River basalt have been normalized to the average composition of the upper and lower crusts. Numbers on each curve indicate the fraction of liquid remaining at the time of segregation. Ti is first enriched and then depleted when Fe-Ti oxides crystallize. Sc, Cr, and Ni are depleted in all liquids because of crystallization of augite and olivine. (After Enc Sonnenthal, 1990, PhD Thesis, University of Oregon.)

growing in a well-stirred liquid. Instead, a boundary layer develops in which the liquid is locally depleted or enriched, and the liquid that the crystals "see" differs in composition from that of the larger reservoir some distance away.

The effect of these kinetic limitations is that the concentrations of excluded elements in a growing crystal tend to be more and those of included elements less than they would be under conditions of perfect equilibrium. With increasing rates of crystal growth or slower diffusion, the effective distribution coefficient—that is, the ratio of the concentration in the crystal to the concentration in the main mass of liquid far from the crystal—approaches a value of 1.0 for both included and excluded elements. The

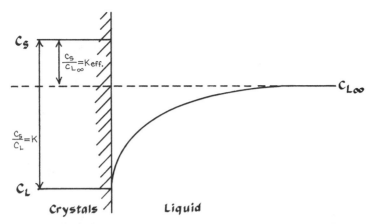

Figure 5-23 Advancing crystallization produces compositional gradients in the layer adjacent to the growing crystals, with depletion of elements that are preferentially included in the solid phases. As a result, the liquid next to the crystals is impoverished in these elements relative to the concentration at a greater distance in the interior of the static liquid. The concentration of an element in the crystals is still determined by the distribution coefficient, K, but because the concentration in the liquid at the crystal face is lower, so too is the concentration in the growing crystals, and the effective distribution coefficient is correspondingly lower. The greater the rate of crystallization or the slower the rate of diffusion, the larger this effect will be and the closer the effective distribution comes to a value of 1.0.

result is that magmas crystallizing rapidly remain undifferentiated, and the composition of the final crystalline assemblage is essentially the same as that of the starting liquid.

Even with relatively slow cooling, kinetic effects may cause the concentrations of trace elements to differ from what they would be in a perfectly equilibrated system. The form these effects may take in a multicomponent liquid is illustrated by a hypothetical three-component system in **Figure 5-24.** With perfect equilibration, if the temperature of a liquid X drops to that of the temperature contour shown by the curved line, crystals of C begin to grow, and the course of the liquid composition X would be directly away from C in the direction of the broken line C–Y. When the liquid reaches point Z, it is just saturated at the prevailing temperature, and the crystal ceases to grow. If we consider the role diffusion could play, we see that the actual course of the liquid may be somewhat different. As the growing crystals remove component C from the adjacent boundary layer, the components A and B are residually enriched (**Fig. 5-25a**). As C diffuses down the compositional gradient to feed the growing crystals, A and B must diffuse in the opposite direction. If A has a greater diffusivity than B, its concentration gradient will be less steep, and the ratio of A to B will vary with distance away from the crystals. As a result the course of the liquid during crystallization will deviate

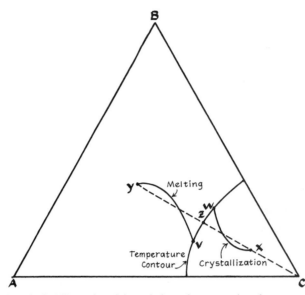

Figure 5-24 A hypothetical illustration of the variations of concentration of two components with differing diffusivities (A diffusing more rapidly than B) during melting and crystallization in a three-component system (see the text for an explanation). (After Y. Oishi, A. R. Cooper, and W. D. Kingery, 1965, *Jour. Amer. Ceramic Soc.* 48:88–95.)

from a straight line and follow a curved path such as that shown by the line X–W in Figure 5-24.

A similar but opposite effect would accompany melting. If a crystal of C were added to a liquid of composition Y and the temperature remains as before, the crystal would melt, and under conditions of perfect equili-

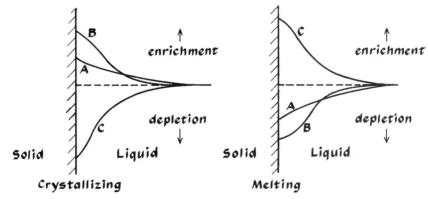

Figure 5-25 Schematic traces of compositional gradients for crystallization and melting in the three-component system in Figure 5-24.

bration, the course of the liquid would be a straight line toward C. If the liquid is not stirred and the components diffuse through the boundary layer at different rates, the effect would be to produce differing diffusion gradients of A and B toward the crystal–liquid interface (Fig. 5-25b), and the course of the liquid would follow a curved path, such as Y–V in Figure 5-24.

The diffusivities of trace elements in silicate liquids vary inversely with their ionic radius and the square of their charge. Because cations may range in charge from +1 in the case of alkalies, such as lithium, cesium, and rubidium, to as much as +6 in certain large-ion lithophile elements, such as uranium, molybdenum, and tungsten, the diffusivities of elements may differ by orders of magnitude, and the degrees to which they are fractionated may differ accordingly. Even though an element may have a very large or very small distribution coefficient, if it also has a low diffusivity, its effective distribution coefficient may be closer to 1.0, and fractionation will be less effective than would be anticipated from its equilibrium distribution coefficient alone.

The effective distribution coefficient is related to the equilibrium value, K, and the ratio of the rate of linear crystal growth to diffusivity, R/D, according to the equation

$$K_{eff} = \frac{1}{2}\left\{1 + erf\left[\frac{\sqrt{(R/D)X}}{2}\right] + (2K-1)\times \exp\left[-K(1-K)\left(\frac{R}{D}\right)X\right] erfc\left[\frac{(2K-1)}{2}\sqrt{\left(\frac{R}{D}\right)X}\right]\right\} \qquad (5\text{-}14)$$

in which X is the linear dimension of the solid. This rather formidable-looking equation can be solved quite easily by use of the table of error functions in Appendix C. In doing so, one finds that, for excluded elements for which K is small (less than about 0.3), the effective distribution coefficient will be within about 20 percent of the equilibrium value so long as the value RX/D is less than about 0.2. For rates of growth less than about a tenth of a millimeter per year, diffusivities of 10^{-8} cm^2 sec^{-1} or more, and crystal sizes of less than 5 cm, the effective distribution coefficient does not differ greatly from the equilibrium value, but for elements with large values of K the problem is more serious and can result in an effective value that is only 30 percent or so of the equilibrium one. Thus, excluded trace elements can be modeled with reasonable accuracy, especially in slowly cooled magmas, but included elements of low diffusivities may deviate widely in their partitioning from the patterns predicted by equilibrium models.

In summary, then, we see that the limitations imposed by slow diffusion can have an important effect on distribution coefficients, especially for included elements. In systems in which individual components have differing diffusivities, the variations of concentration during crystallization or melting may result in the differentiated liquid following a path of evolution that is a function, not just of the distribution coefficients, but of the rates of crystallization and diffusivities as well.

Selected References

Arth, J. G., 1976, Behavior of trace elements during magmatic processes: A summary of theoretical models and their applications, *Jour. Res. U.S. Geol. Surv.* 4:41–47. A very concise and well-written explanation of the basic relations in trace-element fractionation.

DePaolo, D. J., 1981, Trace element and isotopic effects of combined wall-rock assimilation and fractional crystallization. *Earth Planet Sci Ltrs.* 53:189–202. A theoretical model for calculating the effects of simultaneous assimilation and crystal fractionation.

Greenland, L. P. 1970, An equation for trace element distribution during magmatic crystallization, *Amer. Mineral.* 55:455–465. A basic treatment of the equations for partitioning of trace elements during crystal fractionation.

Hart, S. R., and C. J. Allegre, 1980, Trace-element constraints on magma genesis, in *Physics of Magmatic Processes.* R. B. Hargraves, ed., 121–160 p. An excellent summary of the principles of trace-element fractionation with somewhat more detail than the reference by Arth listed previously.

Helz, R. T., 1987, Differentiation behavior of Kilauea Iki lava lake, Kilauea Volcano, Hawaii: An overview of past and current work, in *Magmatic Processes: Physicochemical Principles.* B, Mysen, ed., The Geochemical Society, pp. 241–258. A summary of studies of the most important of the Hawaiian lava lakes. Contains much data on the effects and mechanisms of differentiation.

Hoffmann, A. W. 1980, Diffusion in natural silicate melts: A critical review, in *Physics of Magmatic Processes.* R. B. Hargraves, ed., pp. 385–418. A summary of both the theory and experimental data bearing on diffusion in silicate melts.

O'Hara, M. J., and R. E. Mathews, 1981, Geochemical evolution in an advancing, periodically replenished, periodically tapped, continuously fractionated magma chamber, *Jour. Geol. Soc. London.* 138:237–277. A theoretical analysis of the evolution of a magma chamber from which magma is periodically withdrawn and replenished. The basic principles are laid out in Chapter 7.

Marsh, B. D., 1996, Solidification fronts and magmatic evolution. *Mineral Mag.* 60:5–40. A very thorough review of the physical mechanisms of magmatic differentiation.

Shaw, D. M., 1970, Trace element fractionation during anatexis. *Geochim. Cosmochim. Acta.* 34:237–243. A basic treatment of the equations for fractionation of trace elements during partial melting.

Watson, E. B., 1990, Chemical diffusion in magmas: An overview of experimental and geochemical applications, in *Physical Chemistry of Magmas,* L. Perchuk and I. Kushiro, ed., Springer-Verlag: Berlin. A comprehensive review of the role of diffusion in petrologic processes.

6

Magmatic Differentiation: Basic Intrusions

Having considered the basic principles of crystallization and differentiation, we can now turn to geological examples in which these processes have operated under natural conditions. The most instructive examples of this kind are found in large intrusions that have cooled slowly at moderate depths and have later been exhumed by uplift and erosion. Although these bodies represent only the shallowest kinds of differentiation and must differ in many ways from what would be seen at greater depths in the lower crust or mantle, they are the only ones accessible to our view, and, as such, they provide the best direct evidence we have.

Two broad groups of bodies can be distinguished: basic intrusions, mainly of gabbroic or doleritic rocks, and granitic plutons of more felsic composition. They differ not only in the composition of the rocks but in their form, tectonic setting, and the way they crystallize. Here we consider only the former and defer the discussion of granitic intrusions until a later chapter.

Petrologists have compiled a vast amount of structural, mineralogical, and geochemical information on differentiated basic intrusions, but explaining the compositional evolution of these bodies has been a much more difficult task than describing them. Anyone attempting to understand these cold, dissected intrusions without ever having observed a large body of slowly crystallizing magma is in much the same position as a paleontologist trying to interpret fossil remains without ever having seen the living organism. Nevertheless, much has been learned, and our understanding is slowly advancing.

◼ Doleritic Sills

Although they differ widely in detail, basic intrusions fall into two general groups that differ conspicuously in their dimensions, rates of solidification, and the textures and compositions of their rocks. As noted in Chapter 2,

Figure 6-1 One of a large number of Mesozoic sills forming a belt extending from Australia to South Africa is exposed in steep, glaciated cliffs in the Taylor Glacier region of Antarctica.

hypabyssal bodies are distinguished from *plutonic* intrusions mainly by their smaller size and shallower depths of emplacement (**Fig. 6-1**). Whereas felsic plutons are more common than mafic ones, the reverse is true of hypabyssal intrusions. The mafic rocks of basic dikes and sills are usually referred to as *dolerites* or *diabases* because their grain sizes are intermediate between those of basalt and gabbro. The felsic differentiates are normally *granophyres*, consisting chiefly of fine intergrowths of quartz and alkali feldspar. Unlike the coarse, hypidiomorphic, granular textures that are so typical of larger, slowly cooled bodies, dolerites are typically subophitic (Fig. 4-21c), with a wide range of grain sizes and a relatively restricted range of compositions and lithologies. Most commonly, their trends of differentiation resemble those of the Hawaiian lava lakes mentioned in earlier chapters.

Many varieties of doleritic sills have been described, but most have broadly similar characteristics that can be illustrated with a few examples, such as the well-studied Palisades Sill of New Jersey. This intrusion is part of a large group of Triassic sills, dikes, and subaqueous lava flows that were emplaced in a subsiding basin during an episode of continental rifting. Exposed for 80 km along the western banks of the Hudson River, it reaches thicknesses of more than 300 m. The vertical section illustrated in **Figure 6-2** is probably typical of most of the sill. Above the chilled base and the conspicuous olivine-rich layer a few meters higher, the dolerite becomes coarser and

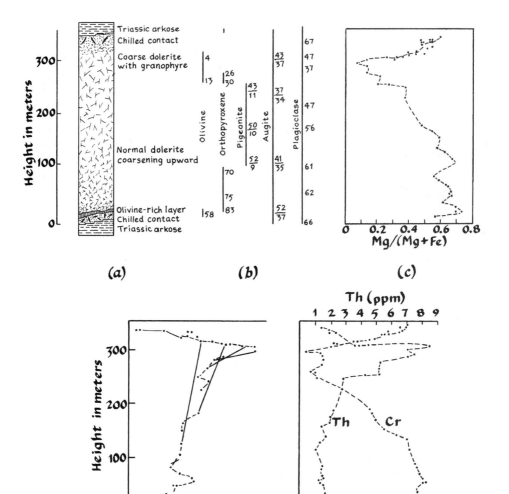

Figure 6-2 (a) A schematic section through the Palisades Sill of New Jersey and New York. (b) Mineral compositions vary systematically, with height as mafic minerals become more iron-rich and plagioclase more sodic. The section from the floor up to about 300 m has a compressed, inverted counterpart in the rocks that crystallized from the roof downward. (c) The ratio of Mg to Mg + Fe is a good measure of differentiation. Its value declines from the floor up and from the roof down to reach a minimum at the 300-m level. Several reversals mark places where new batches of less differentiated magma were injected and reset the composition of the remaining liquid. (d) The sizes of mineral grains increase toward the 300-m level where the last liquid crystallized. Lines join corresponding stages in the mineralogic sequence shown in (b) and illustrate the somewhat coarser texture of the roof rocks at any given stage. (e) Concentrations of chromium and thorium illustrate the variations of included and excluded trace elements. The scale for Cr is logarithmic, and Cr is depleted more rapidly than Th is enriched. (Adapted from K. R. Walker, 1969, *Geol. Soc. Amer. Spec. Pap.* and D. N. Shirley, 1987, *Jour. Petrol.* 28:835–865.)

more felsic up to the 300-m level where granophyre accounts for more than 50 percent of the rock. It was in this "sandwich horizon" that the last crystallizing liquid was sandwiched between the dolerites accumulating simultaneously on the floor and under the roof. The sequence of minerals in the roof rocks is a compressed mirror image of the one on the floor. It is also coarser and has suffered more hydrothermal alteration.

Because it has a 10- or 20-meter thick olivine-rich layer near the base (Fig. 6-2a), the sill was often cited in the early literature as an example of magmatic differentiation by crystal settling. More recent work has shown that the story is not so simple. Much of the compositional variation is a result of initial differences in three or four successive pulses of intruded magma. Although compositional breaks marking these separate injections are difficult to detect in outcrops, they are evident in chemical variations, such as the ratio $Mg/(Mg + Fe)$, which is a good measure of the degree of differentiation of basic magmas. When plotted against height in the Palisades Sill, this ratio decreases upward from the floor and downward from the roof to reach a minimum value about 300 m above the base (Fig. 6-2c). In detail, however, the trend has small but distinct reversals that are thought to result from new injections that set the magma back to a slightly more primitive composition.

Most of the range of differentiation is characterized by a steady enrichment of iron (**Fig. 6-3**). Only in the very late stages does this trend reverse when silica and alkalies are greatly enriched in the granophyres. The silica content, which averages about 54.6 percent for the sill as a whole, exceeds 62 percent in the most differentiated rocks. Plagioclase increases, both in modal abundance and albite content from Ab_{34} in the chilled base to Ab_{63} at the 300-m level, while olivines and pyroxenes, including both the Ca-rich and Ca-poor varieties, become increasingly iron-rich. After the initial segregation of olivine phenocrysts, which were probably in the magma at the time it was intruded, olivine ceased to crystallize. The rare grains found in the main body of the sill are rounded and rimmed with pyroxene as a result of reaction with the liquid. In the late stages of differentiation, however, the magma became rich enough in iron to precipitate fayalitic olivine, which, as we saw in Chapter 4 (Fig. 4-14), is stable in the presence of quartz.

Mechanisms of Differentiation

Because of its shallow depth and rapid rate of cooling, the sill crystallized in much the same way as the Hawaiian lava lakes. Despite its apparent simplicity, however, the manner in which it differentiated is far from obvious.

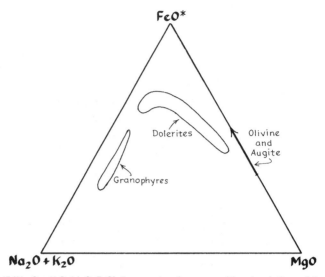

Figure 6-3 AMF ($Na_2O + K_2O$, MgO, FeO) diagram showing compositional variations of the Palisades Sill and Red Hill dolerites. The original compositions of the magmas were rich in MgO and became relatively more iron rich in the main stages of differentiation. At later stages, however, alkalies increased at a greater rate as crystallization of magnetite removed iron. At that point, the trend turned sharply toward the composition of granophyres along the left side of the triangle. (After K. R. Walker, 1969, *Geol. Soc. Amer. Spec. Paper* 111, and I. McDougall, 1962, *Bull. Geol. Soc. Amer.* 73:279–316.)

Crystal settling seems not to have played a major role. The rate of advance of the front of crystallization under the roof was so much faster than the settling velocities of crystals that nucleated and grew there that few crystals escaped entrapment in the growing crust. Clusters of crystals may have been able to break loose from the mush under the roof and because of their greater aggregate size settle to the floor, but the importance of this process is difficult to assess. The preservation of sharp discontinuities marking the boundaries of different pulses of magma clearly rules out vertical movements of large amounts of material.

A more important effect was probably gravitational segregation of liquids from the interstices between crystals growing on the floor and under the roof. As long as they were denser than the liquid, the crystals may have compacted under their own weight, but with time, compaction would become less important as the liquid became denser by virtue of its increasing iron content. Instead, it would drain from the crystals growing under the roof and descend into the main mass of underlying magma in the manner illustrated in the previous chapter (Fig. 5-2). Similarly, when the composition of the liquid passed the point of maximum iron enrichment and became increasingly enriched in silica and volatiles, the resulting density inversion would enable a buoyant, evolved liquid to rise and collect under the roof.

Buoyant liquids that infiltrate upward through the porous network of crystals may collect in discrete bodies where their rise is impeded by the reduced permeability and temperature of the crust. They then inflate the crystal mush and either dislodge slabs of the roof or form irregular lenses that solidify in place. Pods of granophyre formed in this way are especially common in doleritic sills and thick lava flows. The process has been observed in the Hawaiian lava lakes, where plumes of volatile-rich residual liquid became buoyant with respect to the rest of the crystal mush and rose through it to form pipes and lenses of siliceous pegmatite.

A spectacular example is seen in the Red Hill intrusion, one of a large group of doleritic sills in Tasmania. Differentiation followed a trend very similar to that of the Palisades Sill, but in places, the Red Hill intrusion has much larger accumulations of granophyre (**Fig. 6-4**). The normal thickness of the sill is about 400 meters, but the roof was uplifted locally by as much as 300 meters to form a cupola where disproportionately large amounts of granophyre collected. The up-faulted roof seems to have provided a structural trap into which late-crystallizing liquid migrated from a broad lateral extent.

Upward migration of late-stage liquids produces a redistribution, not only of silica and volatiles, but of most excluded trace elements as well (Fig. 6-2e). For example, the concentration of thorium, which is effectively excluded from all minerals of the Palisades dolerite, reaches a maximum at the sandwich horizon where the growing crust of the roof and floor converged, while Cr, which enters readily into magnetite and pyroxene, is progressively depleted until it reaches a minimum concentration at the same 300-meter level.

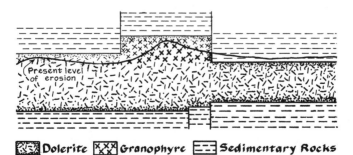

Figure 6-4 A schematic section through the Red Hill Sill of Tasmania. The maximum thickness is about 700 m where granophyre has accumulated under an upfaulted block of the roof. (After I. McDougall, 1962, *Bull. Geol. Soc. Amer.* 73:279–316.)

If the crystals nucleated and grew *in situ* and most of the heat loss was through the roof, why is the sequence on the floor so much thicker than the one that formed under the roof? Comparisons of the mineral assemblages and compositions indicate that solidification advanced about seven times faster on the floor than under the roof. This large difference seems inconsistent with the way the sill cooled. Heat lost through the floor could not have been greater than that lost through the roof. If anything, it was probably less because a vigorously convecting hydrothermal system in the overlying sediments would have enhanced cooling of the roof.

One possibility is that the magma became density stratified, and the liquid under the roof was richer in excluded components, especially volatiles. As we saw in Chapter 3, the addition of another component to a crystallizing liquid tends to lower its liquidus temperature and extend the range of crystallization. This effect is especially pronounced for late-stage residual components, particularly water. Thus, the accumulation of volatiles and other excluded elements in the upper part of the magma could have caused it to lag behind the front of crystallization on the floor.

Alkaline Dolerites

Basic sills of more alkaline composition have many of the features of tholeiitic dolerites, but instead of quartz-rich granophyres, they differentiate to felsic rocks devoid of quartz and rich in alkali feldspar and silica-poor minerals, such as analcite and, more rarely, nepheline. A particularly well-studied example is the Prospect intrusion near Sydney, Australia. Part of the same large group of Mesozoic sills as the Tasmanian dolerites, the Prospect body is exposed in quarries where extensive sections reveal complex lithologic variations, both vertically and along strike (**Fig. 6-5**).

The magma must have been very inhomogeneous with large concentrations of olivine crystals in the last pulse of intruded liquid. No other explanation can be found for the very olivine-rich compositions in the middle of

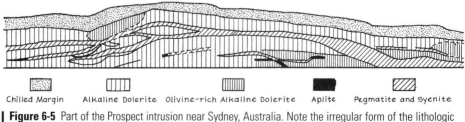

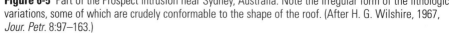

Chilled Margin   Alkaline Dolerite   Olivine-rich Alkaline Dolerite   Aplite   Pegmatite and Syenite

Figure 6-5 Part of the Prospect intrusion near Sydney, Australia. Note the irregular form of the lithologic variations, some of which are crudely conformable to the shape of the roof. (After H. G. Wilshire, 1967, *Jour. Petr.* 8:97–163.)

the sill, precisely where one would expect to find the most differentiated rocks. Although some olivine settled and accumulated near the floor, olivine-rich rocks are also found at almost all levels, especially near irregularities and elevated parts of the roof where crystals seem to have been concentrated by flow segregation at the time of emplacement.

The rocks and minerals have a broad range of composition. The most differentiated rocks are in long, thin lenses of very coarse-grained syenite that cross the sill at odd angles. They are thought to have formed when late-stage liquids, rich in alkalies and volatiles, moved into fractures, cutting the viscous, partly crystalline magma. In places, this caused extensive replacement of plagioclase by alkali feldspars to form syenites rich in anorthoclase, oligoclase, and analcite.

With the effects of all these processes of differentiation—crystal settling, flow segregation, volatile transfer, and porous flow—superimposed on one another, the end result has been a complex vertical and horizontal distribution of the chemical components and lithologic units.

■ Layered Intrusions

Form and Mode of Emplacement

Basic intrusions that reach thicknesses of more than 400 or 500 meters tend to be notably more coarse grained than thinner dikes and sills, and almost all have some form of layering. The critical thickness that separates basic intrusions into two distinct groups, one nearly structureless and the other layered, must be related in some way to the manner in which magmatic bodies of differing dimensions crystallize while losing heat to their surroundings. It seems that above some critical size the mechanisms of crystal–liquid fractionation are much more effective. Whatever the reason, most large, basic intrusions have undergone more extensive differentiation and contain more rocks of extreme compositions than smaller bodies with similar initial compositions.

The largest layered intrusions measure tens or hundreds of kilometers across and several kilometers in thickness. Most of the largest bodies are of Precambrian age; younger intrusions are, on average, much smaller. Unlike granitic plutons, which tend to be tall and steep sided, basic intrusions are generally flatter. Many early descriptions referred to the latter as *lopoliths*, which are defined as large, basin-shaped bodies with thicknesses less than 5 to 10 percent of their diameters. Few, however, are depressed in their centers. Instead, they tend to be funnel shaped or trough-like with a narrow feeder-dike. Although intrusions are only rarely eroded in ways that expose their roots, at least two, the Muskox intrusion of Canada and the Great Dyke of Zimbabwe, have feeder-dikes that extend for tens or hundreds of kilometers. They seem to have been intruded passively into zones of tension and dilation, and as in the case of sills, the magma probably rose and spread laterally at a level determined by its density relative to that of the surrounding rock.

Special mention should be made of intrusions related to meteorite impact. We have already noted (in Chapter 1) the Sudbury Complex of Canada. The distinctive thing about these bodies is that the magma comes in at temperatures well above their liquidus. They are the only known exception to the rule that basic magmas reaching shallow levels of the crust are never superheated. All others have already begun to crystallize by the time they are intruded or erupted at the surface.

Most shallow intrusions have a narrow, fine-grained margin resulting from chilling of the magma against cold wall rocks. This quenched selvage may preserve the initial composition of the undifferentiated magma, but in many cases, its composition has been altered by reaction with the walls or by transfer of components across the steep thermal and compositional gradients at the contact. Moreover, the first magma to intrude often has a composition somewhat different from that that follows.

The calculated cooling times of large masses of magma range from a few tens to hundreds of thousands of years. These estimates are uncertain, however, because they are very sensitive to how much heat has been carried away in meteoric water circulating through the surrounding rocks. The heat of the intrusion tends to set up a convective hydrothermal system that can greatly enhance the transfer of heat.

Layering and Zoning

Three main types of layering have been recognized. The most conspicuous is *modal layering* in which the proportions of minerals differ over vertical distances ranging from a centimeter or so to a few meters (**Figs. 6-6a and 6-6b**). Layers of this kind may have a uniform spacing, in which case they are said to be "rhythmic" (Fig. 6-6c), but many are far from regular and are more properly referred to as "intermittent." The modal differences may be either sharp or gradational, and in extreme cases, alternating layers may be nearly monomineralic (Fig. 6-6b). Layering can also be defined by differences of grain size (Fig. 6-6d). In some instances, the sizes are graded, as they are in many kinds of sediments, but more commonly they are not.

The term *phase layering* is used to denote intervals defined by the presence or absence of a particular mineral in the crystallization sequence. Thus, a distinctive mineral can serve to delineate units of a differentiated intrusion, just as diagnostic fossils are used to identify sedimentary horizons in a stratigraphic sequence.

Finally, the compositions of minerals, such as plagioclase, olivine, and pyroxene, change in systematic ways during progressive differentiation. When recorded in a continuous series of rocks, these changes are usually referred to as *cryptic layering* because they produce no marked contrast in the appearance of rocks in outcrops. They are, however, a basic feature of the compositional variations of all rocks produced by the evolution of a differentiating magma.

(*a*)

(*b*)

Figure 6-6 Types of modal layering. (a) Layering in the Skaergaard intrusion results from differing proportions of plagioclase, magnetite, pyroxene, and olivine, which vary in spacing and proportions from one layer to the next. (b) Nearly pure chromite forms layers within very plagioclase-rich rocks of the Bushveld intrusion. Both rocks are nearly monomineralic.

Figure 6-6 (Cont.) (c) Fine-scale alternations of plagioclase and pyroxene in the Stillwater Complex form a regular pattern on a scale of inches. (d) Differing sizes and proportions of olivine and pyroxene emphasize the layering and cross-bedding of peridotites in the Duke Island intrusion of southeastern Alaska.

The Origins of Modal Layering

Most early studies of igneous differentiation assumed that gravitational settling and various mechanisms of sedimentation were responsible for modal layering. It is now recognized, however, that layering may have a number of different origins.

Certain intrusions of peridotite have well-developed layering with many of the characteristics of sedimentary deposits. The best example, by far, is the peridotite intrusion of Duke Island, Alaska. **Figures 6-6d and 6-7a** illustrate some of the features, such as size sorting and cross-bedding, that closely resemble alluvial deposits of coarse sand and gravel. The crystals were probably laid down by swift currents of exceptionally fluid magma sweeping across the floor of the intrusion.

Gabbroic intrusions show similar evidence of magmatic currents, but the graded layers (Fig. 6-7b) bear only a superficial resemblance to beds laid down by density currents during rapid sedimentation. Minerals of higher density may be concentrated at the base, but their grain sizes are not systematically greater than those deposited at higher levels in the same layer. It will be recalled from the preceding chapter (Equation 5-1) that because settling rates vary with the square of the radius of the particle, they are much more sensitive to size than to density contrast. Any assemblage of minerals, such as magnetite, olivine, and plagioclase, that is segregated by gravitational forces should therefore have a consistent size–density relationship. For these three minerals to settle at the same rate from magma with a density of 2.65, olivine crystals would have to have almost twice the diameter of magnetite grains, and the plagioclase would have to be at least seven times as large. Few of the graded layers in gabbroic intrusions show this relationship. Moreover, graded layering is not confined to the floor; it can be found in rocks that crystallized under the roof or on near-vertical walls. Dark minerals tend to be concentrated in the outer, cooler part, regardless of how these layers are oriented with respect to gravity. The density relationships of the crystals and liquid pose an added problem. Magnetite, olivine, and pyroxene, which must have been denser than the liquid from which they grew, have accumulated under the roof, while plagioclase, which was probably lighter than the iron-rich magma, is an abundant constituent of layered rocks on the floor. It is clear that some mechanism other than gravitational segregation must be capable of producing layering of these types.

Several alternative origins have been proposed. Most involve some form of in situ crystallization and compositional stratification. This must certainly be true of rocks with crescumulate textures, in which slender crystals have grown perpendicular to the plane of layering (Fig. 4-26b). The unusual, elongated form of crystals in rocks of this kind results from nucleation and

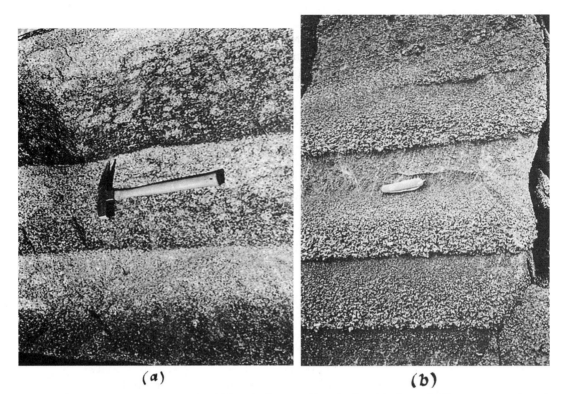

(a)   (b)

Figure 6-7 Two types of graded layers. One (a) from the Skaergaard intrusion is mainly the result of larger concentrations of dark minerals near the base but lacks the size grading seen in the Duke Island ultramafic intrusion (b) where the crystals are larger at the base as would be expected from gravitational settling.

rapid growth in the manner explained in Chapter 4 (Fig. 4-27). Under some conditions, a process of this kind may be repetitive and produce a succession of layers, just as it does in oscillatory zoning of plagioclase (**Fig. 6-8**).

Another mechanism governed by compositional gradients and diffusion has been proposed to explain regular rhythmic layering of the kind illustrated in Figure 6-6c. We saw in Chapter 4 that because of their greater surface energy large crystals have an advantage over smaller ones whenever

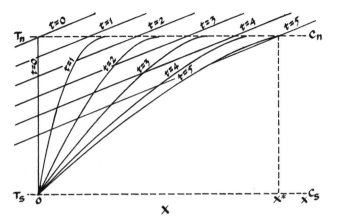

Figure 6-8 Oscillatory crystallization may result from differing rates or diffusion of heat and chemical components through the liquid adjacent to a front of nucleation and crystal growth. Nucleation requires a combination of temperature, T_n, and concentration, C_n (both on the vertical axis). The liquid is supersaturated at time $t = 0$ when crystals begin to grow at $X = 0$. Instantly, the concentration drops to the saturation level, C_s, and a concentration gradient (curved lines) is set up. This diffusion-controlled gradient retreats rapidly at first but gradually becomes slower until at time $t = 5$ the temperature gradient (straight lines) overtakes it. A new layer forms at X^* where the combination of concentration and temperature is again appropriate for nucleation.

crystals with a range of sizes are in contact with the same liquid (**Fig. 6-9**). During very slow cooling, an aging process results in large crystals growing at the expense of small ones. Because of small differences in surface energy, a homogeneous interstitial fluid cannot be in perfect equilibrium with both small and large crystals. It may be slightly undersaturated with respect to the small crystals, which will tend to dissolve, and oversaturated with respect to larger ones, which will tend to grow. As a result, compositional gradients are set up between the liquids adjacent to crystals of differing sizes. Whether the initial differences in grain size result from irregularities of crystallization or from inhomogeneities of some other origin, they may lead to small-scale cyclical variations in the sizes and distribution of crystals. It is doubtful, however, whether they can be responsible for large-scale layering because the rates of diffusion are too slow to be effective over distances greater than a few centimeters.

■ The Bushveld Complex

Any discussion of layered intrusions must begin with the Precambrian Bushveld Complex of South Africa. Not only is it the largest intrusion

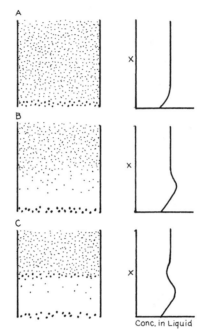

Conc. in Liquid

Figure 6-9 Layering can also develop at a late stage of cooling and recrystallization. As shown in A any initial condition that gives the crystals at one level an advantage over others and causes them to grow larger will produce a compositional gradient through the intercrystalline liquid. Because of their lower surface energy, large crystals grow and extract components from the adjacent liquid, while crystals of smaller size dissolve and cause a local enrichment of the liquid (B). This sets up a compositional gradient allowing components to be transferred from the dissolving crystals to those that are growing. As this process advances, the differences of crystal sizes and liquid compositions become greater, and with time, the small crystals may be completely dissolved. As shown in C the process can be repetitive because the dissolving crystals are also at a disadvantage with respect to other crystals at a higher level, and the latter will begin to grow and generate a new layer. (After Alan Boudreau, 1982 Master's Thesis, University of Oregon.)

known, but in many ways, it is one of the most remarkable bodies of igneous rock on Earth. Apart from its geological interest, it has great economic importance as a rich repository of platinum, chromium, vanadium, and other valuable metals. Thanks to extensive mapping, underground mining, and exploratory drilling, a wealth of data is available for the intrusion, but because of its enormous size and complexity, much still remains to be learned.

The intrusion was emplaced as a nearly flat, tabular mass of gabbroic magma that spread laterally from at least two separate feeders beneath a thin cover of sediments and felsic volcanic rocks. The latter have about the same age as the intrusion and could have been part of the same mag-

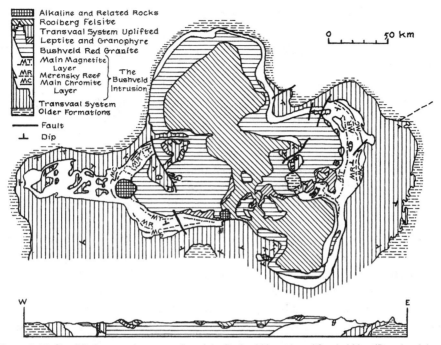

Figure 6-10 Simplified map and cross-section of the Bushveld intrusion of South Africa. (Based mainly on a compilation by J. Willemse.)

matic event. The present area of outcrop (**Fig. 6-10**) measures about 270 km north–south by 450 km east–west, and its original thickness must have exceeded 8 kilometers. The time required to crystallize this huge mass is thought to have been on the order of 200,000 years. About 50 to 100 million years later, a granitic pluton rose through the gabbro and spread laterally over much of its top.

The Bushveld layered gabbros are an extreme example of the difference between the thicknesses of rocks that formed from the floor up and from the roof down. In this case, there is no upper series whatever, and instead of crystallizing, the magma actually melted parts of its roof. As a general rule, the proportion of rocks that have crystallized under the roof of an intrusion is an inverse function of the thickness of the intrusion—the thicker the body, the smaller the roof series. In the case of sills, we noted that the upward concentration of volatiles could account for at least part of this difference. In a large body such as the Bushveld, a more important factor may have been the convective transfer of heat from the zone of crystallization on the floor. Because the pressure at the base was substantially greater than that under the roof, the temperature of crystallization at the base was also higher, and when magma from the floor was carried to the roof, adiabatic cooling was not enough to offset this difference. The rising magma that reached the roof was not only above its crystallization temperature but even had enough heat to melt some of the overlying rocks.

The igneous rocks of the complex resemble a thick section of stratified sediments. They are well layered, and individual horizons have remarkable continuity over vast distances. The best-known unit, the Merensky Reef, is a complex assemblage of orthopyroxene, olivine, chromite, and sulfides, from which much of the world's platinum is mined. Although only 1 to 5 meters thick, it has been traced for distances of more than 300 km.

The gabbroic and ultramafic rocks making up the Layered Series have been divided stratigraphically (**Fig. 6-11**) on the basis of distinctive mineral assemblages that crystallized with progressive cooling and differentiation. The Lower Zone is divided into two parts by the Main Chromite Seam, below which olivine is an essential component of the rocks and above which it is absent. The platinum-bearing Merensky Reef occupies a similar stratigraphic position close to the level where chromite ceased to crystallize at the base of the Main Zone. The top of the Main Zone is marked by the first appearance of magnetite as a primary phase, and the Upper Zone is subdivided at the level where olivine reappears and again at the first crystallization of apatite as a primary phase.

Superimposed on these major divisions is a cryptic zoning resulting from the compositional changes of individual minerals precipitated from the magma as it steadily evolved. The earliest pyroxenes and olivines are magnesium-rich but become more iron-rich upward until, in the most extreme differentiates, they approach the pure iron end-members, hedenbergite and fayalite. The first plagioclase has the composition of bytownite (about An_{80}) and becomes more albitic until it reaches calcic oligoclase (An_{30}) in the uppermost horizons. As the differentiating magma was steadily enriched in ferric iron, phosphorus, potassium, and silica, it eventually became saturated with these components and precipitated them as magnetite, apatite, potassium feldspar, and quartz.

It is difficult to conceive of a body of molten magma with the enormous dimensions of the Bushveld Complex solidifying with such regularity that crystal assemblages maintained nearly uniform compositions over great distances, but for the most part, this was the case. Apart from minor differences between the western and eastern parts of the intrusion, certain of the units consist of layer upon layer of crystals, varying little in composition from one part of the intrusion to another. Many of these rocks are essentially monomineralic and have monotonously homogeneous compositions.

The lower parts of the intrusion contain a complex series of orthopyroxenites, gabbros, and anorthosites, as well as a number of economically

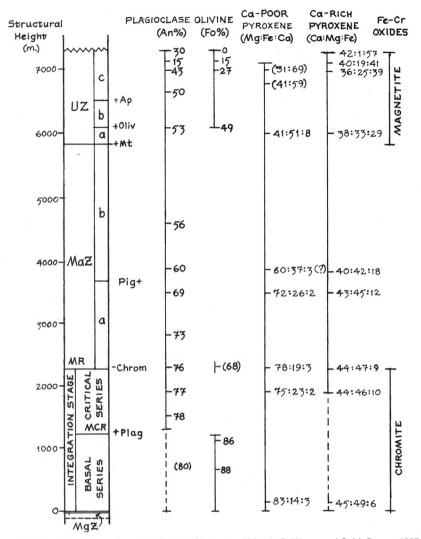

Figure 6-11 Stratigraphic section of the Bushveld intrusion. (After L. R. Wager and G. M. Brown, 1967, *Layered Igneous Rocks,* Oliver & Boyd.)

important horizons rich in chromite. Some of these units are repeated in cyclical sequences in which the modal variations and cryptic zoning of minerals are repeated in regular sawtooth patterns. The base of each cycle is marked by an abrupt step to a more olivine-rich assemblage, often with interlayered chromite in its lower level, followed by greater proportions of orthopyroxene that first become more magnesian and then progressively more iron-rich upward. The cyclic units as a whole become more iron-rich upward and at the same time become progressively thinner until they eventually disappear near the base of the Main Zone.

The origin of these units is still a matter of dispute. It was long assumed that they resulted from sudden introductions of new magma that mixed with the earlier magma and then evolved as separate units. More recently, petrologists have begun to question the plausibility of this explanation because, as they point out, it requires an unrealistic regularity of the spacing of new injections of magma, each with the precise volume and degree of differentiation needed to produce the orderly repetition of mineralogical assemblages. Among the alternative explanations that have been suggested are sudden changes of water pressure or oxygen fugacity, possibly as a result of volcanic eruptions or periodic convective overturn that brought fresh magma to the floor. Whatever their origin, patterns of this kind are by no means confined to the Bushveld rocks. They are common in almost all large basic intrusions.

■ The Muskox Intrusion

Most of the basal horizons that crystallized during the earliest stages of cooling of the Bushveld magma are buried deep below the present level of erosion. If they could be seen, they might resemble the root zones of more deeply eroded bodies, such as the Muskox intrusion in northern Canada. This long, trough-shaped body has an exposed length of 120 km, and geophysical surveys indicate that it has at least that much additional length beneath its roof rocks. It has been tilted slightly toward its upper end so that the exposed section is equivalent to an original vertical dimension of over 6 km (**Fig. 6-12**). A thin feeder-dike of olivine-rich gabbro can be seen passing upward into a prism-shaped body of more differentiated rocks.

The main part of the intrusion is made up of numerous cyclical sequences that follow a regular order of crystallization before reverting to a more primitive composition. The cyclic units consist, from their base upward, of dunite, olivine clinopyroxenite, and olivine gabbro so that the order of appearance of crystallizing minerals was first olivine and then clinopyroxene, followed by plagioclase. Orthopyroxene appears late in the lower units, but higher in the section, it enters at an earlier stage of the cycle of crystallization. The nearly monomineralic character of the dunites testifies to efficient segregation of olivine crystals with only small amounts of interstitial pyroxene or plagioclase. Overlying layers of olivine clinopyroxenite become increasingly rich in clinopyroxene upward and end abruptly

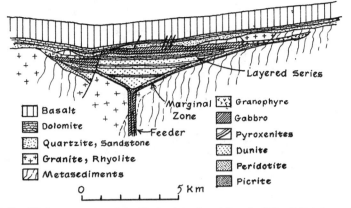

Figure 6-12 Simplified cross section of the Muskox intrusion of Canada. (After T. N. Irvine and C. H. Smith, 1967, in *Ultramafic and Related Rocks,* P. J. Wyllie, ed., John Wiley & Sons.)

with the appearance of plagioclase. From that level upward, the gabbroic rocks become proportionately richer in plagioclase.

The cyclical series may have resulted from new injections of fresh magma that reset the bulk composition and gave rise to repeated sequences of crystallization. The effects of these influxes are clearly shown by the nickel content of the rocks, which have a remarkably regular sawtooth pattern (**Fig. 6-13**). The order of crystallization of minerals was slightly different in later cycles, possibly because the new liquid differed in composition from the earlier one or, more likely, the magma became contaminated with greater amounts of siliceous roof rocks.

The sequence of crystallization corresponds well to one predicted by experimentally determined phase relations. **Figure 6-14a** shows the modal compositions of the early series of dunites, clinopyroxenites, and olivine gabbros, and **Figure 6-14b** shows how crystallization in the system olivine-clinopyroxene-silica could have produced each of the observed rocks from an evolving liquid as it first crossed the olivine field and then descended along the pyroxene boundary until it reached the plagioclase field.

The gabbroic section of the intrusion is capped by a thick layer of granophyre immediately under the roof. The volume of the siliceous liquid from which these rocks crystallized was too great to be a simple product of differentiation of the main magma. Instead, it must have come from melting of sediments in the roof where the low-density melt remained as a segregated body in the uppermost levels of the chamber.

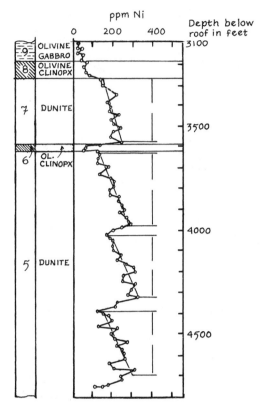

Figure 6-13 Cyclical variations of nickel in the Muskox intrusion. These saw-tooth patterns result from new injections of fresh, undifferentiated magma that replenish the concentration of nickel in the magma. Strongly included elements of this kind are useful for detecting a new influx of more primitive magma because their concentrations are rapidly depleted in the early stages of differentiation and are therefore more sensitive to a fresh liquid that resets their abundances. (After T. N, Irvine, 1969, *Geol. Soc. South Africa, Spec. Publ.* 1:441–476.)

■ The Stillwater Complex

The Stillwater Complex of southwestern Montana was probably comparable to the Bushveld intrusion in its original thickness but somewhat smaller in lateral extent. It is difficult to be sure because the rocks have been tilted, eroded, and partly buried under sediments. The base and slightly more than 5 kilometers of the lower section are exposed in a 30-kilometer long strip. Like the Bushveld intrusion, the Stillwater rocks can be divided into major stratigraphic units on the basis of distinctive minerals that appeared at successive horizons as the magma evolved (**Fig. 6-15**). Much of the section is poorly exposed, but detailed stratigraphic data are available from drill holes and mining. The intrusion has been mined for chromite, which forms rich

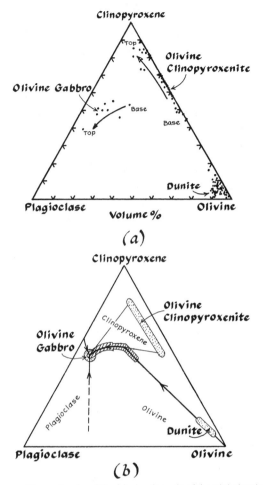

Figure 6-14 Modal compositions of rocks of the Muskox intrusion (a) and their relationships to the course of crystallization of liquids in the system clinopyroxene-plagioclase-olivine (b). (After T. N. Irvine, 1969, *Geol. Soc. South Africa. Spec. Publ.* 1:441–476.)

layers similar to those of the Bushveld and has at least one platinum-bearing horizon having much in common with the celebrated Merensky Reef but a somewhat lower grade.

The Ultramafic Zone

The principal minerals, in order of appearance, are olivine, chromite, orthopyroxene, plagioclase, and augite. The ultramafic series laid down before crystallization of primary plagioclase contains at least 15 cyclic units not unlike those of the Bushveld complex. From base to top, a typical cycle (**Fig. 6-16**) consists of dunite, commonly with a chromite horizon near its top,

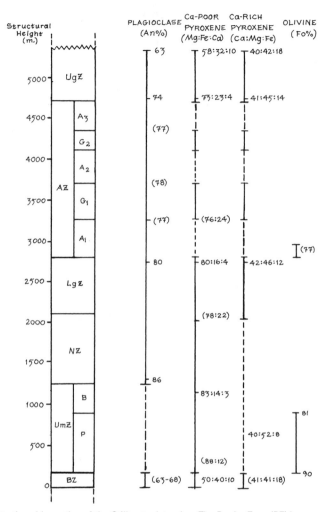

Figure 6-15 Stratigraphic section of the Stillwater intrusion. The Border Zone (BZ) is a group of hornfelsic contact rocks. The Ultramafic Zone (UmZ) consists of peridotite (P) and bronzitite (B) members. The appearance of large crystals of plagioclase marks the base of the Norite Zone (NZ), and in a similar way, the incoming of augite marks the base of the Lower Gabbro Zone (LgZ). The Anorthosite Zone (AZ) consists of alternating zones of anorthosite (A) and gabbro (G). This is overlain by the Upper Gabbro Zone (UgZ) in which the Ca-poor pyroxene is inverted pigeonite. (After L. R. Wager and G. M. Brown, 1967, *Layered Igneous Rocks*, p. 303.)

followed by poikilitic harzburgite (olivine plus orthopyroxene), granular harzburgite, and finally orthopyroxenites. Plagioclase and clinopyroxene are sporadic interstitial phases, and a few cycles lack orthopyroxene. The way in which these units evolved individually and as a whole is illustrated schematically in **Figure 6-17.**

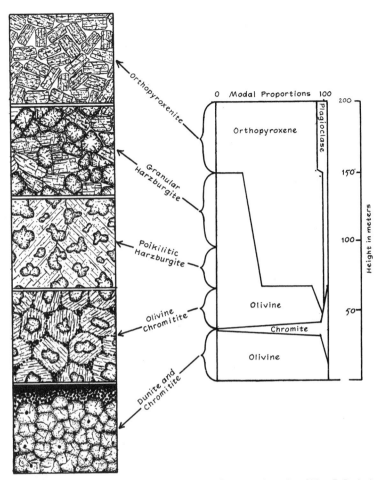

Figure 6-16 Cyclic sequence in the ultramafic zone of the Stillwater intrusion. (After E. D. Jackson, 1969, *Geol. Soc. South Africa, Spec. Pub.* 1:391–424.)

The poikilitic harzburgites consist of grains of partly resorbed olivine surrounded by larger crystals of orthopyroxene that appear to have grown by reaction of the liquid with earlier olivine. This textural relationship is in contrast to that of the granular harzburgites slightly higher in the succession where olivine and orthopyroxene appear to have grown simultaneously from the same liquid. The resorption of olivine in the poikilitic harzburgite is an excellent illustration of the reaction relationship between olivine and orthopyroxene. After first crystallizing olivine, the composition of the cooling liquid reached a peritectic, and part of the olivine was resorbed while orthopyroxene crystallized around it. The granular harzburgites, on the

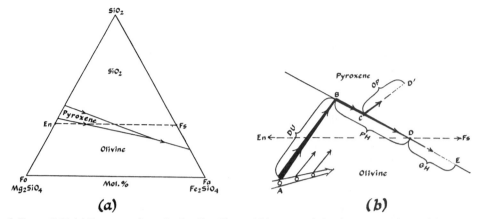

Figure 6-17 (a) The system forsterite-fayalite-silica and (b) an expanded and exaggerated part of the same system showing the evolution of liquids in cyclic sequences in the ultramafic zone of the Stillwater intrusion. DU is dunite, PH poikilitic harzburgite, GH granular harzburgite, and OP orthopyroxenite. The main course of the magma is indicated by the lower open arrow, and the course of a separate unit that differentiated independently is shown by a heavy line of diminishing width. Depending on the degree of reaction, the liquid may at some stage leave the boundary between olivine and pyroxene and embark on a new course (CD') across the field of pyroxene.

other hand, show no sign of this reaction; their textures indicate that olivine and orthopyroxene crystallized together without reaction.

The changing pattern of crystallization in these rocks seems to have resulted from independent crystallization and differentiation of individual magmas, each derived from the main reservoir but evolving as a more or less isolated system (Fig. 6-17b). As noted in Chapter 4 (Fig. 4-14), the reaction relation along the peritectic boundary between olivine and pyroxene changes to a cotectic as the liquid becomes richer in the fayalitic component. (At elevated pressures, this change takes place at somewhat lower degrees of iron enrichment.) Thus, the change from reaction in the poikilitic harzburgites to co-precipitation in their granular equivalents could reflect the enrichment of iron in successive liquids evolving down the olivine–pyroxene boundary. Higher in the cycle, the remaining liquid left this boundary and crossed the pyroxene field, crystallizing only pyroxene to form monomineralic orthopyroxenites.

The Banded Series

The appearance of abundant plagioclase at a structural height of about 2200 meters marks the base of the Banded Series, in which plagioclase joined pyroxene as a primary precipitate. The first rocks in this series are norites. (By definition, it will be recalled, norite is a gabbro in which the dominant mafic mineral is hypersthene.) Augite comes in about 700 meters higher and becomes increasingly abundant so that the norites give way to gabbronorites. Quartz, magnetite, and ilmenite are also present, but only as minor interstitial phases. A remarkable feature of this part of the intrusions is the fine "inch-scale" layering (Fig. 6-6c) in which pyroxene alternates with plagioclase in thin, regularly spaced layers through sections of several tens of meters.

Plagioclase is increasingly abundant in the upper part of the intrusion. The Anorthosite Zone contains three major units, each about 400 to 500 meters thick, consisting of more than 90-percent plagioclase. These thick, nearly monomineralic layers of anorthosite are difficult to explain. At the time they formed, the magma had long since been precipitating pyroxene and should have crystallized that mineral along with the plagioclase. Equally puzzling, the plagioclase was probably lighter than the liquid from which it grew, and there is no apparent reason why it should accumulate on the floor.

One can visualize changes of conditions that might have driven the composition of the liquid temporarily off the cotectic with pyroxene and into the plagioclase field of crystallization, but to precipitate hundreds of meters of plagioclase in this way would require changing the composition of an enormous mass of magma. Certain petrographic observations may have a bearing on the question. First, the grain size of the plagioclase is about twice that of plagioclase elsewhere in the intrusion, and although the crystals have complex oscillatory zoning, their average composition is constant over many hundreds of meters thickness. As we noted in Chapter 4, the oscillatory zoning of plagioclase is commonly attributed to growth in a convecting column of magma. This suggests that the crystals grew while freely suspended in a large volume of convecting liquid, while the denser minerals accumulated on the floor. By this reasoning, the anorthosites represent plagioclase that crystallized along with the orthopyroxene of the ultramafic and lower banded zones and remained suspended in the convecting overlying magma until the latter became choked with crystals and ceased to convect. At that point, the crystals would have settled to form massive layers. This explanation finds support in the distribution patterns of rare-earth elements. As we saw in the preceding chapter, plagioclase takes in such large amounts of europium relative to the other rare-earths that other minerals precipitated from the same liquid normally have anomalously low Eu abundances. The ultramafic rocks have just such a deficiency of Eu. Because the large amounts of plagioclase required to produce such a negative Eu anomaly are not in the ultramafic rocks, putting them into the anorthosites seems like a good way out of the dilemma.

The uppermost gabbronorites of the Upper Banded Series, the highest and most differentiated unit in the exposed section, contain hypersthene with inclined exsolution lamellae of augite like those shown in Figure 4-10, indicating that they must have inverted from pigeonite. The magma at this level was moderately enriched in iron, as well as silica, alkalies, and other residual components, but the amount of differentiation shown by the compositions of the minerals is relatively small compared with that of the upper parts of the Bushveld Complex. We cannot say how the rest of the Stillwater intrusion differentiated because only about 40 percent of the total thickness of the original intrusion is exposed; the rest lies hidden beneath the cover of younger rocks north of the present exposures.

■ The Skaergaard Intrusion

Generations of petrologists have devoted countless years to studies of innumerable layered intrusions of differing forms, compositions, ages, and tectonic settings. No two of these bodies are alike, and each has had something to teach us about magmatic differentiation. There is no way to do proper credit to all this work or even to summarize all the important concepts that have come from it, but if one were forced to select those intrusions that have been particularly instructive, the Skaergaard intrusion of East Greenland would have to be near the top of anyone's list. It is probably the single most intensively studied body of igneous rocks on Earth. Soon after it was first described by Lawrence Wager and Alex Deer in 1939, it became a prime example of igneous differentiation, and in the ensuing years, principles deduced from it had an unparalleled influence on many basic concepts of igneous differentiation. For these reasons, the intrusion merits discussing in some detail.

Form and Structure

The Skaergaard magma was emplaced during the Eocene igneous episode associated with the opening of the North Atlantic Ocean and is closely related, both in time and space, to voluminous fissure eruptions of flood basalts. The magma rose through Precambrian gneisses and amphibolites to enter the base of a thick pile of basaltic lavas and may even have breached the surface in a volcanic vent or caldera. Later uplift, tilting, and erosion have revealed a vertical section of the intrusion measuring about 3200 meters, most of which is now exposed in fresh, glaciated outcrops (**Fig. 6-18**).

Although the body was originally thought to have been a steep-sided pipe, geophysical surveys have shown that it narrows sharply into one or two slender feeders at a shallow depth. The magma seems to have been emplaced in a single surge, displacing a block of the gneissic basement and flaring into an asymmetrical wedge near the gneiss-basalt contact. A dense,

Figure 6-18 Simplified map of the Skaergaard intrusion of East Greenland. Units of the Layered Series are Lower Zone (LZ) a, b, and c; Middle Zone (MZ); and Upper Zone (UZ) a, b, and c. The Marginal Border Series (MBS) and Upper Border Series (UBS) merge in the southern part of the map and are not shown separately. The cross-section below is drawn along the line A–B normal to the southward dip of the intrusion.

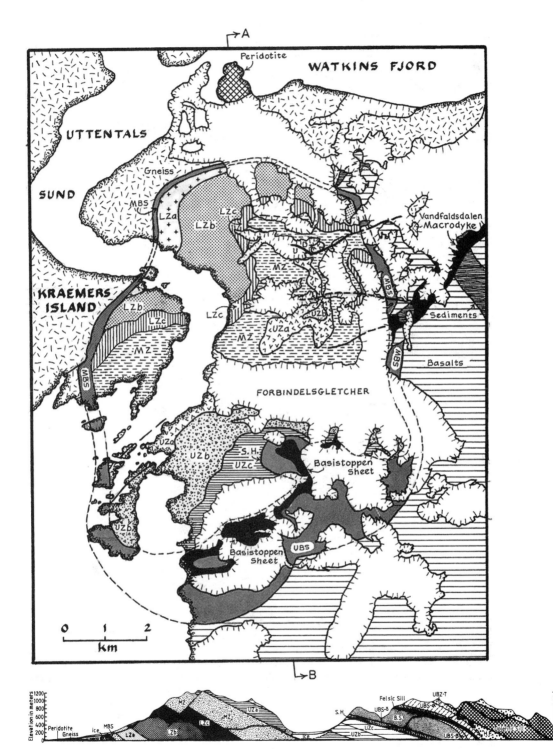

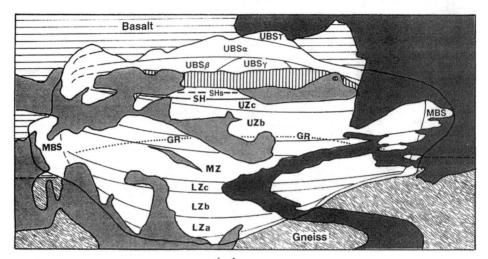

(a)

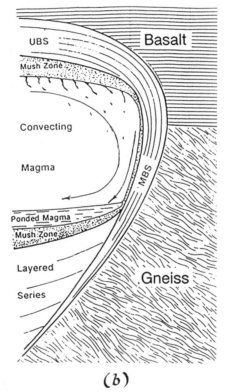

(b)

Figure 6-19 (a) The exposed part of the Skaergaard Intrusion shown in Figure 6-18 has been projected to a plane normal to the layering in order to show an east–west cross-section. The dark shading is water and the lighter shading snow and ice. As the magma crystallized under the roof and along the walls (b) the adjacent liquid was enriched in iron and is believed to have descended to pond on the floor, while some of the heavy liquid and crystals produced under the roof probably sank into the main magma. The fronts of crystallization converged at the Sandwich Horizon (SH), but the interstitial liquid continued to differentiate and migrate upward. As a result, strongly excluded elements reach maximum concentrations at a second Sandwich Horizon (SH$_S$) in the lower part of the Upper Border Series. Interstitial granophyre appears above the dotted line labeled GR. (The lower diagram is schematic and not to scale.)

relatively fine-grained chilled margin, about a meter in width, separates the wall rocks from the coarser-grained gabbro of the interior.

Lithologic Units

The intrusion has been divided into three major series (**Fig. 6-19**), the Layered Series that crystallized from the floor upward, the Upper Border Series that crystallized from the roof down, and the Marginal Border Series that crystallized inward from the walls. As the Layered Series and Upper Border Series converged, they met at the "Sandwich Horizon," where the last and most strongly differentiated magma crystallized. Each of the major units is subdivided on the basis of its phase layering, and although their mineral sequences differ in minor ways, they followed parallel trends of differentiation.

The 2500-meter section of the Layered Series is divided into a lower, middle, and upper zone, and the lower and upper zones are further divided into three subzones (**Fig. 6-20**). The divisions of the two Border Series are similar to those of the Layered Series but more compressed. The order, of course, is inverted in the Upper Border Series because the rocks crystallized from the roof down, and the Marginal Border Series is zoned inward from the wall.

The petrographic basis of the phases layering is illustrated in **Figure 6-21**. Calcium-rich pyroxene is an interstitial or poikilitic mineral in the lowermost unit, Lower Zone A, but became a primary liquidus phase at the base of Lower Zone B. The base of Lower Zone C is marked by the first appearance of abundant iron oxides. Olivine, which was a primary precipitate in the earliest rocks, continued to crystallize up to the base of Middle Zone, at which level it disappeared only to reappear in the rocks of the Upper Zone. Apatite is an abundant primary phase above the base of Upper Zone B, and the base of Upper Zone C is defined by pyroxene that crystallized initially with a structure of wollastonite (or ferrobustamite) and on cooling inverted to hedenbergite.

The mineralogical changes in the Skaergaard rocks broadly resemble those of other intrusions discussed earlier, but the variations are more regular. No firm evidence has been found for repeated injections of fresh magma, and no important cyclical variations have been identified. The exposed section lacks an ultramafic zone of the kind laid down in the early stages of differentiation of the Bushveld and Stillwater magmas. (Blocks of peridotite found in parts of the Marginal Border Group were once thought to come from a deeper, unexposed, ultramafic zone but are now known to be xenoliths from an adjacent body of older and apparently unrelated rocks.)

The trends of cryptic layering are smooth and unbroken. Olivine and clinopyroxene became increasingly iron-rich upward until, in the last liquid to crystallize, they reached the compositions of the pure end-members, fayalite and hedenbergite (**Fig. 6-22**). Calcium-poor pyroxene is much less abundant in the Skaergaard rocks than it is in the other large intrusions we have

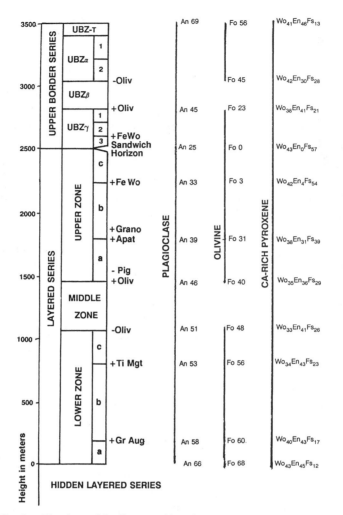

I Figure 6-20   Stratigraphic column of the Skaergaard intrusion.

considered. It may not have been a primary phase because it typically forms reaction rims on earlier crystals of olivine or augite. It ceased to crystallize, except by exsolution from Ca-rich pyroxene, soon after olivine reappeared at the base of the Upper Zone. It never crystallized directly as an orthopyroxene, but only as monoclinic pigeonite that inverted to hypersthene on cooling. (An example of pyroxene exsolution that reflects the change of structural form is shown in Figure 4-10b.) In contrast to the ferromagnesian minerals, which evolved to an extreme degree of iron enrichment, the change in plagioclase is limited to a much smaller range. It advances from An_{66} at the base to An_{30} at the top of the Layered Series (Fig. 6-22b).

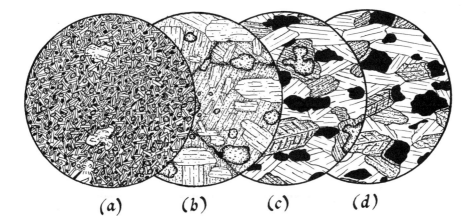

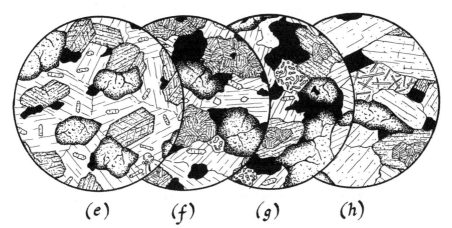

Figure 6-21 Petrographic features of the principal rocks of the Layered Series of the Skaergaard intrusion. (a) Rocks of the chilled margin have crystallized to granular "hornfelsic" rocks containing scattered xenocrysts of reversely zoned and partly resorbed plagioclase. (b) Lower Zone A consists of plagioclase and olivine encased in poikilitic crystals of clinopyroxene. (c) In Lower Zone C clinopyroxene forms separate crystals, some with exsolution lamellae of inverted pigeonite. Magnetite is an abundant primary phase, and in this particular example, olivine is partly resorbed and encased in pyroxene. (d) Rocks of Middle Zone contain no primary olivine, but reaction has produced thin rims of that mineral between pyroxene and magnetite. (e) Olivine returns as a primary mineral in Upper Zone; in Upper Zone B, it is joined by apatite. Upper Zone C (f) is distinguished by the mosaic structure of hedenbergite that has inverted from an original wollastonite structure. Granophyric quartz and alkali feldspar fill interstices between the other minerals. The Sandwich Horizon (g) contains the same minerals as Upper Zone C, but granophyric quartz and alkali feldspar are more abundant. Zircon, sphene, and sulfides are accessory minerals. The rocks of the Upper Border Series (h), represented here by an example from UBS-γ, are somewhat more plagioclase rich than equivalent units of the Layered Series. Some contain elongated crystals of quartz that originally crystallized as tridymite and on cooling inverted to quartz (diameter of fields 3 mm).

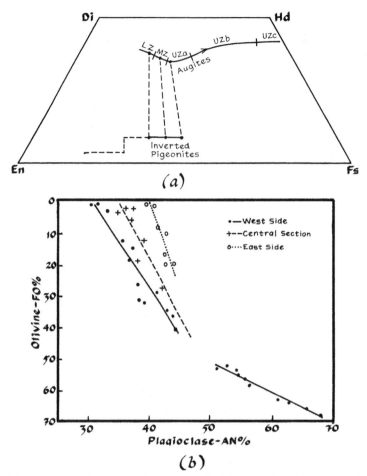

Figure 6-22 (a) Compositional variations of pyroxenes in the Layered Series. Tie lines connect coexisting Ca-rich and Ca-poor pyroxenes. (b) Variations of plagioclase and olivine compositions with height in the center and margins of the Layered Series.

Granophyres

The intrusion contains numerous bodies of granophyre that account for only a small fraction of the total volume but occupy an important place in the differentiation series. They can be grouped into three types. The largest and most common are those that form dikes and sills (**Fig. 6-23**). Because they intrude and cross-cut the other rocks, they are referred to as "transgressive granophyres." Their trace elements and isotopic composition show that they are genetically related to the gneissic country rocks and must have been derived from them by partial melting of the metamorphic wall rocks.

Some of the smaller dikes in the upper part of the Layered Series have a much wider range of compositions, even within a single dike, and unlike the

Figure 6-23 A granophyric sill in the upper levels of the Skaergaard Intrusion is the result of partial melting and mobilization of felsic metamorphic rocks underlying the intrusion. The part shown here is about four meters thick.

transgressive granophyres intruded from greater depths, they are visibly rooted in the ferrogabbro. They appear to have been segregated into pipes and dilational fractures in the crystallizing gabbro and can be interpreted as true differentiates of the Skaergaard magma.

A third group has been called "melanogranophyres" because of the large, irregular clots of ferromagnesian minerals that give them a somewhat more mafic bulk composition. They form pods up to a meter or two across in Upper Zones B and C (**Fig. 6-24**). These granophyres are also differentiates of the iron-rich gabbroic magma, but they separated as immiscible liquids at a more advanced stage of differentiation. Granophyric liquids rich in silica and alkalies have been shown experimentally to be immiscible with the more iron-rich and silica-poor liquid from which the host rocks crystallized (**Fig. 6-25**).

Trends of Differentiation

Differentiation produced more or less parallel changes in the bulk chemical compositions of the rocks in all three series. Most obvious is the steady

Figure 6-24 Irregular pods of granophyre separated as immiscible liquids during crystallization of Upper Zones B and C. The host gabbro must have been crystallized to the extent that it was viscous enough to prevent the buoyant granophyres from rising to a higher level. The pod shown here is about a meter across.

increase in iron and decrease of magnesium throughout the entire sequence (**Fig. 6-26**). Iron enrichment is less pronounced in the Upper Border Series, probably because iron-rich interstitial liquids (along with a certain proportion of the mafic minerals) drained from the zone of crystallization under the roof. Similarly, the Marginal Border Series retained little of the liquid that evolved from crystallization of magma flowing down the walls (Fig. 6-19b). The Layered Series is thought to have crystallized from this liquid after it ponded on the floor.

The distribution patterns of included trace elements reflect the fractionation of minerals in which they are most strongly concentrated. Compare, for example, the distribution of nickel and vanadium (**Fig. 6-27**). The former, because it enters all the mafic minerals, declines steadily in abundance throughout the sequence. Vanadium, on the other hand, does not enter olivine or pyroxene in any substantial amounts but is strongly partitioned into magnetite and ilmenite. For that reason, its concentration in the

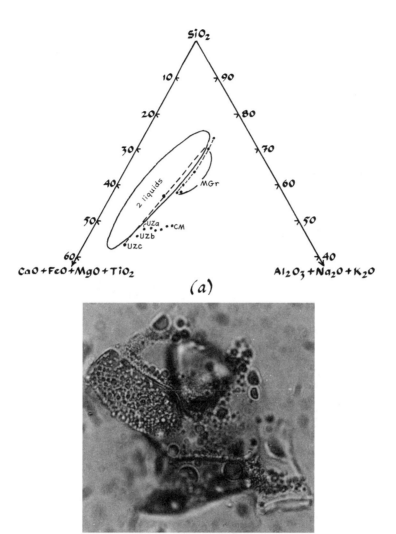

(a)

(b)

Figure 6-25 (a) The compositional relationships of immiscible liquids are illustrated in a diagram plotting SiO_2 versus $Na_2O + K_2O + Al_2O_3$ and $CaO + FeO + MgO + TiO_2$. The course of the main gabbroic magma is shown evolving from the chilled margin (CM) through successive stages until it reaches the boundary of a two-liquid field and then separates into two fractions, one rich in iron, magnesium, and titanium, and the other in silica. The general form of the field of two liquids shown here was first outlined by Edwin Roedder (1951, *Amer. Min.* 36:282–286). The immiscible relationships of granophyric and ferrogabbroic liquids in the Skaergaard intrusion was verified experimentally by holding an intimate mixture of the two compositions [indicated by the dot inside the two-liquid field of (a) at their liquidus temperature and allowing them to separate into two discrete liquids, shown here (b) as blebs of glass of differing refractive indices]. The width of the view is about 0.5 mm.

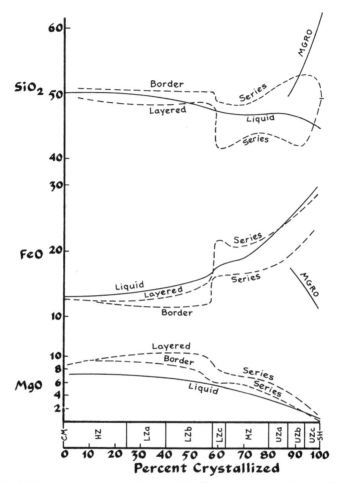

Figure 6-26 Variations of some key major elements of the Skaergaard rocks and magma. Estimated liquid compositions are shown by solid lines, rocks of the Layered and Border Series are shown by broken lines. Two curves are shown for the variations of silica and iron, one for the principal rocks and another for the immiscible melanogranophyre (MGRO) that separated from the iron-rich, silica-poor magma in the upper-most parts of the Layered Series.

Layered Series increased until the oxide minerals came on the liquidus in Lower Zone C, after which it was rapidly depleted.

Excluded elements, such as Rb, Nb, and the rare-earths, have low concentrations in the lower part of the Layered Series but increase sharply near the top of Middle Zone (**Fig. 6-28**). The abrupt change comes at the same level at which the ratio of fluorine to chlorine in apatite suddenly increased. Chlorine tends to follow water when a vapor phase is exsolved, while fluorine is more strongly concentrated in silicate melts. Hence, the sharp increase

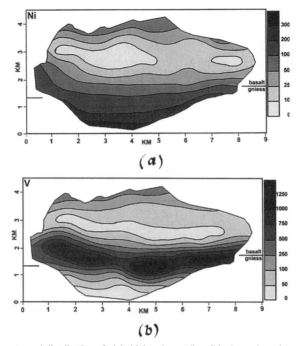

Figure 6-27 The contoured distribution of nickel (a) and vanadium (b), shown here in sections based on the projection in Figure 6-19a, illustrates how strongly the included elements are controlled by the minerals into which they are partitioned. Nickel, which is taken into all the main mafic minerals, declined steadily throughout the crystallization sequence. In contrast, the only minerals that take in significant amount of vanadium are the iron oxides. As a result, the element increased until those minerals began to crystallize in abundance in Lower Zone C and Middle Zone and then declined sharply.

of F/Cl must indicate that the magma became saturated with water at this level and exsolved a vapor phase that depleted the chlorine content of the melt. Sulfides and chalcophile metals such as copper, gold, and the platinum-group elements were deposited at this same level. If the strongly excluded elements were carried upward in this vapor phase, it would explain why they do not reach their highest concentrations in the Sandwich Horizon but about 100 meters higher in the Upper Border Series.

Estimating Temperatures and Pressures

We have several methods by which the mineral assemblages of rocks can be used to deduce the temperatures and pressures at which they crystallized. The most reliable estimates for the Skaergaard intrusion are those for levels close to the Sandwich Horizon (**Fig. 6-29**).

Upper Zone C, as we noted earlier, is characterized by an unusual, iron-rich clinopyroxene that first crystallized with the structure of Fe-wollastonite and, after a short interval of cooling, inverted to hedenbergite.

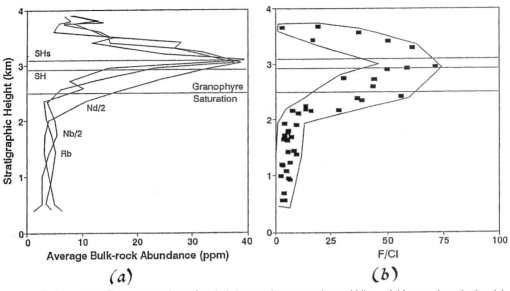

Figure 6-28 The concentrations of excluded trace elements such as rubidium, niobium, and neodymium (a) remain nearly constant up to a level near the top of Middle Zone where they suddenly begin to increase. The change of behavior comes at the same level where the F/Cl ratio in apatite increased as a result of exsolution of a water-rich vapor (b). The excluded elements appear to have been carried in the volatile phase, which infiltrated upward and reached maximum concentrations about 100 meters above the Sandwich Horizon (SH).

A similar polymorphic change from tridymite to quartz occurred in the Upper Border series just above the Sandwich Horizon. Some of the quartz crystals in these rocks have a distinctive, needle-like morphology inherited from an original tridymite structure, while the silica mineral in rocks below that level is more equidimensional because it crystallized directly as quartz and has not inverted from another form.

The experimentally determined conditions of temperature and pressure for the inversion to hedenbergite are plotted in Figure 6-29b. The pressure dependence of the tridymite-quartz inversion has not been determined experimentally but can be calculated by means of the Clapeyron equation. At atmospheric pressure, tridymite inverts to quartz at 867°C (or 1140°K), giving off about 7.43 calories per gram. At that temperature, the density of quartz is 2.536, while that of tridymite is 2.189. Using these values, we obtain

$$\frac{dT}{dP} = \frac{T\Delta V}{\Delta H} = \frac{1140\,(1/2.189 - 1/2.536)}{7.43 \times 42.7} = 0.23°\text{kg}^{-1}\text{cm}^2 \text{ or } 229°\text{kilobar}^{-1} \tag{6-1}$$

for the rate at which the inversion temperature increases with pressure. (The factor 42.7 kg cm cal^{-1} is the mechanical equivalent of heat.) When plotted

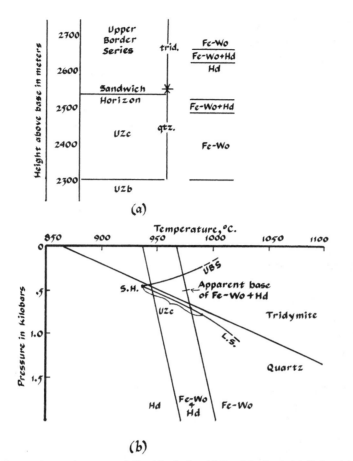

Figure 6-29 Temperature and pressure relationships in the vicinity of the Sandwich Horizon (a) are defined by the stability relations of the polymorphic forms of SiO_2 and $CaFeS_2O_6$ (b).

on Figure 6-29b, the inversion curve for the polymorphs of SiO_2 has a high-angle intersection with those for hedenbergite and Fe-wollastonite and provides an excellent grid of temperature and pressure to which we can relate the mineral assemblages at that level (Fig. 6-29a).

The rocks of the Sandwich Horizon contain quartz and hedenbergite that crystallized originally in those forms, but in rocks only a short distance above that level, these minerals originally crystallized as tridymite and Fe-wollastonite. Rocks just below the Sandwich Horizon are similar to those above it, but the quartz shows no evidence of having first crystallized as tridymite. The curves in Figure 6-29b show how the temperatures and pressures of the Layered Series and Upper Border Group would correspond to these phase relationships. Note that the rocks of the Sandwich Horizon are narrowly confined to a liquidus temperature of about 940°C and a pressure close to 500 bars, the only conditions under which SiO_2 and pyroxene could have crystallized initially as quartz and hedenbergite.

If we assume a density of about 2.8 g cm^{-3} for the overlying rocks, the pressure indicated for the Sandwich Horizon corresponds to a depth of 500 × 1020/2.8 = 182,000 cm, or about 2 km. Deducting the visible thickness of about 1000 meters for the overlying Upper Border Series, the roof of the intrusion could not have been much deeper than one kilometer and might well have been even less.

Again, using the thickness and density of the rocks, we can calculate the pressure at the level in Upper Zone C at which the pyroxenes entered the field of Fe-Wo + Hd, and we find that it falls at the level indicated in Figure 6-29b, well inside the field of tridymite. We know, however, that the silica mineral in these rocks was always quartz. The most plausible explanation for this apparent anomaly is that the rocks crystallized, not at a single temperature, but through a range of 40 or 50 degrees between their liquidus and solidus. By this reasoning, the quartz would have crystallized later and at a lower temperature than the pyroxene. This inference is supported by the textural relationships of the minerals in the rocks of this zone (Fig. 6-21f), which show that quartz was a late-crystallizing interstitial mineral.

Other methods of estimating the temperatures and pressures of crystallization of the Skaergaard rocks are based on the experimentally determined temperature–composition relationships of coexisting iron-bearing minerals. We saw in Chapter 4 that the compositions of coexisting iron-titanium oxides are governed by the temperatures and oxygen fugacities at which they crystallized. If magnetite and ilmenite are both present in the same rock and have not changed composition since they crystallized together from a liquid, then we can use the relationships shown in Figure 4-16 to estimate what those conditions were when the Skaergaard magma crystallized. When this is done, we find that the temperatures for the Skaergaard oxide minerals are as much as 200 degrees below those at which the gabbros would be totally crystallized under any conceivable conditions. The explanation for this inconsistency must be that the minerals re-equilibrated at lower temperatures during slow subsolidus cooling.

Similar methods are based on experimentally determined equilibrium constants for reactions such as:

$$2Fe_3O_4 + 3SiO_2 \leftrightarrow 3Fe_2SiO_4 + O_2 \qquad (6\text{-}9)$$

magnetite   quartz   fayalite   oxygen

and

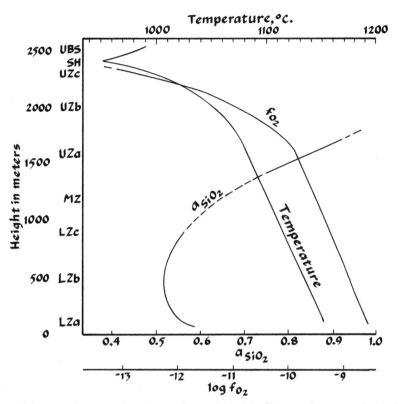

Figure 6-30 Temperature, oxygen fugacity, and silica activity of the Skaergaard magmas calculated from the compositions of coexisting mineral phases and experimentally determined reaction constants. (After R. J. Williams, 1971, *Amer. Jour. Sci.* 270:334–360; S. A. Morse, 1980, *Amer. Jour. Sci.* 279:1060–1069; and S. A. Morse, D. H. Lindsley, and R. J. Williams, 1980, *Amer. Jour. Sci.* 280-A:159–170.)

$$Mg_2SiO_4 + Fe_2Si_2O_6 \leftrightarrow Fe_2SiO_4 + Mg_2Si_2O_6 \qquad (6\text{-}10)$$

forsterite   ferrosilite   fayalite   enstatite

The distribution of iron, silica, and oxygen between a liquid and crystals (6-9) or of iron and magnesium between coexisting olivine and pyroxene (6-10) is a function of the intensive parameters temperature, pressure, and oxygen fugacity. If the reactions are allowed to reach equilibrium under a range of controlled conditions in a series of laboratory experiments, the compositions of coexisting phases can be measured, and in this way, a relationship between compositions and the conditions of crystallization can be established. The technique is subject to uncertainties due mainly to the sensitivity of the reactions to small differences of conditions and to the possible effects of other components of the natural system. It is encouraging, however, that the estimates of temperature and oxygen fugacity obtained in this way (**Fig. 6-30**) are in fairly close agreement with the values obtained by independent experimental studies of the stability relations in the natural rocks.

Subsolidus Re-Equilibration

The discovery that the iron-titanium oxides had re-equilibrated at much lower temperatures naturally raised the question of whether the compositions of other minerals may have been changed as well. The silicate minerals appear to be fresh and unaltered, but on closer scrutiny, it soon became apparent that many of the mineral assemblages had indeed been altered. A good example is the angular blocks that are common in much of the central part of the Layered Series (**Fig. 6-31a**). Large gaps in the Upper Border Series correspond to places where these blocks, some as large as a house, became detached from the roof series and fell to the floor, but their present composition no longer corresponds to that of the units from which they fell. Plagioclase has replaced most of the original mafic minerals, while the latter have been concentrated in the surrounding gabbro. This change must have occurred after the blocks reached the floor, for if they had had such plagioclase-rich compositions at the time they were dislodged, they could not have sunk through the iron-rich magma.

Other examples are found in the lower part of the Layered Series where parts of the original layered gabbro have been replaced by a featureless anorthosite (Fig. 6-31b). The residual islands of gabbro retain the original orientation of their layers, but they have been greatly enriched in mafic minerals—the same components that have been depleted from what is now the surrounding anorthosite. In the same part of Lower Zone A, irregular schlieren of almost pure anorthosite and pyroxenite cut across the layering at all angles (**Fig. 6-32a**). The strontium isotopic ratios of the plagioclase and pyroxene in the two closely associated rocks differ from one another and from those of the host gabbro as well (Fig. 6-32b). Such differences could only be produced by metamorphic differentiation of the rocks during a long period of subsolidus cooling.

Rates of Cooling

As we saw in Chapter 2, the rate at which a body of magma cools is governed chiefly by the mode of conductive heat transfer across its boundary. In the Skaergaard case, the extensive hydrothermal alteration of the roof and upper walls indicates that cooling was enhanced by the circulation of meteoric water through the overlying permeable lavas (**Fig. 6-33**). A convecting hydrothermal system of this kind can greatly increase the rate of cooling if the circulating groundwater penetrates close to the crystallizing

(a)

(b)

Figure 6-31 (a) Fragments of the Upper Border Series became detached from the roof and sank to the floor where they were converted to almost pure anorthosites. (b) Similar metasomatic changes affected parts of the Layered Series. In this example from Lower Zone A, the original layered gabbro has been largely replaced by a more plagioclase-rich, featureless rock while the residual islands of layered rock have been enriched in mafic components.

(a)

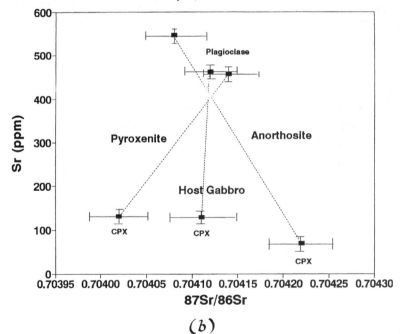

(b)

Figure 6-32 (a) Irregularly shaped schlieren of anorthosite and pyroxenite cut across the layered gabbros of Lower Zone. (b) The strontium isotopic ratios of the co-existing minerals in the same rocks differ from one another and from those of the host gabbro.

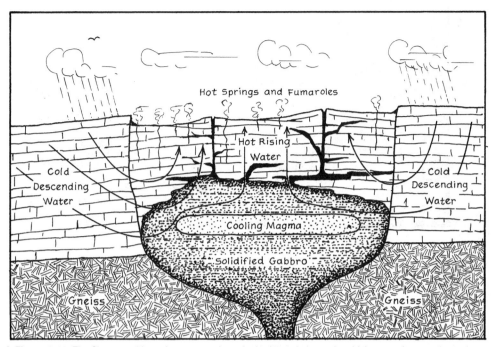

Figure 6-33 The Skaergaard intrusion set up a large hydrothermal system in which meteoric water heated by the cooling gabbro convected through the permeable basaltic rocks of the roof and upper walls. (After H. R. Taylor and R. W. Forester, 1979, *Jour. Petr.* 20:355–419.)

margins and reduces the width of the zone across which heat is transferred by conduction.

The magnitude of this hydrothermal system can be evaluated from its effect on the isotopic composition of the rocks. The two stable isotopes of oxygen, ^{18}O and ^{16}O, have very different proportions in rocks and meteoric water. In igneous rocks they have a characteristic ratio inherited from the mantle, while the oxygen of meteoric water has an isotopic character governed by processes in the atmosphere and hydrosphere. As hot water flows through the rocks it exchanges oxygen with the minerals, and the ratio of isotopes is altered by an amount that is a function of the temperature and quantity of circulating water.

An ingenious mathematical model was devised to analyze the flow pattern of meteoric water and its effect on the rate of cooling of the intrusion. Using the experimentally determined rates at which oxygen is exchanged between water and the major minerals, one can calculate the amount and temperature of the water that moved through the rocks. Knowing from field relationships the permeability of the rocks and the probable form of the

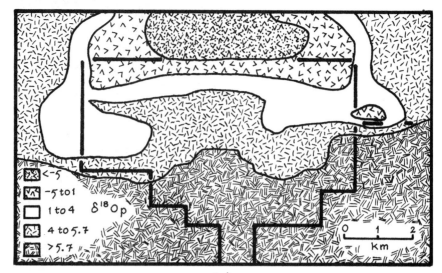

(a)

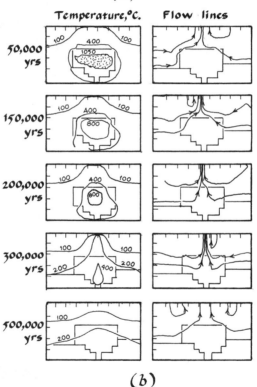

(b)

Figure 6-34 Convecting meteoric water exchanged oxygen with the rocks and reset their isotopic ratios by amounts that were functions of the temperature and amount of water that flowed through the rock. The result is the pattern of zoning shown in the schematic diagram (a) used to model the system. The calculated flow lines and temperature contours of the hydrothermal system (b) provide a way of estimating the cooling history of the intrusion. The speckled pattern indicates magma; all else is solid rock. (After D. Norton and H. P. Taylor, 1979, *Jour Petr.* 20:421–486.)

groundwater system, one can set up a computer model to determine the flow lines and estimate how they affected the magma and surrounding rocks. Results of this calculation (**Fig. 6-34**) show that the magma was totally crystallized after 130,000 years, about half the time that would have been required had there been no hydrothermal system. Another 400,000 years passed before the body reached ambient temperatures.

The manner in which this pattern of cooling affected the interior of the crystallizing intrusion can be modeled in the laboratory by a simple, low-temperature analog (**Fig. 6-35**). Because of the insurmountable problem of scaling the physical parameters to appropriate dimensions, a model of this kind cannot be a precise replica of a large intrusion. A good model need not be an exact replica of nature, however. By judiciously scaling the dimensions and properties of the materials, one can match the fluid dynamic behavior of the model to that of the natural body and thereby produce an analog that is at least qualitatively similar. The model can then be of great value, especially if it generates unexpected insights that can be tested by means of observable geological relationships.

In keeping with the studies of oxygen isotopes, the laboratory experiment was designed so that the "magma" was cooled mainly through the roof and upper walls where it precipitated the Upper and Marginal Border Series (Fig. 6-35b). The combined thermal and compositional boundary layers carried a crystal-laden liquid down the walls and across the floor. Most of the heavier crystals suspended in these currents were dropped near the base of the wall, while the remaining liquid ponded on the floor and crystallized to form the Layered Series. Thus, the liquid from which the Layered Series crystallized was not the main magma but one that had already evolved somewhat by crystallization on the walls. We shall see that this result has an important bearing on our interpretations of how the series of differentiated liquids evolved.

■ Differentiation Mechanisms

Early studies of the intrusion inferred that the principal mechanism of differentiation was gravitational segregation of crystals. Structural features, such as cross-bedding and graded layering (**Figs. 6-36** and 6-7a), were seen as overwhelming evidence for sedimentary deposition. We have since learned that the picture is more complex. Crystals undoubtedly settled out of currents flowing down the walls and across the floor, but for the reasons explained earlier, we now know that little of the layering in the interior of the Layered Series is sedimentary and that most of the crystals nucleated and grew in situ. When this became apparent, we were forced to abandon a con-

(a)

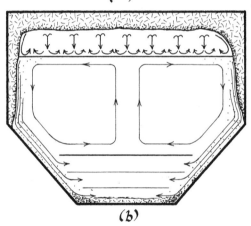

(b)

Figure 6-35 Cooling and crystallization of the Skaergaard magma have been modeled in the laboratory using dilute aqueous solutions of salts, such as sodium carbonate, in a tank with the form of an idealized cross-section of the intrusion (a). By varying the properties of the solution and the ways in which it cools and crystallizes, it is possible to study the nature of convective flow and compositional stratification. The results of one such experiment (b) illustrate a possible form of convection in the major zones. Because the upper part of the intrusion was contaminated with felsic gneiss, its density was reduced to the extent that it convected and crystallized more or less independently to form the Upper Border Series. Meanwhile, the main magma cooled and crystallized against the walls to produce the Marginal Border Series. A heavy compositional boundary layer descended to pond on the floor where it formed the rocks of the Layered Series.

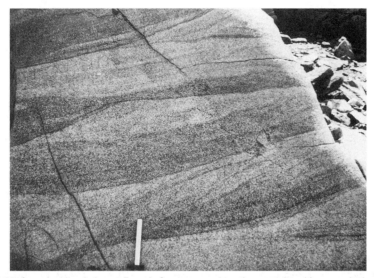

Figure 6-36 Cross-bedding in Lower Zone A. Structures such as these are found in a marginal zone where currents descending along the walls reached the floor and dropped most of the crystals they held in suspension.

cept that for more than a century had been our principal explanation for magmatic differentiation.

Two other mechanisms of crystal fractionation have been proposed as plausible alternatives: convective exchange and compaction. Convective exchange is driven by the density contrast of an interstitial liquid and the main reservoir of magma. In the case of the Skaergaard Layered Series where the liquid became more iron-rich and hence denser as crystallization advanced, the liquid between crystals growing on the floor would be gravitationally stable, but it would be reversed when iron enrichment gave way to increasing silica and volatiles. In the roof series, the relationship would, of course, be the reverse of those on the floor. A gravitationally unstable liquid would infiltrate up or down by porous flow through an interlocking network of crystals.

The factors governing the rate of flow of the liquid, q, can be seen from the Darcy equation:

$$q = \frac{kdP}{\eta dZ} \tag{2-16}$$

where k is the permeability of the crystals and η the viscosity of the liquid. The pressure gradient, dP/dZ, includes a buoyancy term $\Delta\rho g$ for the difference of the liquid densities. Any differential motion of the crystal framework can also be added to this equation. Thus, the effectiveness of the process depends not only on the density contrast of the liquids but also on their viscosities and the permeability of the crystals through which they move. The properties of the liquids can be estimated from their compositions and temperatures and seem to be within the ranges required for effective porous flow of most basaltic liquids, but the permeability of the crystal mush is more difficult to assess because it is linked to other processes, such as rates of nucleation and growth of crystals, that are a function of the system as a whole. The same is true of thermal properties. The liquids in both limbs of a convective system are moving in a thermal gradient that is buffered by the heats of reactions between liquids and crystals. This reduces both the thermal and chemical gradients and tends to compress the zone of crystallization and reduce permeability.

The second alternative mechanism, compaction, has long been seen as a potentially important process of crystal–liquid fractionation, but the process has only recently been examined quantitatively. The studies of the Palisades Sill mentioned earlier were the first to demonstrate how compaction of a crystallizing mush could lead to expulsion of residual liquids and evolution of an overlying magma. They also indicated how a pore liquid might re-equilibrate with the crystals through which it rises. The effectiveness of this process, like that of convective exchange, is critically dependent on the permeability of the crystal mush, which, in turn, is a function of the rate of cooling and crystallization.

Compaction involves a variety of mechanisms. If the volume proportions of crystals are still small, the grains can compact by mechanical reorganization, but with increased crystallinity, pressure solution and re-precipitation become more important. Crystals dissolve at points of contact and grow on nearby crystal faces that are not under stress. Further compaction normally requires very slow cooling because rates of pressure solution may be limited by the slow diffusion of material from grain contacts through the interstitial melt to free crystal faces. The process can be greatly enhanced however, if a fluid is infiltrating through the remaining pore spaces.

■ Reconstructing a Liquid Line of Descent

It seems ironic that the more efficient the process of crystal–liquid fractionation the more difficult it is to define the differentiated liquids it produced. To the extent that components excluded from the principal minerals are effectively removed, the bulk compositions of the rocks diverge from those of the liquids from which they crystallized. Moreover, the mineral assemblages may re-equilibrate during slow cooling so that their original compositions and textural relationships are no longer preserved. As a result, even though we have a wealth of detailed information on the crystalline assemblages, there is no simple way of tracing the evolution of the magma from which they grew.

Several different methods have been used to attack this problem. The earliest and most direct approach is based on a simple mass balance. When Lawrence Wager studied the Skaergaard rocks half a century ago, he reasoned that, whatever the exact mechanism of differentiation, the entire intrusion was the sum of its parts and the composition of the liquid at any stage was the result of separating from the initial magma all the rocks that had crystallized up to that point. Taking the chilled margin as the initial composition of the magma, he subtracted from it the volume-weighted composition of each successive rock unit to obtain the compositions of a series of derivative liquids ending with the Sandwich Horizon. The liquid line of descent he arrived at was not very different from our best estimates today, but we now know that his method had several limitations.

First, the composition of the chilled margin no longer matches the original bulk composition of the magma (if it ever did). It differs from place to place because it was altered to various degrees by reaction with the wall rocks. Second, the mass-balance calculation is no better than the values used for the volume of each rock unit. Although it is a simple matter to measure the stratigraphic thicknesses of the individual units, it is much more difficult to estimate their original lateral extent, and any calculation based on volumetric proportions is subject to large errors. Third, the calculation does not use realistic constraints from crystal–liquid equilibria, in particular the relevant compositions and proportions of co-precipitating minerals.

A variation on Wager's summation method has been devised that surmounts the problem of uncertain volumetric relations. It uses known crystal–liquid relationships compiled from experimental studies of phase equilibria to formulate empirical equations for the temperatures at which minerals begin to crystallize from a liquid of a given composition. For example, the temperature at which olivine crystallized is determined by an equation such as

$$T_{olivine} = C + a\ Si + b\ Fe^{3+} - c\ Fe^{2+} + d\ Mg + e\ Ca + f\ (Al + Na + K)^{0.5} - g\ H_2O \qquad (6\text{-}2)$$

where C is a constant; Si, Fe, etc. are mole fractions of the oxides; and a, b, c, etc. are empirical constants reflecting the effect of each element on the liquidus temperature.

With a set of equations of this kind, one can start with an assumed initial composition and calculate the temperatures at which the principal minerals would crystallize. The temperatures for the minerals known to have been on the liquidus at that stage should have the same high value, while those of all other minerals should be lower. One subtracts from the liquid composition the proportionate amounts of elements that would be fractionated by each liquidus mineral during a specified interval of cooling. This gives a derivative liquid that is then treated in the same way, and the calculation is carried forward step by step until the entire magma is crystallized. The amount of material removed in fractionated solids is determined, not in proportion to stratigraphic measurements but from the proportions predicted by phase relationships.

The results obtained in this way are similar to other estimates for the early stages of differentiation, but after large amounts of iron-titanium oxides are removed at Lower Zone C and beyond, silica and alkalies begin to increase, and a large amount of granophyric liquid is predicted—much more than is seen in the field. These results had a certain appeal because they predicted a trend of differentiation similar to that of many tholeiitic volcanic series. The missing granophyre was easily explained as having been expelled in eruptions of rhyolitic tephra, a suggestion consistent with the presence of a few siliceous ash beds in the overlying volcanic series. Nevertheless, the scheme was unanimously rejected by geologists who had seen the Skaergaard rocks in the field and found it inconceivable that the large mass of ferrogabbro in the Upper Zone of the Layered Series could come from a silica-rich, iron-poor magma. The controversy was effectively resolved when it was noted that the iron content of plagioclase, which is a direct function of the activity of that element in the liquid, continues to increase throughout the entire sequence. The immiscible separation of melanogranophyres and ferrogabbros near the end of the crystallization sequence was equally fatal because these relationships develop only in liquids with very high iron contents.

Several reasons have been proposed to explain why the liquid trend predicted by this method diverged toward high silica contents. Part of the problem may be in the imperfect nature of the crystal fractionation. The material removed from the liquid included, not only the liquidus phases used in the calculation but a certain proportion of residual liquid as well, and this may have been enough to impede silica enrichment. As we noted in the preceding chapter, a small difference in the relative concentrations of an element in the liquid and crystallizing minerals can set a trend toward enrichment or depletion that becomes magnified with progressive differentiation (Fig. 5-13a).

If fractionation was less than perfect and the rocks retained some finite amount of interstitial liquid that crystallized in place, it seemed logical that one could heat the rocks to a temperature slightly above the solidus, restore this liquid, and analyze it. Suppose, for instance, one finds that a representative specimen contains a certain amount of an element that is not included in any of the minerals that were liquidus phases at that stage. The abundance of that component should be proportional to the amount of liquid trapped between the main crystallizing minerals. Phosphorus in rocks below Upper Zone B would be an example of such a component. If the rock contains 0.2-percent P_2O_5 and the magma at that stage contained on the order of 2.0 percent, then about 10 percent of the rock would be the product of crystallization of trapped liquid. One could then reheat the rock under appropriate conditions of pressure and oxygen fugacity (**Fig. 6-37**) until it has the appropriate proportion of melt, quench the sample, and analyze the resulting glass. To confirm the validity of this composition, one can prepare a synthetic sample with this same composition and test it to see whether the correct minerals are on the liquidus at an appropriate temperature. If not, the composition is adjusted until they are.

This technique also has its weaknesses. First, the amount of trapped liquid indicated by the abundance of excluded elements differs from one element to another, simply because they are not all retained to the same degree. On cooling, some elements will go into minerals crystallizing from the interstitial liquid, while others are removed to differing degrees by compaction or diffusion. As noted in the preceding chapter, the composition of the liquid from which crystals are growing depends on the rates at which excluded components are expelled and included components are replenished at the crystal–liquid interface. This involves a host of kinetic processes, and there is no simple way to design a laboratory experiment that would take all these complexities into account.

A third method that has been used to decipher the evolving liquid is based on the principles of fractional crystallization outlined in our discussion of the partitioning of trace elements (Chapter 5). We saw that the compositions of minerals are determined by the activities of their constituent elements, and if one knows the crystal–liquid partition coefficient, the concentration of an element in one phase is a simple function of its concentration in the other. For example, the composition of olivine has been shown to be proportional to the mole fractions of Mg and Fe^{2+} in the liquid and can be predicted by a coefficient, K_D, defined as

$$K_D = \frac{(Fe^{2+}/Mg)_{olivine}}{(Fe^{2+}/Mg)_{liquid}} \tag{6-1}$$

The value of K_D has been found to be between 0.30 and 0.35 for a wide range of intermediate liquids and temperatures. The corresponding K_D for clinopyroxene is 0.25 ± 0.02. Knowing the composition of the minerals at a

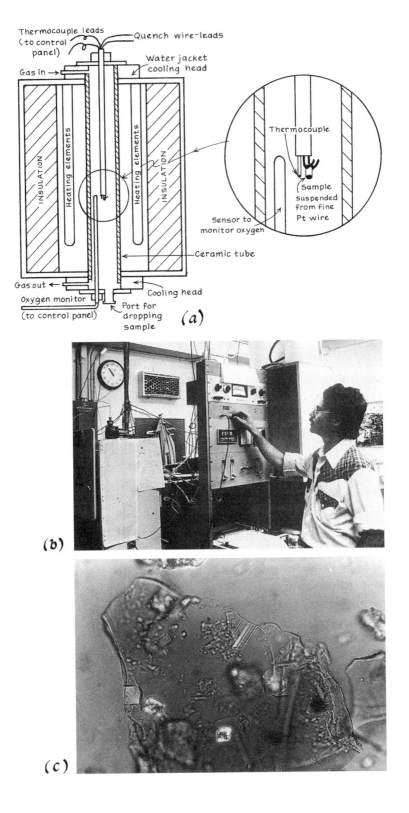

Thermocouple leads
(to control panel)

Quench wire-leads

Gas in →

Water jacket
cooling head

INSULATION

Heating elements

Heating elements

INSULATION

Thermocouple

Sample
suspended
from fine
Pt wire

Sensor to
monitor oxygen

Ceramic tube

Gas out ←

Cooling head

Oxygen monitor
(to control panel)

Port for
dropping
sample

(a)

(b)

(c)

Figure 6-37 The liquid compositions that would crystallize the observed minerals of the Layered Series at the conditions shown in Figure 6-30 can be tested experimentally using a controlled-atmosphere furnace of the kind illustrated here. (a) A sample of each composition is held at its appropriate temperature for periods of 5 hours or more to reach equilibrium. The oxidation state of the sample is controlled by passing a mixture of carbon dioxide and carbon monoxide through the furnace and around the sample. At the end of the run, electric current is passed through the fine platinum wire suspending the sample. When the wire is fused, the sample drops from the furnace and is quenched in water. In this way, the liquid is converted to glass, and phases that were in equilibrium with the liquid can be identified under the microscope. (c) A photomicrograph of a typical quenched charge with small crystals of plagioclase and pyroxene in glass.

given stage of crystallization, it is possible to calculate the iron–magnesium ratio of the liquid with which it was last in equilibrium. One can then turn to the relationships for crystal fractionation discussed in Chapter 5 and use the equation

$$\frac{C_L}{C_0} = F^{(D-1)} \tag{5-11}$$

to determine, F, the fraction of liquid remaining. This approach was used with some success to trace the differentiation of the Kiglapait Intrusion, a well-studied body in Labrador composed of troctolites and plagioclase-rich gabbros. The method was well suited for these rocks because iron-oxide minerals did not become abundant until very late, and throughout most of the course of differentiation, the iron-magnesium ratio was determined almost entirely by olivine and augite The results were in good agreement with the observed compositions (**Fig. 6-38**), but unfortunately, the

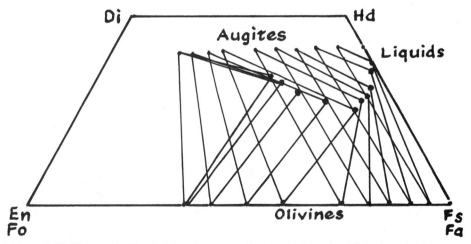

Figure 6-38 Olivine-augite-liquid relationships were calculated for the Kiglapait Intrusion using the experimentally determined exchange coefficients for these three phases. Compare the calculated trend of the liquid with that shown in the experimental system shown in Figure 4-11. (After S. A. Morse and M. Ross, 2004, Kiglapait mineralogy IV, the augite series. *Amer. Min.* 89:1380–1395.)

method was found to be less precise once iron-oxide minerals begin to crystallize.

It is now possible to define the major-element compositions of crystals and liquids from experimentally determined thermodynamic data for virtually all elements of interest. This has been a major advance, for it enables us to calculate the course of liquids during crystallization, but in order to interpret the differentiation of a solidifying body of magma we must also be able to interpret the physical conditions under which crystallization takes place, and this is a much more complex problem. In fact, it is one of the major challenges of igneous petrology today.

Numerical modeling

The time is rapidly approaching when we shall be able to use numerical analogs to model the crystallization and differentiation of large intrusions. This approach has already been used in simplified, one-dimensional models, but the results are questionable if they do not incorporate the body as a whole. For example, a one-dimensional model of the Skaergaard Layered Series alone has limited meaning, because it fails to take into account simultaneous crystallization at the roof and walls and precludes, among other things, a proper analysis of the role of convection.

In defining the geometrical form of a numerical model, the boundaries must be broad enough to treat the system in thermal and compositional isolation from its surroundings. The boundary conditions need not be fixed, of course; they can be open to a specified flow of heat or fluid coming from an outside source. A grid system is set up within these boundaries so that the spatial evolution of the system can be tracked through repetitive calculations for appropriate time steps. Because the processes being evaluated range from small-scale crystal-liquid reactions and diffusion to much larger effects, such as convection and heat transfer, the spacing of the grid points will depend on the conditions in a given part of the system and on the nature of the process under consideration.

Physical properties, such as the densities of the liquid and solids and the viscosity of the melt, must be calculated at each time step as cooling and crystallization advance. The set of equations for a given grid point will depend on the nature of the processes taking place in the local regime. They are relatively simple in those parts of the system that are either totally solid or totally liquid. In these regions the main processes of interest are the transfer of heat and mass by diffusion and, in the case of the liquid, thermal, and compositional convection. The calculations for the zone of solidification are more complex, for they include a greater array of factors.

The temperature at any point is a function of the net heat flux, as well as the thermal effects of crystallization. The temperature changes determine the rates of nucleation and crystal growth, which in turn govern the proportions and sizes of crystals. At the same time, the composition, temperature, density, viscosity, and thermal properties of the liquid are constantly changing. When crystallization is proceeding on the floor, the effective stress from the overlying crystals causes compaction and expulsion of liquid, while the opposite conditions prevail under the roof. In either case, any density contrast between the interstitial liquid and adjacent magma may cause convective exchange and compositional changes within the main body of magma. The ability of the liquid to percolate through the crystals is governed by Darcy's Law (eq. 2-16) and must be considered in terms of the gradients of density, pressure, and permeability through the zone of solidification. Relative motion of the liquid and crystals entails some degree of reaction and re-equilibration, and this will be a function of the temperature and compositional differences as well as diffusion of the components within the crystals.

Setting up a model of this complexity is obviously a daunting task that requires a thorough understanding of a wide range of physical and geochemical processes. With our increasing knowledge of magmatic systems, it is only a matter of time until we shall be able to employ numerical analogs to achieve a realistic picture of magmatic differentiation.

Selected References

Cawthorn, R. G., ed., 1996, *Layered Intrusions,* Elsevier, 531 p. A compilation of papers describing the most recent state of knowledge for most of the major layered intrusions.

Jackson, E. D., 1961, Primary textures and mineral associations in the Ultramafic Zone of the Stillwater Complex, Montana, *U. S. Geol. Surv. Prof. Paper 358*, 106 p. A classic work on chromite-bearing ultramafic rocks with a number of important insights into their origins.

Parsons, I., ed., 1987, *Origins of Igneous Layering*, NATO ASI Series C, Vol. 196, 666 p. A compilation of papers representing research on basic intrusions with special emphasis on the origins of layering.

Shirley, D. N., 1987, Differentiation and compaction in the Palisades Sill, New Jersey, *Jour. Petr.* 28:835–866. The most complete study of differentiation of a classic doleritic sills.

Wager, L. R., and W. A. Deer, 1939, Geological investigations in East Greenland, Part III. The petrology of the Skaergaard Intrusion,

Kangerdlugssuaq, East Greenland, *Meddelelser om Grønland* 105:1–352. Although somewhat out of date, this remains a major classic of petrological literature.

Young, D., 2003, *Mind over Magma: The Story of Igneous Petrology,* Princeton University Press, 686 p. A comprehensive history of the evolution of ideas with emphasis on the concepts of crystal fractionation.

7 Basalts and Magma Series

In 1892, J. P. Iddings first proposed the concept that certain groups of igneous rocks have common genetic lineages stemming from different parents. Recognizing that their chemical nature is more important than their mineral assemblages, Iddings proposed that all igneous rocks fell into one of two "consanguineous" series, one alkaline and the other subalkaline. This basic division is still recognized today, even though several subsidiary series have been defined within the two major branches.

All but a few igneous rocks fit into one of a small number of types of sequential series, each having its own particular mineralogical and compositional trends—hardly any stand in isolation from other rocks. A set of rocks that is perceived to be linked through a process of differentiation, however obscure or tenuous, is said to be co-magmatic and is referred to as a series. Groups of rocks may be closely associated in the field and may even form an eruptive sequence from the same volcano without necessarily having a common magmatic origin. An assemblage of this latter type is usually called a suite to indicate that it has a spatial association whether or not the rocks have any direct genetic coherence. Thus, a suite of rocks from a given igneous center may include rocks from various sources and more than one series, while lavas of many different volcanoes may be assigned to the same series, if they have similar compositional trends.

The principal magmatic series recognized today were first established through geological studies of the natural associations of rocks in the field, and they were defined mainly on the basis of their modes of occurrence and mineralogical characteristics. Experimental studies later showed that the trends predicted by phase relations correspond quite well to those established from field and petrographic evidence.

It may seem odd that most of these series are based on petrographically diverse volcanic rocks rather than on plutonic rocks in which the assemblages of minerals are closer to equilibrium and less dependent on the vagaries of rapid crystallization. The reason effusive rocks are emphasized is that their trends and ranges of differentiation are more easily related to liquid lines of descent. There can be no doubt that their composition, at least that of their groundmass, corresponds to a liquid and that phenocrysts were mineral phases residing in that liquid before it was erupted. In this way, an eruptive sequence can provide a way of tracing the trend of evolution of liquids and their coexisting minerals. Plutonic rocks with longer histories of crystallization and re-equilibration are more difficult to interpret in this way.

Basalts and other mafic rocks having high liquidus temperatures are normally thought to be the original magmas from which more evolved rocks are derived. For this reason, they are assigned a special place as parental magmas, even though they may not be the most voluminous or even the most distinctive members of their particular series. Most parental magmas are thought of as primary in the sense that they are direct products of melting. They need not be basaltic as long as they are not derived from another magma. The process of producing a magma by total or partial melting is known as *anatexis*, and magmas produced in this way are said to be *anatectic*. The term is used most often to imply melting of crustal rocks, but its definition does not limit it to that sense.

The principal types of magma series and their relationships to large-scale tectonic features were mentioned briefly in the opening chapter. Here we examine the two principal divisions, alkaline and subalkaline rocks, and explore some of the ways in which their differentiated series evolve.

■ Chemical and Petrographic Characteristics

Alkaline and Subalkaline (Tholeiitic) Basalts

Before proceeding, we should review the differences between alkaline and subalkaline rocks and elaborate on the distinguishing criteria cited in Chapter 2. We focus first on the main tholeiitic branch of the subalkaline group, deferring the other major divisions to separate chapters of their own.

The *basalt tetrahedron* (**Fig. 7-1**) illustrates the petrologic basis of the distinction between basic alkaline and tholeiitic magmas. Its components—diopside, olivine, hypersthene, nepheline, plagioclase, and quartz—correspond to the principal normative minerals used to characterize basalts. The tetrahedron is divided into three volumes by two interior planes, one a "plane of silica saturation" that separates compositions with normative quartz from those with normative olivine and hypersthene and another a "critical plane of silica undersaturation," separating the latter from compositions with normative nepheline. Thus, the three volumes correspond to three principal types of basalt—quartz tholeiites, olivine tholeiites, and alkaline basalts.

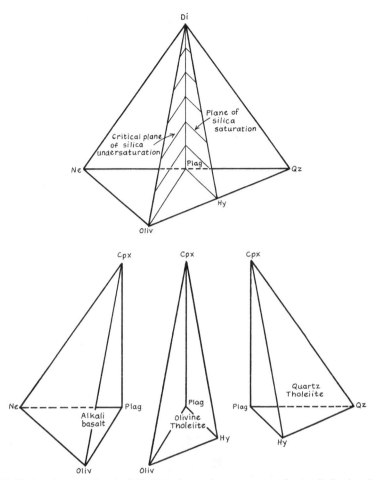

Figure 7-1 The basalt tetrahedron is divided into three volumes corresponding to alkaline basalt with normative Ne (or basanite with modal Ne), olivine tholeiites with normative Hy and OLIV, and quartz tholeiites with normative Hy and Q. The first two are separated by the "critical plane of silica undersaturation" between alkaline rocks with normative Ne and tholeiitic rocks with normative Hy. Olivine tholeiites are separated from quartz tholeiites by the "plane of silica saturation."

    The division between critically undersaturated basalts (i.e., alkaline basalts with normative nepheline) and olivine tholeiites (with normative olivine and hypersthene) is more important than that between saturated and oversaturated tholeiites. The reason for this can be seen from the simple system plagioclase-olivine-silica (**Fig. 7-2**) that corresponds to the base of the tholeiitic side of the tetrahedron. The broken line crossing the interior indicates the trace of the plane of silica saturation in Figure 7-1. Because of the incongruent melting of hypersthene, the boundary between the fields of olivine

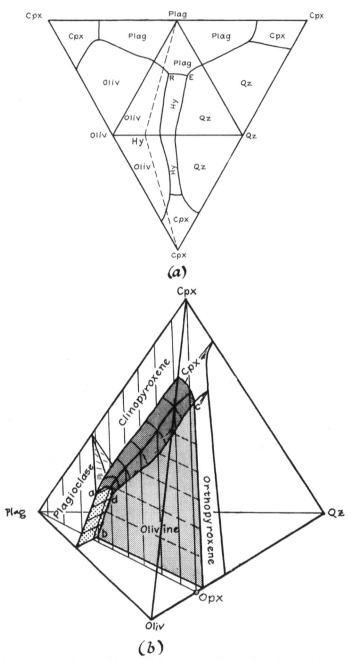

Figure 7-2 (a) The sides of the tholeiitic part of the basalt tetrahedron have been unfolded to show the systems making up the individual faces. The experimental systems on which this part of the simplified tetrahedron is based are shown in (b). The system differs in that the phases are not plagioclase, augite, and hypersthene but the high-temperature end-members anorthite (An), diopside (Di), and enstatite (En). (After D. C. Presnall, et al., 1979, *Jour. Petrol.* 20:12.)

and hypersthene is a peritectic, and any liquid with both normative olivine and hypersthene (i.e., any composition to the left of the broken line) evolves by crystal fractionation to the same end point, E, as one with normative quartz (to the right of the broken line). Thus, both olivine tholeiites and quartz tholeiites evolve toward greater degrees of silica enrichment. This course of evolution is in contrast to compositions with normative nepheline, which evolve toward more silica-deficient compositions.

The triangle forming the back face of the principal tetrahedron (Fig. 7-1) corresponds to the system Di-Ne-SiO$_2$ (**Fig. 7-3a**), in which albite represents plagioclase, and the intersection of the planes of silica saturation with that face corresponds to the diopside-albite join. Albite is an intermediate compound between nepheline and SiO$_2$ and melts congruently to a liquid of its own composition, and thus, the three-component system must have a thermal divide separating it into two subsystems, Di-Ab-SiO$_2$ and Di-Ab-Ne, each with an end point at a eutectic.

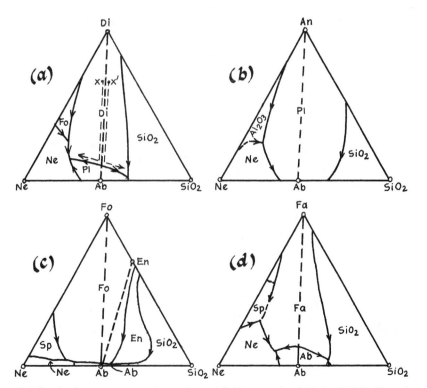

Figure 7-3 Four basic ternary systems have a thermal divide separating fields in which liquids evolve to different end-points. (From work of J. F. Schairer, H. S. Yoder, Jr., and N. L. Bowen, compiled by S. A. Morse, 1980, *Basalts and Phase Diagrams*, Springer Verlag, p. 249.)

This system can be visualized as a simplified analog for the different courses of evolution of alkaline and subalkaline magmas. If the points X and X′ are thought of as two parental liquids, the first alkaline and the second subalkaline, we see that, even though their initial compositions are very close and the minerals they crystallize are the same throughout most of their course of differentiation, they lie on opposite sides of a thermal divide and are destined to evolve with increasingly divergent degrees of silica saturation until one crystallizes nepheline and the other a silica mineral, such as quartz. A thermal divide persists through all the systems governing crystallization of nepheline-normative and hypersthene-normative liquids (Fig. 7-3) and becomes especially marked in the later stages of differentiation. We examine the end products of these two lines of descent in Chapters 10 and 11.

The divergent trends of these synthetic systems correspond closely to those of many alkaline and subalkaline series in which felsic differentiates have compositions close to the eutectics and thermal minima. This is not to say, however, that all magmatic series follow one or the other of these two lines of descent. We shall see shortly that some oceanic series follow an intermediate trend toward trachytes with neither strong enrichment nor depletion of silica. The factors governing this seemingly anomalous but geologically important group of rocks are discussed in a later chapter.

Chemical Characteristics

When dealing with the multicomponent compositions of real magmas, the character of a magma series can usually be defined by the abundances or ratios of certain diagnostic elements. The tholeiitic and alkaline rocks of Hawaii, for example, can be distinguished by their ratios of $Na_2O + K_2O$ to SiO_2 (**Fig. 7-4a**). An empirical dividing line corresponds to the critical plane of silica undersaturation in the basalt tetrahedron, but its position for natural rocks is affected by the relative abundance of mafic and felsic minerals and is not an infallible criterion for all rocks. A sounder approach is to compare the proportions of the key normative minerals, hypersthene and nepheline.

A common source of difficulty is the oxidation state of iron, which can have a pronounced effect on the amount of normative nepheline or hypersthene in basalts and can even affect their trends of differentiation. In calculating the norm for an oxidized rock, more iron is assigned to magnetite (as opposed to FeO-bearing silicates), leaving more silica to be assigned to other minerals. In this way, oxidation of iron may reduce and even eliminate normative nepheline that would otherwise be present. If oxidation results from surface weathering, it can be allowed for by recalculating the analysis with a more appropriate ratio of FeO to Fe_2O_3. If the oxidation state is inherent in the magma, however, it may have greater petrologic significance because

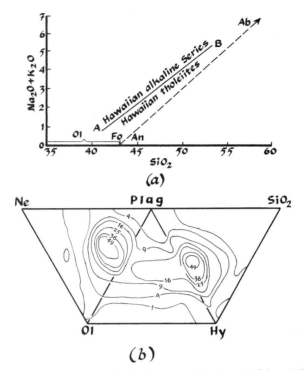

Figure 7-4 (a) When plotted on a diagram for alkalies versus silica, the tholeiitic and alkaline rocks of Hawaii are separated by an empirical boundary parallel to the alkali-silica variation of the plagioclase feldspars. In a rock without mafic minerals, the plagioclase line would separate compositions with normative nepheline (above) from subalkaline compositions (below). Thus, the division for Hawaii, or any other province, will have the same slope, but because of the presence of other minerals, such as olivine, which contain silica but no alkalies, the line has a different origin on the silica axis. Depending on the proportions of feldspar and mafic minerals, the boundary will differ for each rock series. (b) When the normative components of basalts are plotted in a diagram corresponding to the base of the basalt tetrahedron, they fall in the corresponding fields of alkaline basalts, olivine tholeiites, and quartz tholeiites. When large numbers of samples from many tectonic settings are plotted in this way they tend to define two fields of maximum populations.

the effect seen in calculating normative components reflects the silica enrichment in real magmas when equivalent minerals, such as magnetite, are fractionated at early stages of differentiation.

Although a common feature of alkaline basalts is an abundance of MgO and TiO$_2$, these components, like alkalies and silica, must also be considered in the context of the total composition of the rock. A *picrite* (basalt with abundant olivine phenocrysts) may have a very high magnesium content, even though the rock as a whole has normative hypersthene. The same relation holds for TiO$_2$. The early stages of differentiation of tholeiitic basalts may result in strong enrichment of titanium, which normally increases along with iron in ferrobasalts. Such rocks have large amounts of normative and modal ilmenite but none of the other essential features of alkaline rocks.

Petrographic Characteristics

Petrographic criteria can often be just as useful as chemical analyses in characterizing basalts (**Table 7-1**). Rocks containing a modal feldspathoid, such as nepheline or leucite, are obviously alkaline, but, unfortunately, these *basanites,* as alkaline basalts with modal feldspathoids are called, are much less common than alkaline basalts that lack these diagnostic minerals. The latter are often called *basanitoids* and in most cases are poorer in alkalies and richer in silica than basalts with modal nepheline. Whether they have a modal feldspathoid or not, most alkaline rocks contain a late-crystallizing alkali feldspar, normally anorthoclase, which can be recognized from its low refractive index and small negative optic angle.

With few exceptions, nepheline-normative basalts have olivine in their groundmass; tholeiitic basalts do not. The latter may contain stable phenocrysts of olivine, but if they do the olivine tends to be rounded and shows evidence of reaction with the groundmass. A Ca-poor pyroxene, either hypersthene or pigeonite, is, of course, highly diagnostic of tholeiites. The reddish brown titaniferous pyroxenes that are so characteristic of alkaline basalts are rarely seen in subalkaline rocks, even when the latter have bulk compositions rich in titanium. The amount of titanium in the pyroxene structure is a function, not only of the titanium content of the rock, but also of silica content. If the activity of silica in the magma is low, greater amounts of aluminum (+3) enter the tetrahedral site of pyroxenes in place of silica (+4), and the resulting charge deficiency causes a paired substitution of titanium for cations of lower valence. For this reason, titaniferous pyroxenes are a sensitive sign of low silica activity in the melt from which they have crystallized.

■ Examples of Tholeiitic and Alkaline Series

Two oceanic examples will suffice to illustrate the salient features of the two principal series. The differentiated rocks of the Galapagos Islands and Tahiti are ideal for this purpose because both have volcanic and plutonic counterparts of the full range of differentiates, and they are in a setting where the absence of continental crust makes their genetic evolution by crystal fractionation unambiguous.

It is appropriate that we begin our survey of basalts and their differentiates with the Galapagos Islands where Charles Darwin first proposed the concept of crystal fractionation. The islands have since become the focus of

Table 7-1   Characteristics of alkaline and tholeiitic basalts.
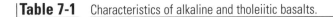

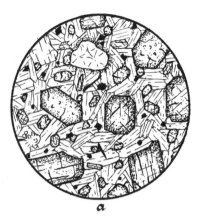
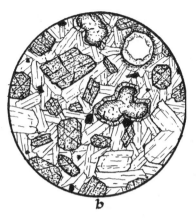

a   *b*

Petrographic Features

Alkaline Basalt	Tholeiitic Basalt
Stable groundmass olivine	Hypersthene or pigeonite
Nepheline or leucite (in basanites)	Interstitial quartz
Reddish titaniferous pyroxene	No olivine in the groundmass
Interstitial alkali feldspar	Phenocrysts of olivine may be rounded or
Plagioclase with sodic runs	rimmed with pyroxene
Vesicles lined with analcite or zeolites	Vesicles lined with a silica mineral

Note: Though common in their respective types of basalts, it is rare to find all these features in a single rock.

Chemical Features

	Normative Ne	*Normative* Hy
SiO_2	46–48	48–52
TiO_2	2–3	1–2
Al_2O_3	14–16	14–16
FeO*	4–7	5–7
MgO	5–8	4–6
CaO	8–10	7–0
Na_2O	2–3	1.5–2.5
K_2O	0.3–1.5	0.1–1

*Total iron as FeO.

intense study because, as we shall see in the chapter that follows, they are well suited for testing may fundamental aspects of oceanic magmatism.

The Galapagos Tholeiitic Series

The active volcanoes of the Galapagos Islands are products of a hotspot in the eastern equatorial Pacific. The differentiated lavas are concentrated in the center of the archipelago immediately downstream from the core of the hotspot, and the plutonic rocks are found as xenoliths among the pyroclastic ejecta erupted from cinder cones in the same region.

The most basic analyzed rocks of the series are picrites and olivine tholeiites with as much as 8.65-percent MgO. This is somewhat less than the value of at least 10 percent that is usually considered a minimum MgO content for primary melts of the mantle and indicates that even these olivine-rich basalts must already have differentiated before erupting at the surface. This illustrates an important point: the most basic basalts that can be sampled at the surface may not be the true parental magma of the series. The primary mantle-derived magma may never be erupted in its original form, possibly because it is too dense to rise through lighter crustal rocks or because it cools and differentiates while rising through the crust.

The series of differentiated volcanic rocks, consisting of olivine tholeiites, ferrobasalts, icelandites, dacites, and sodic rhyolites, is matched by a parallel suite of coarse-grained olivine gabbros, ferrogabbros, diorites, and quartz syenites. Typical compositions are given in **Table 7-2**, and the more notable petrographic features are illustrated in **Fig. 7-5**.

The differentiated series follows a typical tholeiitic trend (**Fig. 7-6**). In the early and middle stages, MgO declines steadily, while iron first increases to a maximum in ferrobasalts and then declines in icelandites, dacites, and rhyolites. Later stages are dominated by an increase of silica and alkalies. At the same time, plagioclase changes from labradorite to oligoclase, olivines, and pyroxenes become richer in iron, and hypersthene gives way to pigeonite. Amphibole and biotite are common in the middle and late members of the plutonic series.

As Figure 7-6b illustrates, these changes can be related to fractionation of the minerals seen in plutonic xenoliths or as phenocrysts in the lavas. The mass-balance comparisons used to test these relationships are based on the principles described in Chapter 5, but, for ease of calculation, are carried out by a computer. The calculation, a sample of which is shown in **Table 7-3**, compares the observed composition of each member of the differentiated series with the composition that would result from removal of a specified set of minerals from their parental liquid. Taking each step of differentiation in

Table 7-2 Compositions of representative lavas of the differentiated tholeiitic series of the Galapagos Islands.

	1.	2.	3.	4.	5.	6.	7.	8.
SiO_2	49.39	48.73	49.72	49.21	52.02	54.82	64.71	71.61
TiO_2	0.69	2.52	2.81	3.55	2.92	2.43	0.93	0.38
Al_2O_3	17.91	15.08	14.56	13.47	14.17	14.62	15.35	13.39
Fe_2O_3	2.38	1.87	1.96	2.18	1.91	1.72	0.87	0.59
FeO	5.26	10.64	11.13	12.36	10.79	9.71	4.96	3.35
MnO	0.12	0.16	0.17	0.20	0.13	0.14	0.09	0.12
MgO	8.82	6.85	5.53	5.18	4.43	3.38	1.57	0.21
CaO	12.34	10.49	9.90	9.42	8.13	7.06	3.71	1.51
Na_2O	2.77	2.96	3.17	3.26	3.76	4.07	5.35	5.65
K_2O	0.17	0.43	0.69	0.76	1.15	1.39	2.17	3.10
P_2O_5	0.14	0.27	0.36	0.41	0.59	0.66	0.28	0.09
Molecular Norms								
Ap	0.29	0.58	0.77	0.89	1.26	1.40	0.59	0.19
Im	0.96	3.54	3.98	5.07	4.14	3.44	1.29	0.53
Mt	2.45	1.97	2.09	2.34	2.03	1.83	0.91	0.61
Or	1.00	2.59	4.13	4.59	6.93	8.36	12.83	18.32
Ab	24.54	26.86	28.99	29.96	34.39	37.11	47.96	50.78
An	35.47	26.87	23.88	20.39	18.71	17.81	11.44	2.01
Di	19.24	19.17	18.96	19.80	14.72	10.70	4.11	3.92
Hy	4.01	8.86	13.32	13.63	16.82	14.98	8.21	3.06
Ol	12.03	9.54	3.88	3.33	–	–	–	–
Q	–	–	–	1.00	4.35	12.66	20.59	
Trace Elements								
Ba	78	80	130	147	160	185	320	441
Zr	97	157	223	262	344	445	617	673
Rb	22	6	12	11	21	25	43	62
Sr	347	320	310	290	296	287	194	74
Cr	596	60	46	25	15	14	8	4
Ni	166	93	39	29	13	7	8	4
La	8	12	7.1	19	25.9	31	41	52
Sm	3.4	5.1	6.9	9	10.1	11.9	11.7	17
Lu	0.29	0.43	0.61	0.65	0.77	0.9	1.0	1.6

1. Picrite gabbro   5. Basic icelandite
2. Olivine tholeiite   6. Icelandite
3. Tholeiite   7. Dacite
4. Ferrobasalt   8. Rhyolite

Analyses are recalculated to 100-percent water free with a uniform ratio of Fe_2O_3 to FeO^* of 0.15.
(Original data from M. Lindstrom, PhD thesis, University of Oregon, with corrections and additions.)

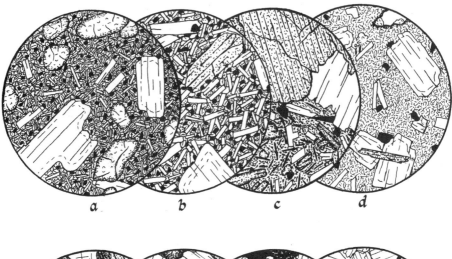

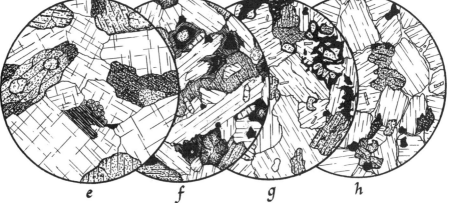

Figure 7-5 Petrographic features of tholeiitic lavas and plutonic rocks of the Galapagos Islands. (a) A typical tholeiitic basalt with phenocrysts of plagioclase and rounded crystals of olivine. (b) A ferro-basalt with abundant groundmass magnetite. (c) An icelandite with large phenocrysts of ferro-augite and labradorite. (d) A siliceous dacite with hedenbergitic pyroxene and vesicles rimmed with cristobalite. Equivalent plutonic rocks are (e) olivine gabbro, (f) ferro-gabbro with magnetite rimmed with biotite, (g) hornblende-bearing leucodiorite, and (h) quartz-syenite with potassic oligoclase, apatite, and sphene.

Figure 7-6 (a) The trend of differentiation of tholeiitic Galapagos lavas shows a strong enrichment of iron and titania in the early stages followed by a decline of those components as silica and alkalies are more strongly enriched. Variations are plotted against the fraction of remaining liquid (F) calculated by the method described in the text. (b) The proportions of minerals that are fractionated at each stage of differentiation have been calculated in the manner illustrated in Table 7-3. (c) The concentrations of trace elements predicted by these calculations (broken lines) are compared with those observed in the rocks (solid lines). See text for explanation. (d) A normalized-element diagram showing the increase of excluded trace elements in the same six rocks relative to their abundances in the most primitive liquid. Although most of the elements are enriched with differentiation, Sr remains almost constant until a late stage, when it is depleted. Ti first increases and then decreases when Fe-Ti oxides start to crystallize; P does the same when apatite is removed.

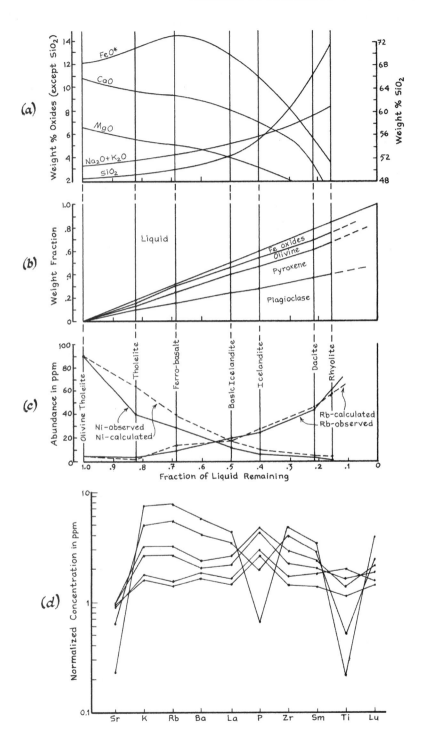

An example of a mass-balance calculation for differentiation of a Galapagos olivine tholeiite (OLIV-TH) to a more evolved tholeiite (TH) by fractionation of plagioclase (AN75), augite (AUG), olivine (F085), and an iron-oxide (MT-ILM), which occur as phenocrysts in the parent. Compositions of the components are listed under INPUT DATA, and the calculated composition of the derivative liquid is compared to the observed composition under SOLUTION. Weight proportions of the minerals subtracted and the remaining liquid are listed under PROPORTIONS OF PHASES. A measure of the quality of the calculation is given by the SUM OF SQUARES OF RESIDUALS, which is a total of the squared differences (DIFF) between the true analysis of the daughter and calculated best-fit composition. By squaring the difference, the effect of the positive and negative signs of these differences is removed. Depending on the reliability of the input data, a value of less than 1.0 is considered satisfactory.

Table 7-3

INPUT DATA

	OLIV-TH	AN75	AUG	F085	MT-ILM	THOL
SIO2	48.73	48.82	51.30	40.01	0.00	49.72
TIO2	2.52	0.00	0.00	0.00	15.80	2.81
AL2O3	15.08	32.21	0.00	0.00	0.00	14.56
FE2O3	1.87	0.00	0.00	0.00	36.84	1.96
FEO	10.64	0.00	6.10	14.35	47.36	11.13
MNO	0.16	0.00	0.00	0.00	0.00	0.17
MGO	6.85	0.00	21.10	45.64	0.00	5.53
CAO	10.49	15.19	21.10	0.00	0.00	9.90
NA2O	2.96	2.80	0.00	0.00	0.00	3.17
K2O	.43	0.00	0.00	0.00	0.00	.69
P2O5	.27	0.00	0.00	0.00	0.00	.36

PARENT - FRACTIONATED MINERALS = DAUGHTER

SOLUTION

	PROPORTIONS OF PHASES	% FRACTIONATED MINERALS
OLIV-TH	1.000	--
F085	-0.032	18.601
AUG	-0.037	21.048
AN75	-0.092	53.151
MT-ILM	-0.013	7.201
THOL	0.826	--

	DAUGHTER ANALYSIS	DAUGHTER CALC	DIFF
SIO2	49.72	49.63	-.09
TIO2	2.81	2.81	.00
AL2O3	14.56	14.62	.06
FE2O3	1.96	1.71	-.25
FEO	11.13	11.33	.20
MNO	0.17	0.19	.02
MGO	5.53	5.55	.02
CAO	9.90	10.04	.14
NA2O	3.17	3.27	.10
K2O	.69	.53	-.16
P2O5	.36	.32	-.04

SUM OF SQUARES OF RESIDUALS = 0.172

TRACE ELEMENTS:

	BULK D	PARENT	DAU OBS	DAU CALC
CR	1.62	60	46	53
NI	2.96	93	39	64
BA	0.05	80	130	98
RB	0.11	6	12	7
SR	1.14	320	310	310

turn, successive compositions can be related in this way to a diminishing volume of remaining liquid.

Having determined the proportions of crystals and liquids consistent with the major-element variations, the relationships can be further tested by use of a number of trace elements that are sensitive measures of crystal fractionation. Taking appropriate distribution coefficients and the equations in Chapter 5, the predicted effect of fractionation is compared with the known trace-element concentrations in successive liquids. The calculated and observed trends for two key elements, Ni and Rb, are shown for the entire series in Figure 7-6c. Another way of showing the behavior of excluded trace elements is shown in a normalized-element diagram (Fig. 7-6d) in which the concentrations in the differentiated lavas have been normalized to those in the most basic parent from Table 7-2.

The question of how and where the Galapagos magmas differentiated will be taken up in the next chapter when we consider the volcanoes in the context of their tectonic setting.

The Tahitian Alkaline Series

One of the most remarkable suites of oceanic alkaline rocks is that of Tahiti in the Society Islands of the central equatorial Pacific. Two large, steep-sided volcanoes, long inactive and deeply incised by erosion (**Fig. 7-7**), have been cut by steep canyons that expose their plutonic cores.

The most common lavas are alkaline basalts, basanites, and ankaramites (basalts in which phenocrysts of both olivine and augite are especially abundant), but a wide range of more differentiated lavas can be found among the products of later eruptions. Two diverging trends can be distinguished (**Table 7-4 and Fig. 7-8**). One is a good example of the type of intermediate

I Figure 7-7 A topographic model showing the twin volcanoes of the island of Tahiti.

| Table 7-4 | Major-element compositions of some typical members of the phonolitic suite of Tahiti. |

	Volcanic Rocks				Plutonic Rocks			
	1.	*2.*	*3.*	*4.*	*5.*	*6.*	*7.*	*8.*
SiO_2	43.26	44.13	42.53	55.81	43.79	43.17	46.18	53.23
TiO_2	3.40	4.02	4.78	0.26	2.57	4.92	2.30	1.10
Al_2O_3	9.69	15.36	16.61	22.00	9.43	14.71	19.32	20.54
Fe_2O_3	3.66	3.08	3.80	1.68	3.47	4.53	4.11	2.86
FeO	8.97	9.09	7.84	0.89	8.80	7.94	4.41	1.81
MnO	0.16	0.14	0.13	0.06	0.14	0.15	0.13	0.07
MgO	12.64	6.37	5.56	0.21	15.72	5.39	3.13	0.77
CaO	12.10	10.61	10.03	0.93	12.46	10.64	8.47	3.26
Na_2O	1.59	2.56	3.85	9.26	1.59	3.43	4.71	7.09
K_2O	1.18	1.77	1.96	5.45	0.72	1.84	2.83	5.40
H_2O+	1.79	1.50	1.27	2.17	1.25	1.13	3.49	3.45
H_2O-	0.67	0.63	0.45	0.46	0.03	0.22	0.07	0.39
P_2O_5	0.61	0.55	0.90	0.02	0.26	1.71	0.89	0.13
Total	99.63	99.81	99.71	99.20	100.26	99.78	100.04	100.10
Molecular Norms								
Ap	1.32	1.20	1.94	0.04	0.54	3.71	1.92	0.27
Ilm	4.89	5.82	6.85	0.35	3.58	7.10	3.30	1.54
Mt	3.95	3.35	4.09	1.49	3.63	4.90	4.43	1.92
Hm	–	–	–	0.15	–	–	–	0.73
Or	7.20	10.87	11.91	31.54	4.26	11.26	17.22	32.08
Ab	10.83	17.50	14.25	33.95	6.63	22.10	22.39	27.66
An	16.35	26.21	22.90	2.33	16.48	20.01	23.93	8.33
Di	33.23	19.83	17.73	1.32	34.93	18.46	10.69	4.50
Ol	19.88	11.39	7.52	–	25.35	6.59	3.41	–
Ne	2.35	3.84	12.78	28.50	4.59	5.87	12.70	21.81

1. Ankaramite
2. Alkaline basalt
3. Basanite
4. Phonolite
5. Olivine-rich gabbro
6. Biotite-rich gabbro
7. Kaersutite essexite
8. Nepheline syenite

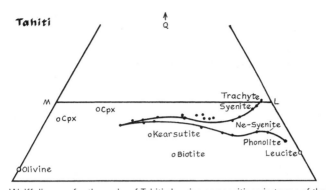

Figure 7-8 Von Wolff diagram for the rocks of Tahiti showing compositions in terms of their normative components. On the left side are the mafic components, pyroxene, olivine, and iron oxides, with "M" at the composition of pyroxene and the more silica-deficient minerals below. Similarly, the leucocratic minerals, feldspar and feldspathoids, are plotted on the right side, with feldspars at "L" and leucite and nepheline below. The degree of silica saturation is plotted as the distance from the apex "Q" so that rocks with normative quartz are above the line ML and those with normative nepheline and leucite below. Note the two trends diverging from a common basaltic composition toward more felsic liquids with differing degrees of undersaturation of silica. One series terminates at trachyte and syenite, the other at phonolite and nepheline syenite. Minerals that appear as phenocrysts are clinopyroxene (Cpx), kearsutitic amphibole, olivine, and biotite.

series mentioned earlier. It consists mainly of marginally alkaline hawaiites and mugearites and culminates in trachyte and its plutonic equivalent, syenite. Some members contain small amounts of normative nepheline, others normative hypersthene. The second series is more strongly undersaturated and includes a variety of feldspathoidal lavas and plutonic rocks (Table 7-4 and **Fig. 7-9**) that have differentiated to nepheline syenites and their volcanic equivalents, phonolites (a fine-grained felsic rock similar to trachyte but with modal nepheline).

Strongly alkaline rocks, including those of Tahiti, have been given an inordinate number of special names in keeping with their great mineralogical and petrographic diversity. These names are useful when describing the series in detail, but for our present purposes, it will suffice to mention some of the typical minerals that characterize rocks of this kind without dwelling on the nomenclature the mineral assemblages define. Most of the pyroxenes are rich in titanium, which gives them a deep, reddish brown color. Some are strongly zoned with green sodic rims. The dark brown amphibole, kaersutite, is also titaniferous, as is the biotite, which is particularly abundant in the more felsic members of the plutonic suite. The alumina-silicates include, in addition to sodic plagioclase, nepheline, analcite, and brilliant blue hauyne.

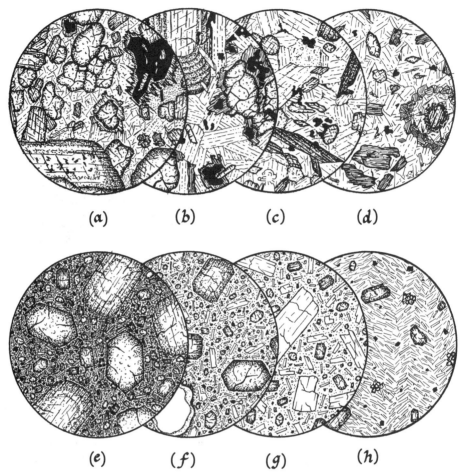

Figure 7-9 Petrographic features of the alkaline plutonic (a–d) and volcanic rocks (e–h) of Tahiti. (a) An olivine gabbro with zoned Ti-augite and titaniferous magnetite rimmed with biotite. A more evolved gabbro (b) has alkali feldspar between the earlier-crystallizing plagioclase crystals. Apatite is an abundant accessory mineral. A nepheline-bearing, two-feldspar monzonite (c) contains a Ti-rich amphibole, kaersutite. (d) A nepheline syenite containing reddish biotite, green Na-augite, sphene, and clots of magnetite, augite, and biotite. The most mafic lavas are ankaramites (e) with abundant phenocrysts of olivine and augite. They grade into intermediate compositions (f and g) with increasing amounts of alkali feldspar, analcite, and feldspathoidal minerals, such as hauyne. The most differentiated lavas are phonolites (h) composed of alkali feldspar, nepheline, sphene, and lesser amounts of amphibole and magnetite in a glassy trachytic groundmass.

Unlike the Galapagos series, the Tahitian alkaline rocks are difficult to explain in terms of their observed mineral assemblages. Although both the trachytic and phonolitic series seem to stem from similar parental compositions, each has followed a separate course of differentiation. The von Wolff diagram in Figure 7-8 illustrates this divergence by emphasizing the amount of silica enrichment relative to the proportions of mafic ("M") and leuco-cratic components ("L"). A fractionation scheme of the kind described for the Galapagos tholeiites cannot account for either series by removal of the minerals seen in the plutonic rocks, namely olivine, pyroxene, feldspar, and amphibole. The problem is particularly acute for the phonolitic series because none of the minerals that crystallized from the basic magmas contain enough silica to drive the composition toward strongly silica-deficient felsic liquids.

The most plausible explanation for this apparent inconsistency is that the magmas differentiated at greater depths where the mineral assemblages differed from those that crystallized near the surface. The high-pressure assemblage of minerals may have had a somewhat greater silica content, causing that component to be relatively depleted in the evolving liquids. The role of pressure will be considered in more detail shortly. The important point here is that the minerals seen in the rocks at the surface are not necessarily the ones responsible for differentiation. Magmas can differentiate under a variety of conditions, and even in a single volcano differentiation may proceed independently at more than one level.

■ The Origins of Magma Series

Modern theories for the origins of the various types of magmatic series have evolved as a result of many years of research, much of it stimulated by intense debates. Although distinctive types of magmas were recognized as early as the 1880s, the causes of their differences were not given serious consideration until the first decades of the last century, when the Scottish Geological Survey turned its attention to the Inner Hebrides and the Eocene volcanic centers on the small islands of Skye, Rhum, Eigg, Muck, and Mull (**Fig. 7-10**). These "monosyllabic appendages of western Scotland" comprise a remarkable assemblage of differentiated volcanic and subvolcanic rocks eroded to levels exposing the inner structures of what were once large volcanic complexes.

Geologists working with the rocks of these centers, particularly on Mull, recognized two principal genetic lineages, to which they gave the names "Plateau Magma Series" and "Main Magma Series." The former is represented by a voluminous sequence of flood basalts and lesser amounts of differentiated rocks discharged from fissures related to the rifting that accompanied opening of the North Atlantic Ocean. It was the first magma erupted, and it reappeared repeatedly at intervals throughout much of the subsequent igneous history of the province. The Main Magma Series, which formed large, central-vent volcanoes, evolved during a later stage and

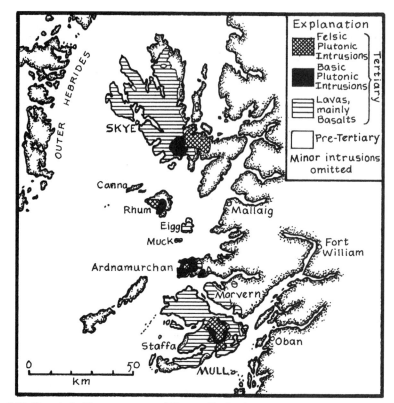

I Figure 7-10 The principal igneous centers of Western Scotland.

though volumetrically subordinate had a greater proportion and range of differentiated rocks. The basic magma of the Plateau Series has all the chemical and petrographic attributes of alkaline basalts. It differs from the tholeiitic basalts of the Main Magma Series in being poorer in SiO_2 and richer in MgO, TiO_2, alkalies, and olivine.

Each of these basalts has its own series of differentiates. The alkaline basalts of the Plateau Series are associated, both in space and time, with small volumes of mildly alkaline hawaiites, mugearites, benmoreites, and trachytes, not unlike Hawaiian rocks of the same names. In a similar way, the tholeiitic basalt of the Main Magma Series was the forerunner and presumed parent of a series of progressively more differentiated rocks that culminated in granites and rhyolites. The local names originally assigned to the intermediate rocks correspond closely to what are now called ferro-basalts and icelandites.

It is instructive to consider the reasoning that led different geologists to arrive at directly opposing interpretations of the relations between these two series. The original workers who studied the field and petrographic relations concluded that the voluminous alkaline basalt of the Plateau Series was a primary mantle-derived magma and that the tholeiitic basalt of the Main Magma Series was the product of contamination of the original magma with continental crust. This inference seemed logical because siliceous granitic and metamorphic rocks visibly underlie the volcanic centers and are found as inclusions in the lavas. Because olivine-rich alkaline basalts are widespread on oceanic islands as well as continents, they were deemed the best candidates for a primary, worldwide, basaltic magma derived directly from an olivine-rich mantle. Tholeiites, it was thought, were confined to continental settings, where basaltic magma rose through thick sialic crust. It seemed reasonable, therefore, to attribute them to contamination of alkaline basaltic magma with siliceous metamorphic and granitic material.

An American petrologist, Norman L. Bowen, demonstrated the weakness of this interpretation. If the tholeiites were the result of contamination of the Plateau Magma with continental crust, that could explain their greater silica contents, but crustal rocks are also rich in alkalies, especially potassium, so the tholeiites should also be richer in the alkalies—which they are not. He showed by a simple mass-balance calculation of the kind illustrated in Figure 5-14 that the composition of any material that must be added to the alkaline basalts to produce tholeiites is quite unlike that of any plausible continental rock. He also pointed out that the compositional variations are consistent with crystal fractionation of the kind predicted by laboratory experiments and outlined a sequence of crystallization by which he thought the tholeiites of the Main Magma Series could evolve from the Plateau Magma. His basic idea was that if olivine were fractionated in a manner that precluded reaction, the remaining liquid would evolve through the reaction point into the range of more silica-rich liquids (Figs. 4-13 and 4-14).

Bowen's work had a logic and rigor that gave it great intellectual appeal. His explanation of the Scottish magmas, however, was not without its own problems. Indeed, the geologists who originally studied the rocks had already considered and rejected the possibility that parental magmas of the two series were related by crystal fractionation. They reached this conclusion mainly from petrographic evidence that was in clear conflict with Bowen's explanation. Look again at the two basalts shown in Table 7-1, and note that the last interstitial liquid to crystallize from the alkaline magmas was not a silica-rich differentiate, as Bowen said it should be if olivine crystallized in abundance and without reaction. Instead, alkali feldspar and silica-poor minerals, such as analcite or natrolite, occur as late-crystallizing phases in the interstices between earlier crystals of plagioclase, pyroxene, and olivine. Where the alkaline basalt contains interstitial minerals enriched in alkalies and depleted in silica, however, the tholeiite contains quartz. In other words, each rock contained within its groundmass a late differentiate with the essential mark of its derivative rock series—one enriched in silica the other poor in silica and rich in alkalies. Neither Bowen nor the Scottish geologists knew at that time that a thermal divide separates the liquid lines of descent of the two contrasting magmas, but if they had, this simple observation, based solely on petrographic evidence, could have provided the clue needed to explain the basic relationships between the two major magma series, not only in Western Scotland but throughout the world.

It was impossible to reconcile the difficulties inherent in any mechanism, whether it be contamination, crystal fractionation, or any of the various ingenious schemes invoked to make one basalt from another, as long as the Plateau Magma was thought to be the only primary mantle-derived magma and, hence, the parent of all others. In Scotland, it was logical to assume this to be so. The Plateau Lavas had almost all the attributes of a primary magma—it was the earliest, most frequently recurring, most voluminous, and most basic rock in the province—but it was not necessarily the only primitive undifferentiated basalt; the belief that it was proved to be the basic fallacy underlying the entire controversy.

Debate continued for a decade or more, until the hypothesis of crustal contamination was dealt a fatal blow by the simple observation that tholeiites are not confined to continents. Alfred Lacroix was the first to point out, in 1928, that subalkaline (tholeiitic) basalts are also found in the oceans, notably on the island of Hawaii where they make up the main mass of the huge shield volcanoes, Mauna Loa and Kilauea. Other workers reached similar conclusions, and as samples from the deep ocean floor became available in increasing numbers, it was soon apparent that tholeiites are by far the most common basalt in the oceanic basins. Because continental crust is absent from the deep oceans, it is obviously impossible for tholeiites to be products of assimilation of silica-rich granitic and metamorphic rocks by a primary alkaline basalt.

It did not follow, however, that Bowen's scheme offered a better solution. The Hawaiian volcanoes discharge enormous volumes of tholeiitic basalt throughout the main shield-building period, and alkaline basalts appear only in small volumes in the late stages of declining growth. Thus, the order of appearance and relative volumes are the reverse of those in Scotland, and if one uses the accepted criteria for a parental magma and holds to the concept of a single primordial basalt, a tholeiite must be the primary magma of Hawaii and alkaline basalts the products of its differentiation!

With the problem now reversed, several schemes were devised to explain these inferred relationships. One of the most durable of these was a process of assimilation of limestone and "de-silication" proposed by Reginald Daly as early as 1910. Daly was an unusually perceptive geologist with wide experience in the field. He documented many occurrences of strongly alkaline rocks closely associated with limestones, and he described how reaction with carbonate xenoliths caused lime to combine with silica and other components of the magma to precipitate Ca-rich pyroxene and plagioclase, leaving a more silica-poor liquid. These effects have been described in terms of simple chemical reactions, such as

$$CaCO_3 + 3(Mg,Fe)SiO_3 \rightarrow Ca(Mg,Fe)Si_2O_6 + (Mg,Fe)_2SiO_4 + CO_2\uparrow$$

calcite   hypersthene   augite   olivine   gas

and

$$CaCO_3 + 2(Ca,Na)(Al,Si)Si_2O_8 + Al_2O_3 \rightarrow 2CaAl_2Si_2O_8 + NaAlSiO_4 + CO_2 \uparrow$$

calcite   plagioclase   melt   anorthite   nepheline   gas

The first equation expresses the way in which addition of lime might convert a hypersthene-normative magma to an olivine-normative one, and the second shows how the melt could become nepheline normative. If the assimilated material also contained aluminous clay, Al_2O_3 would not be depleted from the melt. Daly argued that the limestone would also act as a flux to reduce the viscosity of the magma and facilitate crystal settling and differentiation. At the same time, CO_2 released in this process would account for the more explosive nature of many alkaline magmas. It seemed quite reasonable to postulate coral reefs buried within the internal structure of oceanic volcanoes where they could be assimilated in a shallow reservoir. A lively debate raged over the merits of this scheme until it was finally put to rest by clear geochemical evidence. If alkaline basalts are richer than tholeiites in certain elements, such as Ni, TiO_2, and alkalies, how could they be produced by assimilation of rocks that are nearly devoid of these components?

Clearly, the solution to the problem is not to be found in processes operating at shallow depths; one must look deeper into the levels of the mantle where the magmas originate.

■ Mantle Origins of Basalts

Having seen that the concept of a single primary basaltic magma from which all others can be derived through shallow differentiation fails to explain the diversity of igneous series, we are left to conclude that the mantle is capable of yielding more than one type of primitive magma. The question then is not how all magmas can stem from a common parent but rather how basalts of differing compositions can be derived from the mantle.

Before addressing this question, we should examine more critically the assumption that the parental basaltic magmas to which magma series have been related are the most basic (or "primitive") visible member of that series and that this parental magma is a primary, mantle-derived liquid. At first glance, it seems inconceivable that basalts could themselves be derivative liquids. They erupt at high temperatures, often in very large volumes of nearly uniform compositions. On closer examination, however, certain characteristics of common basalts are difficult to reconcile with direct derivation through melting of the mantle.

We noted in Chapter 2, for example, that most basalts erupt at temperatures on or close to their liquidus, and all their essential minerals begin to crystallize within a narrow interval of temperature. We speak of such basalts as being "multiply saturated." The basalt of the Hawaiian lava lakes, to which we have referred so often, is a good example. Although its temperature of eruption, about 1200°C, is close to the maximum ever recorded, the magma already contained phenocrysts of olivine. Moreover, the other major minerals, pyroxene and plagioclase, began to crystallize after only 10 to 20 degrees of cooling. Neither of these observations is consistent with what would be expected if basaltic magmas rise directly from their source in the mantle.

Consider first the fact that the temperature of eruption is close to or slightly below the liquidus. We can say at the outset (from the principle of Le Chatelier) that melting, because it entails an increase of volume, must occur at higher temperatures at depth than at atmospheric pressure. The amount by which the melting temperature increases with pressure can be estimated from the Clapeyron equation. If we assume that a gram of basalt with a liquidus temperature of 1200°C (1473°K) at the surface has a volumetric increase, ΔV, on melting of 0.03 cm^3 and a heat of fusion, ΔH_f, of 100 cal g^{-1} and that the mechanical equivalent of heat is 41.84 bar cm^3 cal^{-1}, then

$$\frac{dT}{dP} = \frac{T\Delta V}{\Delta H_f} = \frac{(1200 + 273) \times 0.03}{100 \times 41.84} = 0.01°\text{bar}^{-1}\text{ or } \sim 3°\text{km}^{-1} \tag{7-1}$$

Thus, the liquidus temperature of the basalt at a depth of fifty kilometers would be about 150 degrees above the value at surface conditions. A basalt at a temperature of 1200°C at that depth would be largely, if not totally, crystalline! It is clear, therefore, that a magma of this kind must originally have been hotter and lost heat en route to the surface.

Cooling could result from two effects. A magma loses heat to the rocks through which it rises, and it cools as a result of its expansion with relief of pressure. Just as air cools when it rises and expands, so too does a liquid. The amount of this adiabatic cooling (i.e., cooling at constant heat content) resulting from relief of pressure on a rising magma can be estimated from the relation

$$\frac{dT}{dP} = \frac{T\alpha}{\rho C_p} \tag{7-2a}$$

where α is the coefficient of thermal expansion (about 3×10^{-5} deg^{-1}), and C_p is heat capacity (about 0.3 cal g^{-1} deg^{-1}). Thus, for mantle temperatures and densities, the effect of adiabatic cooling would be

$$\frac{(1200 + 273) \times 3 \times 10^{-5}}{2.7 \times 0.3 \times 41.84} = 0.001°\text{bar}^{-1} \tag{7-2b}$$

or between 0.2 and 0.3 degrees per kilometer of rise. This value is only a tenth or so of the melting-temperature gradient estimated from the Clapeyron equation and is negligible compared with the difference between the temperatures of eruption and that of melting. Hence, the fact that basalts erupt with little if any superheat and at temperatures well below those at which they formed cannot be attributed to relief of pressure alone. This must mean that rising magmas are cooled even more, either by losing heat to their surroundings, reacting with the walls of their conduit, melting of crystals carried in suspension, or by some combination of these.

Consider next the observation that many basaltic liquids are close to being multiply saturated; they begin to crystallize their three principal minerals—olivine, pyroxene, and plagioclase—over a narrow range of temperature. A liquid close to equilibrium with all these minerals at low pressures (**Fig. 7-11a**) cannot be in equilibrium with the same minerals at high pressure nor can it have the same composition (Fig. 7-11b). The effect of pressure on the liquidus temperatures must differ from one mineral to another. Each mineral has different thermodynamic properties, and the rate of increase of its liquidus temperature with pressure, dT/dP, must also differ. Thus, liquid compositions that crystallize a given set of minerals simultaneously at the surface cannot have been multiply saturated and in equilibrium with those same minerals at some

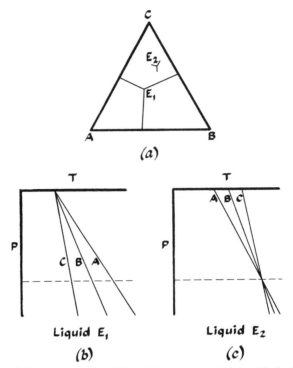

Figure 7-11 Schematic illustration of the stability relations of minerals in basaltic liquids at two different pressures. A liquid in equilibrium with three minerals, A, B, and C, at the surface would have the composition of the low-pressure invariant point E_1(a). When projected to high pressures, as shown in (b), the stability curves of the individual minerals diverge and are no longer in equilibrium with the same liquid at the same pressure and temperature. Conversely, a liquid, E_2, in equilibrium with the same minerals at high pressure (c) will not be in equilibrium at a single temperature with the same minerals at the surface.

greater pressure. By the same reasoning, any liquid produced by melting must have been saturated simultaneously with the crystalline phases of its source rock at the conditions of melting, and a basalt that preserves a primary composition inherited from melting of minerals at high pressures would begin to precipitate these minerals at very different temperatures as it approaches the surface (Fig. 7-11c).

One implication of this conclusion is that few aphyric lavas can be original, mantle-derived liquids. Before this simple fact was recognized, petrologists searching for the elusive "primary magma" shunned porphyritic rocks because they feared that the phenocrysts could have accumulated from other liquids and that the bulk composition of such a rock would not correspond to that of a liquid. It is realized now that porphyritic lavas, such as picritic basalts containing many phenocrysts of olivine, may be closer in composition to original, mantle-derived liquids than are aphyric lavas that must have evolved from other compositions.

If a primary mantle-derived melt were to reach the surface unchanged, how would we recognize it? It should, of course, have a very high liquidus temperature, and it should be rich in those components that are normally removed early in the course of differentiation. However, the most important requirement is that it should have a composition that would be in equilibrium with the principal minerals of the mantle. For example, the ratio of FeO to MgO in the liquid must be one that would be in equilibrium with the olivine in residual mantle peridotites, which is normally between Fo 90 and Fo 92. We saw in the preceding chapter that the experimentally determined distribution coefficient for Fe^{2+} and Mg between olivine and liquid

$$K_D = \frac{(Fe^{2+}/Mg)_{olivine}}{(Fe^{2+}/Mg)_{liquid}} = 0.3 \pm 0.03 \tag{7-3}$$

is nearly constant for a wide range of temperature and pressure. Using this equation and appropriate values for the compositions of olivine in mantle xenoliths, one finds that primary mantle melts should have weight ratios of Fe^{2+} to Mg of 0.425. Thus, basalts with typical FeO contents of about 10 percent should have at least 18 weight percent MgO. The liquidus temperatures of a liquid with the ratios indicated by these relationships would be at least 1400°C at the surface and even higher at mantle pressures. Although no historic eruption has been observed in which temperatures in this range were measured, we noted in Chapter 1 that they do not seem to have been uncommon in the earlier history of the Earth.

Certain Archean ultramafic lavas known as *komatiites* have compositions approaching those postulated for the mantle as a whole (**Table 7-5**). Their curious *spinifex* textures are characterized by highly acicular blades, mainly of olivine but also of pyroxene, that are typical of the crystals in rapidly quenched silicate melts (Fig. 4-19a). Their eruption temperatures were at least 1600°C, and if they rose adiabatically, their temperature at mantle pressures would have been even higher. Their Mg contents are so high that they must have been produced by what was essentially total melting of the mantle.

Because the high degrees of melting required for such extreme compositions are unlikely to have prevailed in post-Archean time, primary melts of the mantle today would probably be porphyritic lavas intermediate in composition between komatiites and common basalts. Picrites and ankaramites are the most common lavas that meet these criteria.

Compositions and petrographic features of komatiites. The appearances of spinifex textures in outcrop and under the microscope are shown in the two drawings.

Table 7-5

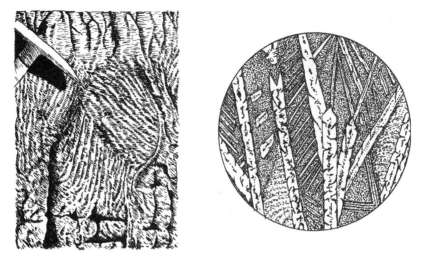

Typical Chemical Compositions

	1.	2.	3.
SiO$_2$	45.9	44.4	45.2
TiO$_2$	0.4	0.3	0.2
Al$_2$O$_3$	8.4	5.4	3.7
FeO*	10.9	11.3	11.0
MnO	0.2	0.2	0.2
MgO	23.5	31.7	32.2
CaO	8.7	4.9	5.0
Na$_2$O	0.8	0.3	0.4
K$_2$O	0.2	0.0	0.2
Ni (ppm)	1049	1930	1931
Cr (ppm)	3250	3100	3190
Ba (ppm)	13	4	11

Molecular Norms

Ilm	0.5	0.4	0.3
Or	1.1	0.0	1.1
Ab	7.0	2.6	3.4
An	18.1	12.7	7.3
Di	18.8	8.3	14.1
Hy	9.0	17.7	14.7
Ol	45.4	58.4	59.2

1. Munro Township, Canada

2. Yilgarn Block, Australia

3. Berberton Mts., South Africa

*Total iron as FeO

Source: From a compilation by R. W. Nesbitt and Shen-Su Sun, 1976, *Earth Planet. Sci. Ltrs.* 431:433–453.

Mantle Source Rocks

If the mantle produced only one primary magma and re-equilibration of that magma was always perfect, all basalts would arrive at the surface with identical compositions and would retain little decipherable evidence of their origin in the mantle. The fact that all basalts are not reduced to a single composition shows that although they may re-equilibrate they do so to differing degrees and may follow somewhat different lines of evolution depending on the composition and depth of their source.

Mafic and Ultramafic Inclusions

The most direct evidence of the mineralogical character of mantle sources comes from coarse-grained rocks brought to the surface as xenoliths in basaltic magmas. These inclusions are especially plentiful in explosive volcanic vents, such as diamond pipes, and in tuff rings or maars formed by magma blasting craters through water-saturated sediments. It is not clear why inclusions should be so much more common in these particular settings. Perhaps the rapid rise enables them to reach the surface in relatively unaltered form. Where the same magmas erupt more quietly as lava flows, inclusions are less common and tend to show more evidence of re-equilibration to low-pressure conditions.

Most inclusions in basaltic lavas fall into one of four groups (Fig. 7-12):

I. Gabbroic rocks, typically *eucrites,* containing calcic plagioclase (bytownite or anorthite), augite, and olivine.
II. Dunites and wehrlites composed of differing proportions of forsteritic olivine and diopsidic pyroxene. The most common accessory minerals are spinel, plagioclase, or both.
III. Lherzolites and harzburgites containing, in addition to the minerals of group II, a magnesian orthopyroxene, usually enstatite. The aluminous phase is spinel.
IV. Garnet peridotites and eclogites consisting of differing proportions of Ca-rich and Ca-poor pyroxene, olivine, and pyrope-rich garnet.

Gabbroic inclusions of Group I are most common; dunites and wehrlites of Group II are slightly less so. In oceanic settings the enstatite-bearing assemblages of Group III are much less common, and the last group of garnet-bearing rocks is known from only a single occurrence at Salt Lake Crater on the island of Oahu, Hawaii. On the continents, however, nodules of Groups III and IV are more widespread. Garnetiferous

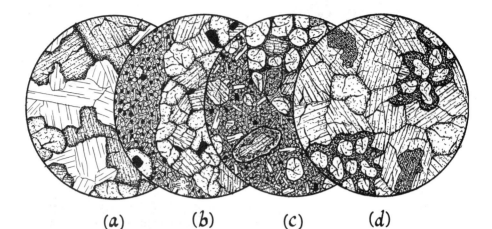

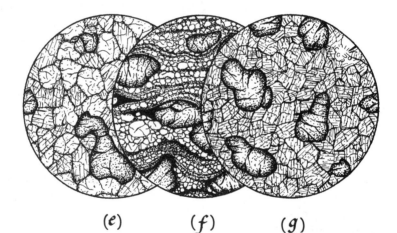

Figure 7-12 Principal types of mafic and ultramafic inclusions found in volcanic rocks. (a) Eucritic gabbro from a tholeiitic volcano in the Galapagos Islands consisting of calcic plagioclase, augite, and olivine. (b) Spinel wehrlite from Hualalai volcano, Hawaii. (c) Spinel lherzolite from a strongly alkaline basanite on Kerguelen Island, south Indian Ocean. Enstatite from the nodule is scattered through the lava. Where it has reacted with the liquid, it is rimmed by a fine-grained aureole made up of the same minerals, olivine and augite, that were crystallizing from the liquid. Enstatite in the outer parts of the nodule also shows evidence of reaction. (d) Garnet peridotite from Kakanui, New Zealand. Dark-brown glass surrounds the partly melted garnet and fills interstices between other minerals. Some crystals of pyroxene contain patches of brown amphibole. (e) A granular garnet lherzolite from a South African diamond pipe consists of nearly equidimensional grains of pyroxene, olivine, and garnet with grain boundaries meeting at angles of approximately 120 degrees. This textural configuration is the most stable one for minerals coexisting at high temperatures and pressures. (f) Similar mineral assemblages may be highly sheared, giving the rock a crushed and highly irregular texture. (g) A garnet pyroxenite or "eclogite" from Salt Lake Crater, Oahu, Hawaii.

inclusions have been divided into two petrographic types (Figs. 7-12e and 7-12f), one with a granular texture and another showing evidence of intense shearing. The latter must result from large-scale plastic flow of the mantle. The name *eclogite* is given to rocks consisting of pyrope-rich garnet and sodium-rich pyroxene (Fig. 7-12g). Orthopyroxene and olivine, if present, are only minor components. The bulk chemical compositions of this type of eclogite are close to that of basalt, of which it is the high-pressure equivalent.

It would be a mistake to take this assemblage of mafic and ultramafic inclusions as a representative sample of the mantle. Some could be the crystalline residue of partial melting, while others are no doubt products of crystallization or reaction at the walls of the conduit through which their host magma rose. They are no less informative, however, because they indicate what minerals are stable in such a system and how those minerals could contribute the essential components of a basaltic liquid. The principal phases, olivine and pyroxene, contain Si, Mg, Fe, and Ca in amounts that are more than adequate to satisfy the requirements of basaltic melts, but it is not so easy to account for the other essential components—Al, Na, K—and a host of minor and trace elements. Some may be in accessory minerals, while others are probably secondary components or impurities in more abundant minerals.

At low pressures, almost all the alumina in mafic igneous rocks is contained in plagioclase, but as already noted (in Chapter 4), at higher pressures, this component enters clinopyroxene as the Ca-tschermak's molecule (CaTs – $CaAl_2SiO_6$) and jadeite (Jd – $NaAlSi_2O_6$), and to a lesser extent, it enters Ca-poor pyroxenes as the Mg-tschermak molecule (MgTs – $MgAl_2SiO_6$). Experiments show that the amount of alumina contained in pyroxene increases with both pressure and temperature. It reaches a maximum of about 4.5 percent for mantle assemblages at pressures of about 20 kilobars and at temperatures approaching those of basaltic melts. Beyond that limit, any additional alumina must be taken up by a separate phase, such as spinel or garnet, and because these latter minerals are denser than pyroxene, they become increasingly important at higher pressures.

The experimentally determined stability ranges of the principal aluminous phases are shown in **Figure 7-13**. The boundaries are only approximate because they are affected by the bulk composition of the rock. For example, rocks rich in alumina may contain both plagioclase and spinel over a range of pressures and temperatures. This is especially true of compositions rich in normative albite, which tends to extend the stability range of plagioclase to higher pressures.

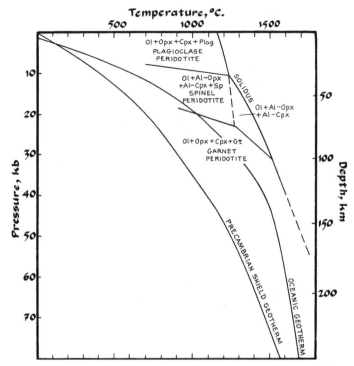

Figure 7-13 Stability ranges of the essential phases of the mantle. (Adapted from D. H. Green, 1970, *Trans. Leicester Lit. Phil. Soc.* 64:28–54 and E. Takahashi and I. Kushiro, 1983, *Amer. Min.* 68:859–879.)

At pressures greater than about 15 kilobars, garnet may be a stable aluminous phase, at least at temperatures well below the melting range of basaltic liquids. At temperatures approaching the solidus, where melting begins, the thermally expanded crystal structure of pyroxene can accommodate more alumina, and the stability of garnet is reduced. For this reason, it is difficult to say from phase relationships alone whether or not garnet is a stable phase in equilibrium with basaltic melts in the mantle. One must also know the bulk composition of the rocks and the position of the melting curve with respect to the stability fields of garnet and aluminous pyroxene.

The succession of reactions for the changes from one aluminous phase to another can be written in terms of plagioclase and olivine in the following simplified sequence:

$$2CaAl_2Si_2O_8 + 3Mg_2SiO_4 \rightarrow CaMgSi_2O_6 \cdot nCaAl_2SiO_6 + 2Mg_2Si_2O_6 \cdot mMgAl_2SiO_6 + MgAl_2O_4$$

An   Fo   Di   CaTs   En   MgTs   spinel and for the albitic component

$$NaAlSi_3O_8 + Mg_2SiO_4 \rightarrow NaAlSi_2O_6 + Mg_2O_6$$
$$\quad Ab \qquad Fo \qquad Jd \qquad En$$

At higher pressures, spinel goes to garnet:

$$MgAl_2O_4 + CaMgSi_2O_6 + 2Mg_2SiO_4 \rightarrow CaMg_2Al_2O_{12} + 2Mg_2Si_2O_6$$
$$\quad spinel \qquad Di \qquad Fo \qquad garnet \qquad En$$

Because the proportion of olivine to plagioclase in the mantle is much greater than that of the reactants in these equations, olivine is also present among the products of both reactions.

Trace elements provide a way of determining whether a given basalt came from a depth where garnet was the stable aluminous mineral. The rare-earth elements (REE) are particularly useful in this regard because their partitioning between the liquid and minerals varies with their atomic radius. Garnet normally contains proportionately greater concentrations of the heavy REE so that liquids produced by small amounts of melting of a garnet-bearing assemblage would have proportions of REE mirroring that of garnet. Amphibole and orthopyroxene have the same effect but to a less pronounced degree.

Figures 7-14a and 7-14b show two sets of curves calculated for different degrees of melting of two source rocks, one with garnet and one without. Note the stronger relative enrichment of light REE in liquids derived from the garnetiferous assemblage and the manner in which the pattern becomes flatter with more advanced degrees of melting. This change results from the fact that as more garnet melts it contributes more heavy REE to the liquid until, at advanced stages of melting, the pattern approaches that of the original rock.

The jadeite molecule of pyroxene is the main Na-bearing constituent of ultramafic rocks at mantle pressures. Ca-rich pyroxene accepts increasing amounts of jadeite in solid solution at high pressures and depths of more than a few tens of kilometers. With appropriate degrees of melting, it could supply the sodium essential to melts of basaltic composition. Minor amounts of sodium may also be contained in amphibole but only under shallow mantle conditions because that mineral is unstable at depths of more than 100 kilometers or so.

Potassium is a different matter. It does not enter pyroxene, olivine, spinel, or garnet in appreciable amounts, and the amounts contributed to basaltic liquids can be contained only in a separate mineral, such as the magnesian

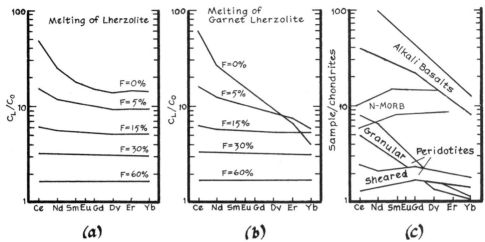

Figure 7-14 Calculated REE distribution patterns for melts derived by batch melting of two lherzolites, one (a) without garnet and another (b) with 5-percent garnet. Note the stronger depletion of heavy REE in small fractions of melt from the garnetiferous lherzolite and the similarity of melts evolved beyond 5-percent melting. Patterns in natural basalts (c) from intraplate volcanoes are strongly depleted in heavy REE as one might expect from small amounts of melting of a garnet-bearing source rock. While most basalts from mid-ocean ridges (N-MORB) have flatter slopes that could result from melting at shallow levels where garnet is not stable or alternatively, by advanced degrees of melting. An enriched type of mid-ocean ridge basalt (E-MORB) has a steeper pattern, suggesting a deeper origin where garnet is stable. Abundances of REE in all three diagrams are plotted on logarithmic scales; those of (a) and (b) are normalized to the abundances in the original peridotite, while those in (c) are normalized to the abundances in chondritic meteorites. The differences between abundances in chondritic meteorites and sheared mantle peridotites are not large. [(a) and (b) after G. N. Hanson, 1978, *Ann. Rev. Earth Planet. Sci.* 8:371–406.]

mica, phlogopite, or the potassium feldspar, sanidine. Small amounts of the former are common in peridotites, especially in continental settings, while the latter is a theoretical possibility but is rare in mantle rocks. Phlogopite is of special interest for other reasons: first, it contains a potentially important radioactive heat source, potassium, and, second, one of its essential constituents is water.

Water is normally the most important volatile component of the mantle. Even though it is not likely to account for more than a fraction of a percent by weight, it plays an important role in magmas because, as we shall see shortly, even small amounts can greatly reduce melting temperatures and affect the stability relations of major minerals. In addition to phlogopite, the magnesium-rich amphibole pargasite is a common hydrous mineral, particularly in ultramafic xenoliths found in continental volcanoes, but it is stable only at relatively shallow depths, certainly less than 100 km.

Carbon dioxide is another important volatile component of the mantle, especially at depths greater than about 70 km. It is contained in calcite, dolomite, and other carbonates that are found as minor but widespread constituents of ultramafic xenoliths of deep origin. Small inclusions of dolomite have been found in diamonds. CO_2 has the opposite effect of H_2O in that it tends to raise rather than lower the melting temperatures of mantle assemblage, but its effect is confined to relatively shallow depths. At depths of more than approximately 100 km, the carbonates that replace it depress the solidus to almost the same extent as water. These relationships are discussed in greater detail in later chapters.

Depths of Origin

By calibrating the compositions of coexisting minerals experimentally, one can assign reasonable values to the temperatures and pressures recorded in the minerals of mantle xenoliths. For example, the compositions of Ca-rich and Ca-poor pyroxenes coexisting with plagioclase, spinel, or garnet are functions of temperature and pressure. Their contents of calcium and aluminum vary in such a way that the abundances of these two components form a grid on a P-T diagram (**Fig. 7-15**). Thus, by determining the composi-

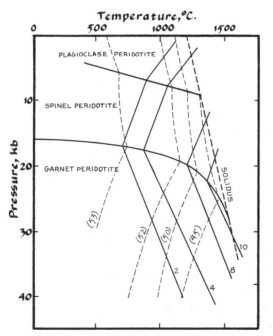

Figure 7-15 T-P-X relations of pyroxenes in equilibrium with olivine and plagioclase, spinel, or garnet. Broken lines are contours of Ca contents (numbers in parentheses), and solid lines are contours of Al contents (numbers without parentheses). (Adapted from compilation by Carmichael et al., 1974, *Igneous Petrology*, Fig. 12-11.) Equations for calculating pressure and temperature are given in Appendix H.

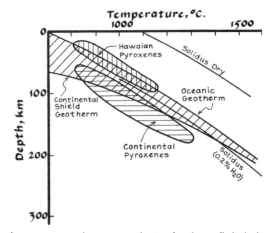

Figure 7-16 Ranges of temperature and pressure estimates for ultramafic inclusions from Hawaii (vertical ruling) and continental diamond pipes (horizontal ruling). A range of estimated temperature gradients is indicated by diagonal ruling between the oceanic and continental geotherms. Two curves for the melting of ultramafic mantle rocks are shown, one for dry conditions and another in the presence of 0.2-percent water. The general concordance of temperatures and pressures estimated from pyroxenes in ultramafic nodules with the range of probable temperature gradients and melting curves indicates that basaltic liquids can be produced by melting in the upper levels of the mantle if small amounts of water are present.

tions of pyroxenes in a particular peridotite, one can estimate the temperature and depth at which they last equilibrated.

Figure 7-16 summarizes the estimated temperatures and pressures for two suites of peridotite nodules, an oceanic one from Hawaii and a continental one, mainly from South African diamond pipes. For comparison, estimated geothermal gradients for oceanic and old continental regions are shown, along with two solidus curves corresponding to the beginning of melting of basaltic compositions. The overlap of these various curves shows that the nodules and their host magmas must be closely related and that the latter could be produced by melting at depths of less than 200 km. The temperatures and pressures recorded by inclusions indicate that magmas are formed in the oceanic mantle at somewhat higher temperatures or lower pressures than those formed under continents. This is in accord with geophysical evidence that indicates a steeper geothermal gradient under the oceans, and it helps explain the differences in suites of ultramafic inclusions found in continental and oceanic volcanoes.

Although many ultramafic inclusions found in continental volcanic rocks resemble those of their oceanic counterparts, the proportions of various types differ in the two settings. As we noted, enstatite-bearing peridotites are much more common in continental suites, and garnetiferous inclusions, which are all but unknown in oceanic settings, are found in many continental localities. In contrast, peridotites consisting mainly of olivine and aluminous pyroxenes, which are so plentiful in oceanic suites, are less common in continental settings. These differences may be related to the differing geothermal gradients of the two types of regions. Referring again to Figure 7-16, note that the steeper oceanic geotherm crosses the stability fields at higher temperatures, where pyroxenes are more aluminous and garnet is less likely to be a stable phase. The lower thermal gradient under the continents, on the other hand, favors garnet-bearing assemblages over a wider pressure range.

Relationships of Ultramafic Inclusions to Basaltic Magmas

It was obvious even to early investigators that inclusions contain valuable evidence of the deep origins of magmas, but the complex associations and great diversity of these rocks were difficult to fit into any simple genetic scheme. At least three possible origins can be imagined. Some inclusions might be interpreted as fragments of the original mantle from which magmas are derived by partial melting, while others are so impoverished in alumina, alkalies, and other essential components of basalt that they could be the depleted refractory residue left after segregation of a partial melt. Still others have textures, compositions, and even layering that give them the appearance of rocks formed by crystal accumulation from cooling magma.

A Japanese petrologist, Hisashi Kuno, was the first to draw attention to a systematic relationship between Hawaiian inclusions and the types of basalt in which they are found. He noted that tholeiitic basalts contain only gabbroic inclusions of Group I, while mildly alkaline basalts contain gabbro, dunite, and wehrlite of both Groups I and II. Enstatite-bearing inclusions of Group III are found along with Groups I and II in basalts of more alkaline composition, and inclusions in the strongly silica-deficient basanites on the island of Oahu include, in addition to all other types, garnet-bearing assemblages of Group IV. Thus, the more alkaline and undersaturated the basalt, the greater the range of inclusions it is likely to contain.

Kuno reasoned that if successively more alkaline magmas come from increasing depths in the mantle, each may bring up fragments of the shallower horizons through which it rises. If gabbroic inclusions are the only type found in tholeiitic basalts, then tholeiites have the shallowest source. More alkaline magmas from greater depths bring with them not only rocks from their own source but those of higher levels as well.

This logical interpretation explained why the sequence of increasingly alkaline basalts and their associated inclusions is also one of progressively greater mineralogical contrasts and disequilibrium. Gabbroic inclusions differ only slightly from their tholeiitic host. Both contain plagioclase, olivine, and pyroxene, all which are stable at the surface. This is in contrast to ultramafic rocks in alkaline lavas, which are much poorer in silica and alumina than their hosts and commonly contain spinels that tend to break down or react with other minerals at low pressures. The enstatite in lherzolites and harzburgites of group III is unstable in silica-deficient magmas, and as we would anticipate from what we know about systems like forsterite-silica, it reacts with the alkaline liquid that has brought it to the surface.

Kuno and his co-worker, Ikuo Kushiro, saw in this relationship evidence that the observed minerals have a pressure-dependent stability relationship to their coexisting liquids. They predicted that experiments would show that enstatite is stable in silica-poor liquids at high pressures, and in due course, when experimental equipment was developed to explore the effects of high pressures, their prediction was fulfilled. This astute deduction, which proved to be a key element in explaining the compositional dif-

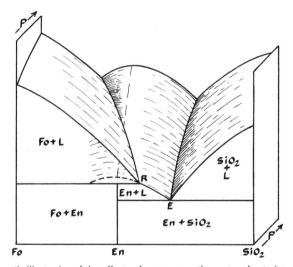

Figure 7-17 Schematic illustration of the effects of pressure on the system forsterite-silica. Pressure elevates the liquidus curves of enstatite more than that of forsterite, causing the stability field of forsterite to become smaller as that of enstatite is enlarged. At the same time, the reaction point, R, retreats to lower silica contents until it passes the composition of enstatite and becomes a eutectic. Melting of any assemblage of olivine and enstatite at high pressures will result in a liquid of lower silica content than one produced at low pressures.

ferences of basalts, was arrived at by a simple projection of the melting temperatures of olivine and enstatite from atmospheric conditions to high pressure.

Although the melting temperature of forsterite is far above that of enstatite at atmospheric pressure (1890° versus 1557°C), pressure has a greater effect on enstatite than on forsterite. The reason for this difference is that the change of volume of enstatite on melting, ΔV, is larger, and its heat of fusion, ΔH_f, is less. Substituting appropriate values in the Clapeyron equation, the effect of pressure on the melting temperature, dT/dP, was found to be about 26° per kilobar for enstatite, compared with only 5° for olivine. Hence, with increased pressure, the stability field of enstatite expands rapidly at the expense of olivine, and the incongruent melting of enstatite disappears at pressures of more than a few kilobars (**Fig. 7-17**).

An effect, such as increased pressure, that reduces or eliminates the incongruent melting of enstatite would account for the lower silica contents of liquids generated at greater depths. At the same time, it would explain why enstatite can be stable in an alkaline magma at mantle depths but react with the same liquid at the surface.

■ Melting at High Pressures

A major advance was achieved when apparatus were developed to carry out petrologic experiments under conditions of temperature and pressure comparable to those of the mantle (**Fig. 7-18**). Before this was possible, one could, of course, use thermodynamic relations to predict the general effects of pressure, just as Kuno and Kushiro did in their elegant analysis of the melting relationships of ultramafic rocks; however, the new experimental techniques made it possible to explore much more complex relationships, and the result has been a wealth of new information on the properties and melting behavior of the mantle.

The effects of pressure on composition are well illustrated by the system forsterite-nepheline-silica (**Fig. 7-19a**) that corresponds to the base of the basalt tetrahedron in Figures 7-1 and 7-3c. The invariant point where melting begins is shown at atmospheric pressure and at 10, 20, and 30 kilobars. Note the steady migration toward liquids of increasingly more silica-deficient compositions. Similar effects are seen in several other systems.

The relevance of these synthetic systems to natural rocks is illustrated by the results of experiments in which samples of mantle peridotite have been melted at high pressures and a series of temperatures. The compositions of melts produced at different pressures and degrees of melting are shown in Figures 7-19b and 7-19c. The diagrams correspond to the front face and

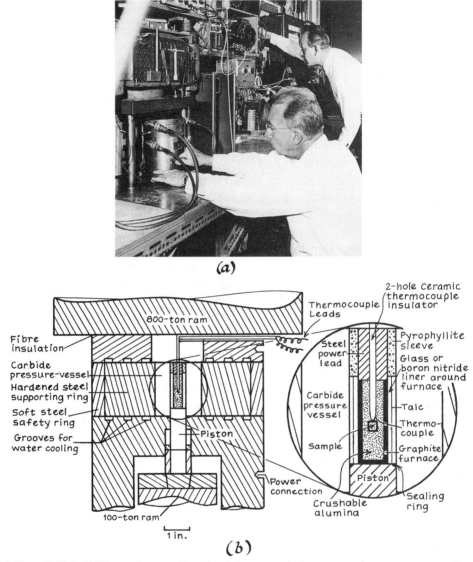

Figure 7-18 (a, b) Using modern experimental techniques, petrologists can reproduce temperatures and pressures equivalent to those at depths as great as 150 km in the earth's mantle. The apparatus illustrated here is one developed at the Geophysical Laboratory, Washington, D.C., by J. L. England (foreground) and F. R. Boyd (rear) (courtesy of H. S. Yoder, Jr., Director of the Geophysical Laboratory, Carnegie Institution of Washington).

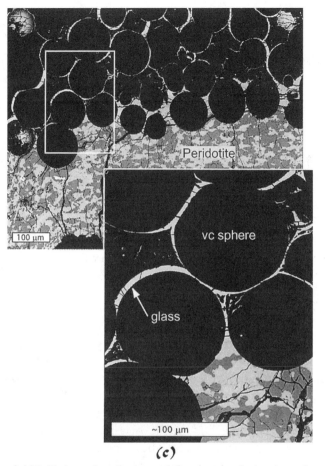

Figure 7-18 (Cont.) (c) A thin layer of small spheres of vitreous carbon is placed over the crystals being melted so that the liquid can be quickly quenched to glass and analyzed under an electron microprobe. Because the carbon spheres are strong enough to withstand high pressures, they are not crushed, and the pressure on the sample, possibly augmented by capillary forces, causes the resulting liquid to move into the open pore spaces.

base of the basalt tetrahedron in Figure 7-2. Again, the chief effect of pressure on liquid compositions is a decrease in silica content, but increased melting has the opposite effect.

These melting relationships are more easily visualized by comparing the compositions of the liquid and residual assemblage produced by melting a mantle peridotite (**Table 7-6**). If we assume that the liquid produced by melting a garnet lherzolite is a picritic basalt, it is a simple matter to calculate, by means of a mass balance (Fig. 5-14), the bulk composition of the refractory minerals that would be left after the maximum amount of melt

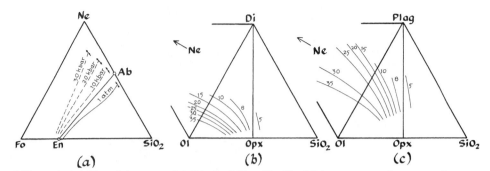

Figure 7-19 (a) Part of the system forsterite-nepheline-silica (Fig. 7-3c) at pressures of one atmosphere, and 10, 20, and 30 kilobars. Only the invariant points and olivine-pyroxene cotectics are shown. Note the migration of the invariant point toward lower silica contents with increasing pressures. The effects of pressure and degree of melting on the compositions of liquids in equilibrium with a mantle peridotite are shown in terms of the normative components olivine-cpx-quartz and olivine-plagioclase-quartz in (b) and (c), respectively. The light numbered lines define the range of liquids produced by different degrees of melting at constant pressure in kilobars. [Sources: (a) F. Schairer and H. S. Yoder, Jr., 1961, *Carnegie Inst. Washington Yrbk.* 60:141–144 and I. Kushiro, 1968, *J. Geoph. Res.* 73:619–634; (b) and (c) E. Takahashi and I. Kushiro, 1983, *Amer. Min.* 68:859–879].

has been removed. The effects are most clearly seen if the compositions of the liquid and residue are expressed in terms of the same set of minerals as the original peridotite. (A procedure by which a chemical composition can be recast into any convenient set of minerals is outlined in Appendix A.)

Note first that the minerals of the peridotite have not been consumed in the same proportions they had in the original peridotite. Instead, garnet, clinopyroxene, and phlogopite have been preferentially incorporated into the liquid, while olivine and enstatite become relatively more important in the residue. Second, the maximum proportion of such a liquid that can be obtained is limited by the K_2O content of the peridotite. It cannot exceed 2.5 percent, the value at which all phlogopite is exhausted, without the amount of K_2O in the liquid falling below its observed concentration. Third, because this amount of liquid is proportionately small, its removal has a correspondingly small effect on the more abundant major-element components of the remaining peridotite.

We should note that the liquids obtained by melting experiments are produced in a closed system at constant pressure, and we have no assurance that the modal proportions we assumed are applicable to a mantle in which mobile components can be collected from a larger volume than that of the rock being melted. If, for example, additional H_2O or alkalies were introduced metasomatically, the volume of the source rock required to produce a given amount of melt would be proportionately less.

Table 7-6	Comparisons of a basaltic magma, postulated to be the parental magma for Kilauea Volcano, and a garnet lherzolite that could be a possible source rock for such a basalt. Also shown is the calculated composition of the residue left on extraction of the liquid after melting 2.5 percent of the peridotite.

	Basalt	Peridotite	Residuum
	Major Elements (wt %)		
SiO_2	46.70	44.50	44.44
TiO_2	1.85	1.30	1.29
Al_2O_3	9.17	2.80	2.64
FeO*	11.90	10.30	10.26
MgO	20.00	37.90	37.90
CaO	7.86	3.30	3.18
Na_2O	1.54	0.40	0.37
K_2O	0.40	0.01	0.00
	Trace elements (ppm)		
Ni	855	2300	2337
Cr	1400	3000	3041
Ba	195	5	< 0.01
U	0.7	0.02	< 0.01
Th	0.3	0.08	< 0.01
	Normative or Modal Minerals		
Olivine	5.1	50.6	54.0
Opx	31.9	20.5	18.0
Cpx	32.4	14.3	20.2
Garnet	25.6	14.4	6.5
Phlogopite	3.2	0.2	0.0
Rutile	1.8	0.0	1.3
Density	2.81	3.47	3.46

*Total iron as FeO.

High-Pressure Crystal–Liquid Relationships

Phase relationships become increasingly complex as the components of natural rocks are added to simple synthetic systems. Each additional component increases the number of degrees of freedom and adds to the difficulty of portraying melting relations in two-dimensional diagrams. The CMAS system, illustrated in **Figure 7-20**, was devised in an attempt to surmount this problem. It groups components that behave similarly in basaltic systems and plots them in a four-component tetrahedron for which the

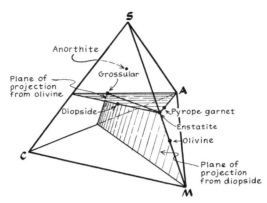

Figure 7-20 The CMAS tetrahedron. Compositions of important minerals are shown at their appropriate locations on or within the tetrahedron. The composition of anorthite falls on the rear face above the plane of the olivine projection.

nominal end-members are CaO, MgO, Al_2O_3, and SiO_2. These end-members are calculated according to the following combinations:

C = (mol. prop. $CaO - 3.33P_2O_5 + 2Na_2O + 2K_2O$) × 56.08
M = (mol. prop. $FeO + MnO + NiO + MgO - TiO_2$) × 40.31
A = (mol. prop. $Al_2O_3 + Cr_2O_3 + Fe_2O_3 + Na_2O + K_2O + TiO_2$) × 101.96
S = (mol. prop. $SiO_2 - 2Na_2O - 2K_2O$) × 60.09

The four factors are then normalized so that they sum to 100.

This rather elaborate arrangement has the advantage that all the major components of natural rocks and all the important minerals—olivine, Ca-rich and Ca-poor pyroxenes, plagioclase, spinel, and garnet—can be plotted conveniently in terms of the composite end-members.

The system can be simplified even more. Because olivine is probably present at all reasonable pressures and stages of melting, liquids can be assumed to be olivine saturated, and their compositions within the CMAS volume can be projected from that mineral to a plane defined by the other phases, pyroxene and garnet or spinel. In this way, melting behavior can be viewed in a simple, two-dimensional diagram, such as the one shown in **Figure 7-21**. The form of the diagram varies with pressure as the fields of individual minerals expand or contract, but for our present purposes, it will suffice to examine a single section corresponding to a pressure at which garnet is the stable aluminous phase.

The first liquid produced by melting of the four-phase assemblage, diopside, enstatite, olivine, and garnet, will have the composition of point B.

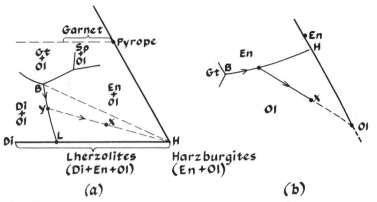

Figure 7-21 (a) Olivine-saturated liquidus field boundaries of diopside, enstatite, garnet, and spinel at 30 kilobars are projected from olivine onto the upper plane shown in Figure 7-20. At this pressure, the point B is just inside the CMAS tetrahedron. At pressures less than 27 kilobars, it lies slightly outside the CAS face of Figure 7-20. (b) A similar projection from diopside shows the field boundaries of garnet, enstatite, and olivine.

(Although B is actually the piercing point of a univariant line passing through the plane, it can be thought of as a eutectic in the projection of Figure 7-21.) With rising temperature, the liquid would follow a path that is governed by the original bulk composition and can be predicted by comparing the melting relations in Figure 7-21 with those that would produce the residual mineral phases of natural mantle rocks.

In progressive melting of a composition, such as X, the liquid would leave B when garnet is exhausted, follow the cotectic to Y until diopside is exhausted, and then cross the enstatite (+ olivine) field. Looking again at the mantle xenoliths described earlier, we can arrange the ultramafic members of the suite in order of decreasing numbers of mineral components:

Olivine + Opx + Cpx + Spinel or Garnet + Phlogopite
↓
Olivine + Opx + Cpx
↓
Olivine + Opx
↓
Olivine

This sequence corresponds to the residual mineral assemblages that would be in equilibrium with successive liquids at each stage of melting of a composition such as X. Lherzolites (with olivine and two pyroxenes but no aluminous mineral) correspond to the area on the front face of the tetrahedron and to the line between Di + Ol along the lower edge of the plane of the projection. Harzburgites (olivine + enstatite) fall along the front edge of the tetrahedron and at point H in the projection (Fig. 7-21b), while pure dunite would be below that point at the composition of olivine.

The order of progressive depletion outlined here may not be the only one. Some petrologists argue on the basis of the abundances of certain trace elements in basalts that clinopyroxene may disappear before the aluminous phase. If point X in Figure 7-21a were slightly richer in garnet, Di would disappear first, and the liquids would follow a different path. In either case, the order of depletion of mineral phases on melting affords a way of tracing the course of liquids in the CMAS system. First to disappear are the accessory phases. These are followed by either garnet or diopside and then by enstatite. Olivine would be the last and most refractory mineral.

We saw in Chapter 5 that liquids produced by progressive melting can be segregated in two ways. In *fractional melting*, small amounts of melt are separated soon after they are formed, while in *equilibrium* or *batch melting*, the liquid accumulates in situ as melting advances, and its composition is the integrated product of a continuous series of liquids. The course is very different if liquid is extracted soon after as it forms. In that case, when the first phase is exhausted, no new liquid can be produced until the temperature reaches the melting temperature of the invariant point L, the composition of the first liquid produced by melting the remaining crystals. Similarly, when diopside disappears, no further liquid is produced until the temperature reaches the eutectic between enstatite and olivine at H. With orthopyroxene exhausted, no additional liquid would appear until the temperature reaches the melting point of pure olivine. Thus, the effect of this type of fractional melting would be to produce a discontinuous series of melts separated in time and with large steps in composition and temperature.

Under the high pressures at depths where most melting occurs, extraction of even small amounts of liquid may be very efficient. Melting of a mantle peridotite normally begins around unstable hydrous minerals or at grain boundaries where incompatible elements tend to be concentrated. Junctions where three or more different minerals are in contact are especially favored because the melting temperature of such an assemblage is lower (**Fig. 7-22**).

The form these first pockets of melt take depends in large measure on the degree to which the liquid wets the surfaces of crystals. If the "wetability" is

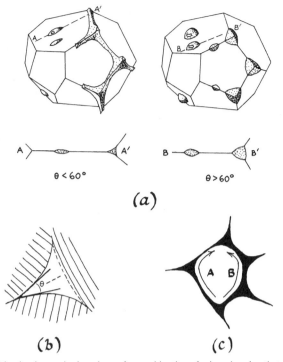

Figure 7-22 (a) Melting begins at the junctions of a combination of mineral grains that corresponds to the lowest temperature melt. The form the melt takes depends on the amount of melting and the degree to which the liquid wets the surface. Pools of melt tend to have concave sides with dihedral angles, θ, less than 60° if they wet the crystals (left) but convex sides (right) if they do not. (After E. B. Watson and J. M. Brenan, 1987, *Earth Planet. Sci. Ltrs.* 85:497–515.) (b) The dihedral angle also varies with the crystallographic orientation of minerals that melt more readily in one direction than another. (c) The permeability of the rock is governed mainly by the narrowest channels, which are usually between the flat surfaces of crystals. As liquid migrates through these spaces it tends to follow a progressively smaller number of preferred channels. The liquid following course A, for example, will be slower than that along course B, if it must follow a thinner or more tortuous channel. The flow tends to widen the preferred channels even more by thermal erosion. With time, course A will be abandoned, and course B will be enlarged.

strong, the liquid penetrates grain boundaries, and pools of melt at triple junctions have the form of prisms or tetrahedra with concave sides and dihedral angles of less than about 60 degrees. The liquid in isolated pockets moves very slowly, if at all, but when these pockets become interconnected and form a continuous, network it can begin to move under any driving force, such as buoyancy or stress, that provides a pressure differential. The amount of melt necessary for this to happen is probably very small, certainly less than 10 percent and possibly as little as 1 percent. More melt is required for pyroxene-rich peridotites than for more easily deformed olivine-rich rocks, such as dunite.

Experiments have been devised to determine the amounts and compositions of melts produced from a natural mantle assemblage by a given increase of temperature. By using a special technique for extracting the small amounts of liquid (Fig. 7-18), it has been possible to do this over a range of possible mantle temperatures and pressures. The results of a set of these experiments are shown in **Figure 7-23**.

Although there are minor differences depending on the proportions of crystalline phases, the amount of liquid increases systematically as long as all the essential minerals are present. Clinopyroxene seems to be the limiting

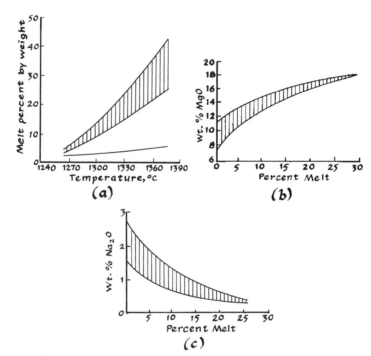

Figure 7-23 (a) Percentage of melt (by weight) as a function of temperature for a number of different proportions of olivine, Ca-rich pyroxene, Ca-poor pyroxene, and spinel. The lowest curve shows little increase in melt fraction until it reached the highest temperatures because it contained only 1-percent Ca-rich pyroxene. (b) The abundance of MgO in liquids increases as the more refractory minerals are melted at higher temperatures. (c) Na_2O has the opposite behavior. It is effectively partitioned into the very first melts, and as melting continues to advance, its abundance in the liquid is reduced by dilution. The diagrams are based on experiments with a range of bulk compositions, and the spread of values at small degrees of melting in b and c reflects differences in the abundances of minor minerals that contribute to the first melt but are soon exhausted. (After J. Pickering-Witter and A. D. Johnston, 2000, *Contrib. Miner. Petrol.* 140:190–211 and B. E. Schwab and A. D. Johnston, 2001, *Jour Petrol.*, 42:1789–1811.)

mineral in this respect because it furnishes at least two essential components, sodium and aluminum, that the pyroxene holds in the form of the jadeite and Ca-tschermak's molecules. This is reflected in a progressive increase of the Na content of the liquid with progressive melting so long as clinopyroxene is still present.

The fact that melts with the extreme compositions that would result from advanced degrees of fractional melting are seldom, if ever, observed among the products of basaltic volcanoes would suggest that fractional melting is less common in the mantle than batch melting. And yet, experimental studies indicate that at these pressures only a percent or two of liquid can be produced before it is squeezed from its crystalline residue. Moreover, we have seen that ultramafic nodules seem to indicate that melting continues until only one or two residual minerals remain. A possible explanation for this apparent inconsistency would be that the liquids are indeed extracted soon after a small amount accumulates but that they collect in intermediate reservoirs before continuing their ascent.

The notion that the mantle produces a single uniform basalt that changes composition only after reaching the crust is long since gone, and we are left with a highly complex system in which the compositions of mantle-derived magmas result from a combination of factors—the depth and degree of melting, the composition of source rocks, and a variety of changes that can occur en route to the surface. How can one sort out all these individual effects and come up with a plausible explanation of why one basalt differs from another? The task is not as hopeless as it may seem. We have seen that basalts preserve distinctive compositional features that provide clues to their previous history, and with the wealth of information gained from experimental and geochemical studies, it has become possible to identify certain systematic variations that reflect their origins and compositional evolution. In the next chapter we see how these principles can be used to interpret the basalts produced in a variety of tectonic settings.

Selected References

Basaltic Volcanism Study Project, 1981, *Basaltic Volcanism on the Terrestrial Planets,* Pergamon Press, 1286 p. A giant volume covering all aspects of basaltic rocks on the earth, moon, and inner planets.

Morse, S. A., 1980, *Basalts and Phase Diagrams.* Springer-Verlag, 493 p. A comprehensive treatment of differentiation of basaltic magmas with emphasis on phase equilibria.

Emeleus, C. H. and B. R. Bell, 2005, *British Regional Geology: The Palaeogene Volcanic Districts of Scotland,* 4th edition. London: HMSO for the

British Geological Survey, 212 p. A condensed summary of the geology of the classical igneous centers of western Scotland.

Yoder, H. S., Jr., 1976, *Generation of Basaltic Magma*, National Academy of Science, 265 p. A comprehensive survey of the state of research into the origins of basalts with emphasis on phase equilibria and physical mechanisms of magma generation. After more than 30 years, this still remains one of the best general treatments of the subject.

8 Oceanic Magmatism and Flood Basalts

If volumetric proportions are any measure of the petrologic significance of any group of rocks, the basalts of the ocean floor are by far the most important on Earth, for they cover about 70 percent of the surface and account for nearly 85 percent of the igneous rocks added to the crust each year. The first evidence that basalt is an important constituent of the seafloor was obtained on the voyage of HMS Challenger, between 1872 and 1876, when Murray and Renard obtained the first samples of deep-sea sediments and noted that they contained large amounts of fragmental volcanic material. Solid basaltic rocks were not recovered until 1905 when Wiseman succeeded in dragging a dredge along the floor of the Indian Ocean. Since then, great numbers of samples have been obtained in this way. Starting in the early 1960s, new drilling techniques made it possible to obtain cores through almost the entire thickness of the oceanic crust, and this, along with refined geophysical studies, now provide detailed information on the subsurface structure of the ocean floor. By comparing these observations with studies of regions where the oceanic lithosphere has been exposed on land, it has been possible to piece together a composite three-dimensional picture of the oceanic crust.

Beneath a thin cover of sediments, the seafloor consists of about 2 kilometers of basaltic pillow lavas that become more compact downward and grade into increasing numbers of basaltic dikes, many of which were feeders of the overlying lava flows. Near the base of the lavas, the dikes extend down into coarse gabbros, which make up the rest of the section down to the mantle 4 to 5 kilometers below the surface.

Most oceanic basalts fall into one of two broad types that are commonly referred to as mid-ocean-ridge basalts (MORB) and ocean–island basalts (OIB). The names are not really appropriate. Oceanic ridges are not necessarily centered in the middle of oceans, and rocks of this type occur in many other settings. Volcanoes that produce basalts of the second type do not necessarily form islands and they too occur in other settings. Nevertheless, the names are now widely used and are not likely to be changed. Although the compositions of these two varieties of basalt are not fixed, the variations within each group are coherent in the sense that they can be related by differentiation of a distinct parental magma. The term MORB is best confined to rocks that demonstrably originated at or near a ridge. They should not be confused with another important variety, oceanic plateau basalts (OPB), that resembles continental flood basalts in that they form large expanses of voluminous basaltic lavas with no apparent relationship to spreading centers.

The average composition given in **Table 8-1**, no. 1, illustrates some of the distinctive features of the normal type of MORB (or N-MORB). They are olivine-normative tholeiites with relatively small amounts of K_2O and TiO_2 (normally less than 0.2 and 2.0 percent, respectively). Because of a somewhat higher alumina content, they are more likely to have plagioclase phenocrysts than would the ocean–island type. Where the two major types of basalt have a close spatial relationship, the ocean–ridge basalts are richer in potassium, titanium, and many lithophile elements (**Fig. 8-1**). They are usually referred to as enriched or "E-type MORB" and seem to result from entrainment of magma of the ocean–island type in normal MORB.

OIBs differ from those of oceanic ridges, mainly in being richer in TiO_2, FeO, and incompatible elements, such as K, Rb, Ba, Zr, and the rare-earth elements (REE) (Table 8-1, nos. 3, 4, and 5). They also differ in their isotopic character (**Fig. 8-2**). MORB and OIB fall at opposite ends of a linear distribution defined by the mantle-derived basalts of different tectonic settings. Normal MORB occupies a relatively restricted range, whereas OIBs tend to have lower Nd and higher Sr ratios with each group of islands having a distinct isotopic character. The variation along this trend, often referred to as a "mantle array," results mainly from differences in the abundances of the parent radioactive elements, Rb and Sm, from which the daughter products, ^{87}Sr and ^{144}Nd, are derived, so that ratios falling along the lower right part of the array come from sources that were relatively richer in these elements. The differences between the basalts of mid-ocean ridges and intraplate volcanoes suggest that the former come from a more depleted and homogeneous source than the latter. On a local scale, islands and seamounts differ from one another over short distances, and even the lavas of a single volcano may differ from one flow to the next. Large volcanoes tend to be more uniform than small ones, probably because their inhomogeneities are better integrated into the larger volume, but even the giant Hawaiian volcanoes have notable geochemical variations.

■ Ocean-Ridge Basalts

The system of spreading ridges extends for 65,000 kilometers through all the principal oceans of the Earth. With an average height of about 2.5 km

Table 8-1

Chemical characteristics of MORBs, OIBs, and ocean plateau basalts (OPB). FeO* is total iron as FeO. Norms are calculated assuming a uniform ratio of Fe_2O_3 to FeO* of 0.15. Data are taken mainly from the references listed at the end of this chapter.

	1.	2.	3.	4.	5.	6.
SiO_2	50.53	49.98	49.28	47.73	43.05	49.60
TiO_2	1.56	1.99	2.61	3.30	2.83	1.08
Al_2O_3	15.27	15.11	12.56	15.53	10.169	14.20
FeO*	10.46	11.04	11.20	10.67	12.34	10.74
MnO	–	–	0.17	0.14	0.19	0.20
MgO	7.47	6.87	9.74	8.37	16.34	8.02
CaO	11.49	11.25	10.80	8.71	10.23	12.44
Na_2O	2.62	2.81	2.10	`2.89	2.08	2.01
K_2O	0.18	0.47	0.48	1.70	1.21	0.06
P_2O_5	0.13	0.32	0.27	0.66	0.48	0.10
Molecular Norms						
Ap	0.27	0.68	0.57	1.39	0.99	0.21
Ilm	2.19	2.80	3.69	4.61	3.91	1.54
Mt	1.65	1.75	1.78	1.67	1.67	1.52
Or	1.07	2.80	2.88	10.08	7.08	0.36
Ab	23.70	25.46	19.13	26.05	4.37	18.45
An	29.59	27.48	23.77	24.48	14.76	30.20
Di	21.58	21.39	23.04	11.66	25.94	25.77
Hy	18.05	13.13	22.17	2.33	–	20.30
Ol	1.89	4.52	2.97	17.74	32.80	1.65
Ne	–	–	–	–	8.48	–
Trace elements (ppm)						
Ba	6	57	125	760	449	2
Zr	74	73	141	255	216	52
Rb	1	5	8	62	35	1
Sr	90	155	368	792	719	88
Cr	960	960	630	–	655	342
Ni	249	487	200	–	467	125
La	2.5	6.3	14.6	41.2	36.0	2.7
Sm	2.6	2.6	5.6	8.1	–	2.3
Lu	0.5	0.4	0.3	0.3	–	0.4
$^{87}Sr/^{86}Sr$	.7028	.7036	.7035	.7053	.7043	.7037

1. Typical N-type MORB (average basaltic glass from the Atlantic, Pacific, and Indian Ocean spreading axes; trace elements taken from Table f, Appendix F).
2. Typical E-type MORB (average basalt from the vicinity of the Galapagos hotspot on the Galapagos spreading axis; trace elements taken from Table f, Appendix F).
3. Olivine tholeiite, 1921 eruption of Kilauea.
4. Potassic tholeiite, Gough Island, South Atlantic Ocean.
5. Nodule-bearing basanite, Tahiti, Pacific Ocean.
6. Ocean plateau basalt, Nauru Basin.

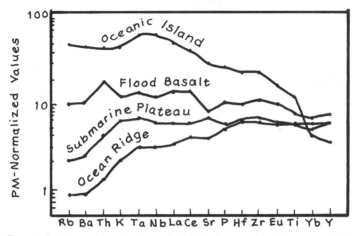

Figure 8-1 Characteristic trace-element patterns for basalts from oceanic ridges, oceanic islands, and flood basalts on the continents (Deccan Traps) and oceans (Ontong-Java Plateau). Values are normalized to the composition of primitive mantle. (From D. A. Wood., 1979, *Geology* 7:499–503.)

and width of about 2,000 km, the total volume of these ridges is of the order of several hundred million cubic kilometers. The ridge axes are characterized by numerous shallow earthquakes, high heat flow, and strong hydrothermal activity. The seafloor gradually subsides with increasing dis-

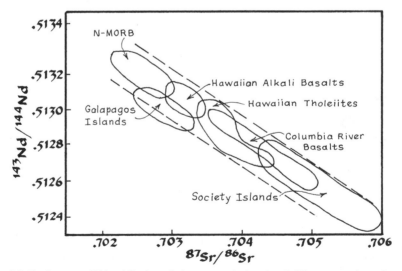

Figure 8-2 The isotopes of Nd and Sr show distinct ranges for basalts of different tectonic settings. Basalts of the mid-ocean ridges have the greatest Nd and smallest Sr ratios, while those of intra-plate volcanoes and flood basalts are more varied and come from less depleted mantle. Although each region and tectonic setting has its own characteristic ratios, most fall within the "mantle array" outlined by dashed lines.

tance from the ridges, at least out to crustal ages of 80 million years or so. This subsidence is thought to be due mainly to conductive cooling of the lithospheric because the rate is directly proportional to the square root of time (Equation 2-13).

The oldest oceanic rocks, which reach Jurassic ages along the western margins of the Pacific Ocean, are chemically and petrographically indistinguishable from the lavas erupted at mid-oceanic ridges today. We can infer, therefore, that basalts of this kind have been produced with little or no change in composition for at least 140 million years. Estimates of the total production of basaltic magma range from 10 to 20 cubic kilometers per year, depending on the assumed thickness of the basaltic layer. Less than half of the new magma added to the oceanic crust is erupted as lava; the larger part is intruded as gabbro, peridotites, and doleritic dikes.

The average rate of formation of oceanic crust is probably close to 10,000 m^3 per kilometer per year but differs widely from one ridge to another. Even on an individual ridge segment, the rates of magmatism vary in both time and space so that only the averages over periods of centuries or more can be correlated with plate motions. If the rate of production of magma is low relative to the spreading rate, the ridge has a lower elevation, and part of the crustal extension is taken up by a combination of extensional fissures and normal faults. Ridges with little magmatism and half-spreading rates slower than about 3 centimeters per year have prominent axial grabens up to a kilometer or more in depth and bounded by multiple, inward-dipping, normal faults. Faster-spreading ridges with greater rates of production lack these features and have higher elevations above the normal seafloor. Most of the dikes feeding lavas erupted at the ridge crest are injected laterally from centralized conduits spaced a few tens of kilometers apart (**Fig. 8-3**).

The volume of rising mantle required to supply this magma is vastly greater than that of the basalts, and yet the magma drawn from this enormous mass is focused in a very narrow axis, commonly less than a kilometer in width. The sharp flexure where the mantle column diverges outward favors a radial orientation of potential fractures crossing the flow lines. This occurs where the viscosity of the mantle increases sharply and deformation changes from plastic flow to brittle fracture. As a result, the melt flows upward and inward through discrete channels that are focused on the axis of the ridge. They feed a narrow, thin lens of magma from which most of the overlying dikes are injected. These melt lenses must be steady-state features because they have been found on nearly every fast-spreading ridge that has been explored seismically.

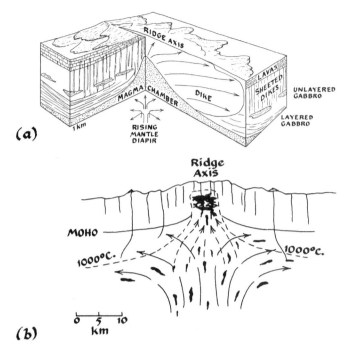

Figure 8-3 (a) Schematic section through a ridge axis based on studies of ophiolites. Magma is supplied mainly by lateral injections from central feeders spaced at intervals along the axis. Note that both the direction and velocity of the mantle differ from those of the lithosphere; the motion of the former is radial from the diapir, while the latter is outward from the ridge axis. (b) Most of the magma produced by passive pressure relief in the rising mantle is channeled into a central zone by the anisotropic stresses where the flow turns from vertical to horizontal. It is focused into a narrow zone under the ridge axis where part crystallizes in a shallow reservoir while another part continues to ascend to the surface. The elevation of the ridge reflects the amount of new material added to the crust, either as lavas or shallow intrusions. (After A. Nicolas, 1989, *Ophiolites, and Dynamics of Oceanic Lithosphere.* Kluwer Acad. Publ., 367 p.)

Although the average compositions of basalts erupted within the axial zones of ridges show no regular differences with time and are only crudely correlated with the rate of spreading, those of fast-spreading ridges seem to be somewhat more varied, possibly because the presence of a melt lens allows for longer term storage and shallow differentiation. Another factor is the rate of melt production. Residual peridotites associated with slow-spreading ridges are mainly lherzolites, while those under more active ridges are more depleted harzburgites. This difference indicates that the rate of magma production is directly correlated with the degree of melting. If melting is the result of decompression, the amount of magma released from a rising column of mantle is a function of the interval between the depth of origin and the level to which it rises (**Fig. 8-4**). The rate of ascent must be fairly

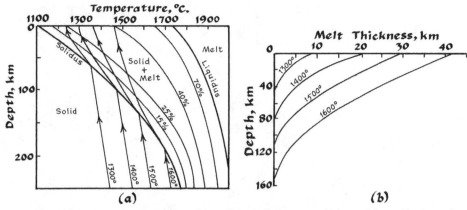

Figure 8-4 (a) A column of asthenosphere rising adiabatically intersects the base of the melting interval at differing depths according to their potential temperatures and compositions. The effect of water and carbon dioxide would be to lower these depths and increase the pressure interval through which melting advances. (b) The total thickness of melt generated by rising dry mantle of different potential temperatures. (b and c are taken from D. P. McKenzie and M. J. Bickle, 1988, *Jour. Petrol.* 29:625–679.)

rapid because the least differentiated basalts are usually charged with numerous phenocrysts of plagioclase and olivine that would be more likely to be segregated from a column that rises more slowly.

Ophiolites

We can see some of the structural and compositional relations of this system in bodies of oceanic lithosphere that have been thrust onto continents and can be observed at close range. Originally defined as an assemblage of mafic and ultramafic rocks in geosynclines, the term *ophiolite* has more recently been applied to an assemblage of pillow lavas, sheeted dikes, gabbros, and residual peridotites typical of the oceanic crust. These bodies were previously considered roots of Alpine-type orogenic belts, but most are now thought to result from oceanic lithosphere being thrust over continental crust.

Ophiolitic complexes have been described from such diverse regions as Cyprus, New Guinea, Newfoundland, and the Coast Range of Oregon and California, but by far the most thoroughly studied and instructive example is in the mountains of southeastern Oman on the Arabian Peninsula (**Figure 8-5**). The body is thought to have been formed at a Cretaceous ridge crest, near what is now the mouth of the Persian Gulf, and thrust westward over late Paleozoic and Mesozoic rocks of the Arabian Shield.

A composite section through the complex, shown in **Figure 8-6**, is probably typical of much of the oceanic lithosphere. The peridotite at the base

(a)

(b)

Figure 8-5 Rocks of the oceanic crust and underlying mantle are magnificently exposed in the desert of Oman. (a) In this view, basaltic pillow lavas rest directly on the lighter-colored rocks of the mantle. (b) A closer view shows well-layered gabbros under what is thought to have been an oceanic ridge. (Photos courtesy of Adolphe Nicolas.)

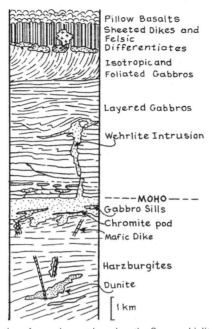

Pillow Basalts
Sheeted Dikes and Felsic Differentiates
Isotropic and Foliated Gabbros
Layered Gabbros
Wehrlite Intrusion
————MOHO———
Gabbro Sills
Chromite pod
Mafic Dike
Harzburgites
Dunite
1 km

Figure 8-6 A composite section of oceanic crust based on the Oman ophiolite. (After R. C. Coleman, 1981, *Jour. Geoph. Res.* 86:2497–2508, with modifications based on A. Nicolas, 1989, *Ophiolites and Dynamics of Oceanic Lithosphere,* Kluwer Acad. Publ., 367 p.)

of the section is interpreted as depleted mantle underlying what was originally a shallow magmatic reservoir. It is composed of highly deformed harzburgite with many dikes and irregular pods of dunite, wehrlite, and gabbro. The overlying gabbros are also deformed but less so than the mantle rocks. Layering is pervasive in the lower gabbro but dies out upward. In most places, the gabbros intrude the overlying dikes and pillow basalts, but a few dikes rise through the gabbro from sources at unseen depths. The contact relationships suggest that the dikes and lava flows were derived from a shallow reservoir at a time when the top of the latter was still largely liquid. Relatively few tapped lower levels. Small pods of plagioclase-rich sodic granophyres, usually referred to as *plagiogranites*, are found in the uppermost levels of the gabbroic section. They must have crystallized during later stages of cooling because they are clearly later than most of the dikes.

Because the magma must come from a source in the mantle focused directly under the ridge, it is logical to interpret the thick, depleted harzburgitic peridotites beneath the gabbros as the residue left after melt was extracted and intruded upward. If so, there is an apparent anomaly, because the gabbros and basalts precipitated olivine and Ca-rich pyroxene long before orthopyroxene. A magma derived from a harzburgitic source rock consisting chiefly of olivine and orthopyroxene would have been saturated with orthopyroxene and, on cooling, would have precipitated that mineral rather than the Ca-rich pyroxene that is much more abundant in the gabbro. A possible explanation is that the basaltic liquids were separated from the harzburgites at greater depths where, as we saw in the previous chapter (Fig. 7-17), the stability of Ca-poor pyroxene is increased relative to that of olivine.

The amount of differentiation shown in cryptic layering of the gabbros is surprisingly small until a late stage of solidification when dioritic and granophyric liquids were segregated and intruded upward (**Fig. 8-7**). This restricted variation in the gabbros is thought to result from frequent injections of new magma that continued to replenish the reservoir until it had receded far enough from the axis to crystallize and differentiate without further mixing.

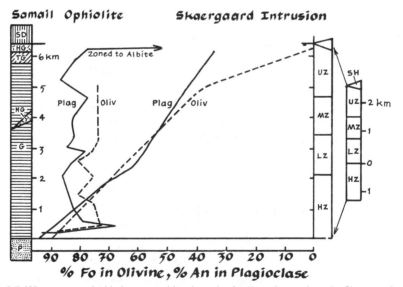

Figure 8-7 When compared with the compositional trends of an intrusion, such as the Skaergaard, which crystallized from a single injection of magma, the gabbroic rocks of the Samail ophiolite show only minor irregular variations reflecting periodic replenishment of the magma and resetting of its composition. Only at the very top where light differentiates accumulated do the rocks show strong changes in the composition of their plagioclase. The vertical scale for the Skaergaard intrusion on the right has been expanded to make it equal to that of the Samail ophiolite on the left. (After J. S. Pallister and C. A. Hopson, 1981, *Jour. Geoph. Res.* 86:2593–2644.)

Shallow Differentiation

The major-element trends of differentiation of ridge basalts differ in minor ways from one locality to another, but they all share certain general features. An unusually complete example, shown in **Figure 8-8**, resembles the Galapagos tholeiitic series described in the preceding chapter (Fig. 7-6). Note that FeO^* first increases and then declines, while SiO_2 increases only slightly until well over half of the original liquid has crystallized. The variations through the early and middle stages of differentiation are consistent with shallow fractionation of the observed phenocrysts—olivine, plagioclase, and augite—which, taken as a whole, cause a depletion of MgO and residual enrichment of titanium, alkalies, phosphorus, and other excluded elements.

The olivine crystals in a given basalt can have a wide range of sizes and compositions. Iron-rich olivines are reversely zones, while forsteritic ones have normal zoning—a clear sign of mixing. The basalts must be blends of two or more magmas, an initial one that was crystallizing and differentiating and a second of less evolved composition. A similar explanation may account for the wide range of trace-element contents in basalts of the same

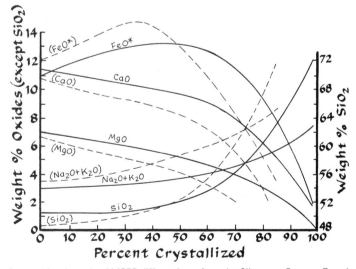

Figure 8-8 Compositional trends of MORB differentiates from the Clipperton Fracture Zone (solid lines) plotted in terms of the fraction crystallized. The trends resemble those of the tholeiitic Galapagos rocks in Figure 7-6a, shown here as dashed lines; however, there is less enrichment of alkalies, and MgO declines more slowly. The amount of differentiation in this example is much greater than normal. (After G. Thompson, W. B. Bryan, and S. E. Humphris, 1989, Axial volcanism on the East Pacific Rise. In *Magmatism in the Ocean Basins*, A. D. Saunders and M. J. Norry, eds., Oxford, UK: Blackwell, pp. 181–200.)

MgO content. If the reservoir under a ridge is periodically recharged with fresh magma in response to spreading and surface eruptions, the volume of liquid would remain more or less constant, and the composition of the differentiating magma would be periodically reset back to a more primitive composition. The effects of differentiation under these conditions differ from those of a closed reservoir. Because the variations during Rayleigh fractionation follow a curved path, while those due to mixing are along straight lines, each cycle of differentiation and replenishment has the effect of increasing the concentrations of strongly excluded trace elements, even though the major elements remain nearly constant.

The concentration of an element in such a reservoir can be calculated from the equations:

$$C_D = C_{N-1} (1 - X)^{D-1} \tag{8-1}$$

$$M_N = M_{N-1} (1 - X - Y) + Z \tag{8-2}$$

$$C_M = C_D (M_N - X - Y) + (C_0 \times Z)/M_N \tag{8-3}$$

where

C_0 = concentration of the element in the primary undifferentiated magma

C_D = concentration after differentiation

C_M = concentration after mixing

C_{N-1} = concentration at the beginning of the current cycle (C_0 for the first cycle)

D = bulk distribution coefficient

X = mass fraction crystallized per cycle

Y = mass fraction discharged per cycle

Z = mass fraction replenished (Z = X + Y for a reservoir of constant volume)

M_N = total mass after the Nth cycle

As shown schematically in **Figure 8-9**, the trend of differentiation follows a curved course defined by the Equation 5-11 for Rayleigh fractionation. When the process is interrupted by discharge of part of the differentiate, C_D, and a corresponding influx of primitive magma, C_0, the composition is set back toward the original composition at a point C_M, determined by the proportions of old and new liquids. This mixture then proceeds to differentiate along a new curved path, and the process continues. With time, the compositions approach a steady state in which the concentrations, C_{SS}, are

$$C_{ss} = C_0 \frac{(X + Y)(1 - X)^{D-1}}{1 - (1 - X - Y)(1 - X)^{D-1}} \tag{8-4}$$

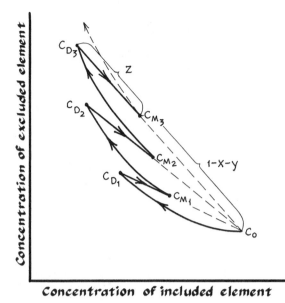

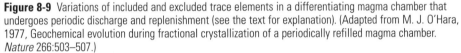

Figure 8-9 Variations of included and excluded trace elements in a differentiating magma chamber that undergoes periodic discharge and replenishment (see the text for explanation). (Adapted from M. J. O'Hara, 1977, Geochemical evolution during fractional crystallization of a periodically refilled magma chamber. *Nature* 266:503–507.)

Thus, the major elements approach constant compositions reflecting fractionation of olivine, plagioclase, and pyroxene, while trace-element fractionation is progressively more exaggerated. This process can account for the remarkable uniformity of major elements in basalts on fast-spreading ridges.

Mantle Sources

Of the three basic forms of magma generation outlined in the opening chapter (Fig. 1-9), the most effective in oceanic systems would be decompression melting. This is the only mechanism that can account for the sustained production of large volumes of nearly uniform basalts at ridges in an extensional tectonic setting. The pressure relief comes about when the asthenosphere rises passively to fill new space created by the spreading lithosphere. On reaching the depth corresponding to the pressure of the solidus, the mantle begins to melt and yields increasing amounts of liquid as it continues to rise to progressively shallower depths (**Fig. 8-10**). Judging from the attenuation of seismic waves passing through the source zones under ridges, the maximum fraction of melt is between 3 and 7 percent.

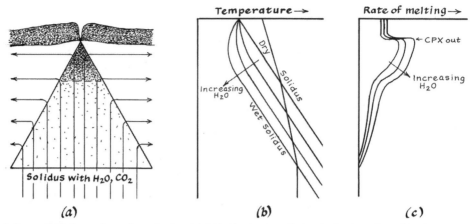

Figure 8-10 As a column of mantle rises (a), it first begins to melt at a depth where the solidus is depressed by strongly excluded components, such as water and carbon dioxide. (b) Because these components are present in relatively small amounts, the amount of melt produced is also small until the column reaches the solidus for the depleted residue, where the amount of melting becomes much greater. (c) As the column approaches the base of the lithosphere, the flow lines diverge outward, while the magma continues to flow into a shallow reservoir. (Adapted from Langmuir, et al., 1992, *AGU Geoph. Monog.* 71:183–280, and Hirschmann, et al., 1999, *Jour. Petrol.* 40: 831–851.)

The composition and quantity of this melt depends primarily on the depth interval through which it rises; the greater the depth at which it crosses the solidus the greater the pressure interval through which melting takes place. The depth at which melting of an upwelling column begins is determined by its potential temperature and composition. A mantle containing excluded components, such as water, that have the effect of lowering melting temperatures would begin to melt at a deeper level than a depleted, anhydrous mantle at the same temperature. As melting progresses and these low-melting components are removed in the fractionated liquid, the rate of melting declines until the column approaches the solidus for anhydrous rocks. Melting then increases sharply and continues at a higher rate until one of the major minerals is exhausted. When the mantle residuum finally reaches the base of the lithosphere and begins to diverge laterally, it remains at a nearly constant pressure and produces little or no additional melt.

Such a process is neither fractional melting nor batch melting but a combination of the two. Although small quantities of melt are probably extracted as soon as they are formed, they are thought to pool in a shallow reservoir so that the lavas reaching the surface are the integrated products of melting and differentiation over a wide range of pressures. Nevertheless, the basalts retain an intrinsic signature from their source,

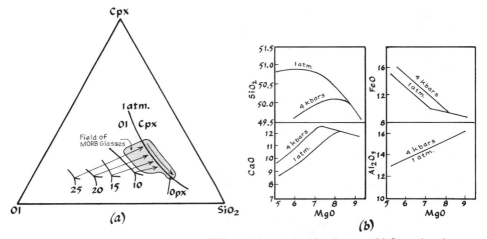

Figure 8-11 When the compositions of N-MORB glasses with more than 9-percent MgO are plotted on an olivine-cpx-silica plane within the CMAS tetrahedron (a), most fall between the olivine-cpx cotectics for one atmosphere and ten kilobars. Those with less than 9-percent MgO are closer to the low-pressure cotectic, as one would expect if they have differentiated in a shallow reservoir. Liquids produced by small amounts or melting at high pressures would crystallize olivine as they rise, owing to the expansion of the olivine field with declining pressure. If they then collect and differentiate at constant pressure, their compositions evolve further (b) by fractionation of plagioclase and augite. (a) From D. Elthon, 1989, in *Magmatism in the Ocean Basins*, A. D. Saunders and M. J. Norry, eds., *Geol. Soc. Spec. Publ.* 42:125–136. (b) Adapted from Yang, Kinzler, and Grove, 1996, *Cont. Min. Pet.* 124:1–18.

and if their compositions can be corrected for the effects of later differentiation, one should be able to determine the conditions under which they were initially formed.

This can be seen in **Figure 8-11a** in which primitive MORBs are compared with the experimentally determined compositions of mantle melts over a range of pressures. The diagram is a section through the CMAS tetrahedron (Fig. 7-20) in which compositions are projected from plagioclase, so that only the fields of mafic minerals, olivine, and the two pyroxenes, are shown. A liquid that is simultaneously saturated with all these minerals (plus plagioclase, spinel, or garnet) would have the composition of a pseudoinvariant point at a given pressure and would correspond to the first melt of a peridotite having those same minerals on the liquidus. (The name "pseudoinvariant point" is used for the eutectic-like junction of the three cotectics because its position varies in another dimension. It is really the point where a univariant line passes through the plane of the diagram.) As we would expect from the experimental systems described in the previous chapter (Figs. 7-17 and 7-19), liquids produced by small degrees of melting at high pressure are poorer in silica than those in equilibrium with the same minerals at shallower depths.

When plotted on the same diagram, most ocean-ridge basalts fall between the olivine-CPX cotectics for one atmosphere and ten kilobars. Because the field of olivine expands with falling pressure, melts rising to shallower levels crystallize that mineral and the liquids evolve in the direction shown by the arrows away from the olivine corner of the triangle. Depending on the pressure at which the melts were first formed, their compositions at lower pressures will fall at differing distances from the low-pressure invariant point. When these liquids pool at a shallow depth of 10 to 15 kilometers, plagioclase begins to crystallize along with olivine and augite, and their compositions take a new course down the low-pressure cotectic. The trends of this shallow differentiation can by calculated from experimental data so that the concentrations of key components can be projected back to the stage at which the magma reached a shallow reservoir. In this way, we can reconstruct its earlier composition and say something about its ultimate origin. An MgO value of 8 to 9 weight percent is commonly used for this because it corresponds to a stage shortly before plagioclase and augite begin to crystallize, and the compositions have a more restricted range (**Fig. 8-11b**).

Correcting for trace-element fractionation is a very different matter. We saw earlier that the effects of differentiation in a shallow reservoir can differ widely depending on the partitioning of a given element between the crystals and liquid and the frequency with which the magma has been replenished. Although their absolute abundances vary widely, the ratios of elements with similar partition coefficients can remain almost constant. This is best seen in isotopic ratios, which are essentially immune to most processes of crystal fractionation or melting and can provide clues to the degree of enrichment or depletion of lithophile elements, such as Rb and Sm, in the source rocks. The relative abundances of REE are also useful in this regard, for they provide a means of identifying the different minerals contributing to the melt (Fig. 7-14).

Having made the necessary adjustments for shallow differentiation, experimentally determined phase equilibria can be used to estimate the conditions under which a basalt of a given composition was last in equilibrium with a mantle source. A simple empirical equation relates temperature to the MgO and SiO_2 content of the liquid:

$$T(^\circ C) = 2000 \frac{MgO}{SiO_2 + MgO} + 969 \tag{8-5}$$

The equation has a relatively large mean error of 40°C, due mainly to the effects of secondary components, such as water, alkalies, and other excluded components that tend to lower the liquidus and solidus temperatures of most anhydrous minerals. Using the temperature obtained in this way, one can then estimate the depth at which a basalt was produced in the mantle. We saw earlier (Figs. 7-17 and 7-19) that increased pressure causes the compositions of melts to migrate toward systematically lower silica concentrations. A simple equation uses this well-established relationship to arrive at an approximate pressure:

$$\text{Ln } P \text{ (kbar)} = .00252 \, t(^{\circ}C) - 0.12 \, SiO_2 + 5.027 \tag{8-6}$$

Again, the value obtained has a rather low precision (a mean error of about 2.7 kilobars) because it is based on a single component. A number of computer-based analogs have been devised to calculate more precise temperatures and pressures by including the effects of a larger number of components.

The wide range of temperatures and pressures estimated in this way indicate that the conditions of melting under ridges are far from uniform. Similar conclusions have been drawn from estimates of the extent to which melting advances at a given temperature and pressure. Experimental studies, such as those described in the preceding chapter (Fig. 7-23c), show how the concentrations of certain components such as sodium are a good measure of the degree of melting. When corrected for shallow differentiation, the sodium contents of basalts correlate remarkably well with the elevations and hence the volumes of ridges, and because the latter are determined by the rates of magma production in the upwelling mantle, the fact that they vary so widely suggests that the degree of melting in the rising mantle column also varies substantially with local conditions.

■ Intra-Plate Volcanism

The magmatism in interior regions of oceanic plates is second only to that of spreading ridges in terms of the amount of basalt it adds to the crust each year. The productivity of these sources differs widely from one region to the next, but the global rate is probably of the order of 1.5 km^3 per year. The most conspicuous expression of this volcanism is seen in long chains of volcanic islands and seamounts extending for thousands of kilometers across the central and northern parts of the Pacific Ocean. The linear age progression in the Hawaiian-Emperor Seamount Chain (**Fig. 8-12**) indicates that the magmas have had a fixed source beneath the drifting Pacific Plate for at least 47 million years. The volcanoes grow rapidly at first then gradually become extinct as they leave their source and move downstream. This is not to say that all intraplate volcanoes are formed in this way. Thousands of seamounts are scattered in seemingly random fashion on the

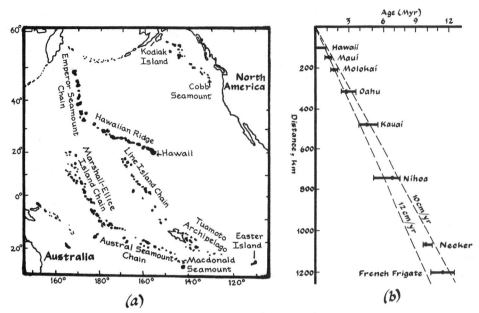

Figure 8-12 (a) Chains of volcanic islands and seamounts in the central Pacific. (b) Relationship between ages of the volcanoes of the Hawaiian Emperor Seamount Chain and distance from the hotspot near the island of Hawaii. (Adapted from G. B. Dalrymple, E. A. Silver, and E. D. Jackson, 1973, *Amer. Sci.* 61:294–308.)

seafloor, and although some are grouped in elongated clusters or chains, their age sequences seldom have any apparent relationship to the direction of plate motion. This is especially true of older volcanoes. Over time, the regions of volcanism have become narrower, more linear, and more consistently age progressive. This trend has been attributed to changes in stress patterns in the oceanic lithosphere as a result of shifts in the directions of subduction and long-term variations in the temperature and density of the lithosphere.

Although each hotspot has its own distinctive basalts and trend of differentiation, a crude relationship has been found between the compositions of lavas on individual islands and their distance from a spreading axis and the thickness of the lithosphere. Many islands have both tholeiitic and alkaline basalts, but those close to spreading axes tend to be predominantly tholeiitic, especially in their differentiated suites. On islands more distant from an axis, alkaline rocks are more abundant and tend to be more strongly differentiated. All islands with true rhyolites, such as Iceland, Easter Island, and the Galapagos, stand close to ridges, while islands with very alkaline differentiates, such as Cook Island and Tahiti in the Pacific and

Trindade and Fernando Po in the Atlantic, are much farther from a spreading axis. This relationship implies that the compositions of the magmas are affected by the age, thickness, and thermal state of the lithosphere through which they rise.

Hawaii

The Hawaiian-Emperor Seamount Chain was the first in which a precise linear relationship was found between the ages of individual islands or seamounts and their distances from a fixed source under the southern end of the chain. The hotspot at the head of the Hawaiian chain has been described as the expression of a mantle plume rooted deep in the mantle, well below the lithospheric plate drifting over it. This is in contrast to the magmatism of spreading ridges, which comes from shallower sources and can migrate laterally along with the overlying plates. In both cases, melting probably results from decompression of an ascending portion of the mantle, but at hotspots, it is thought to be driven by buoyancy effects, whereas the mantle under ridges is simply rising through passive flow into space generated by the diverging plates. The lithosphere drifts downstream at a rate of about ten to twelve centimeters per year until the volcanoes are detached from their source and become extinct. On a long-term scale, the rate of magmatism seems to be accelerating and is now about 10 times greater than it was at the time the trend of the islands took their present course.

The development of the Hawaiian volcanoes can be divided into several stages, each characterized by a distinctive type of activity and group of lavas (**Fig. 8-13**). The youngest in the chain, Loihi, is a small submarine volcano just south of the island of Hawaii. Many of the basalts it erupted initially were distinctly alkaline, but it has also produced tholeiites, and with time, it is expected that the latter will be dominant as the volcano becomes a large, mature shield like Mauna Loa or Mauna Kea. During this main shield-building stage, the growing volcanoes discharge copious lavas, first as submarine pillow basalts and later as subaerial pahoehoe flows, until they attain volumes of 40,000 cubic kilometers or more. Measured from their base five kilometers below sea level, their heights approach that of Everest, and their huge mass is vastly greater than even the largest land volcanoes. It is estimated that the magma required for this scale of volcanism would have to be drawn from a mantle column 600 kilometers in diameter rising at a rate of about 50 centimeters a year.

Later in this shield-building stage, a shallow caldera develops by passive subsidence of a circular or elliptical part of the summit. Nearly all lavas are erupted within the caldera or from one or more rift zones radiating down the flanks. Kilauea, currently the most active of the Hawaiian volcanoes, is

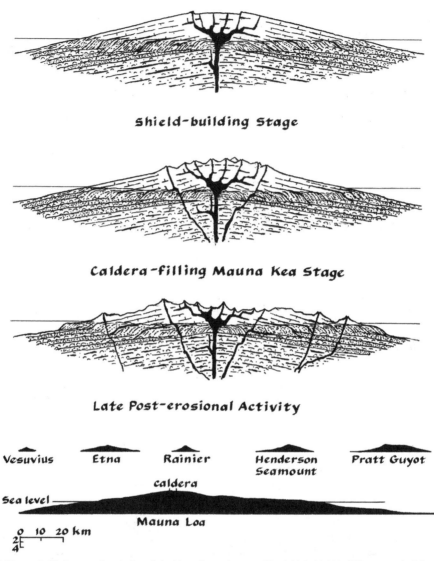

Shield-building Stage

Caldera-filling Mauna Kea Stage

Late Post-erosional Activity

Vesuvius   Etna   Rainier   Henderson   Pratt Guyot
Seamount

caldera

Sea level

Mauna Loa

0 10 20 km

Figure 8-13 Stages of evolution of the Hawaiian volcanoes. The initial shield-building stage is followed by a period in which the caldera is filled and small vents break out in the summit region. Later, after an interval of erosion, a final stage is characterized by small eruptions of very alkaline magmas, mainly on the lower slopes. The enormous size of the volcanoes can be appreciated by comparing a profile through Mauna Loa with others, such as Etna, the largest continental volcano.

in this stage of evolution. For many years, the activity inside its caldera was characterized by persistent lakes of molten lava that, until 1924, occupied a deep pit on the floor of the caldera. More recently, summit eruptions have migrated southward from the caldera.

Eruptions from the summit and flanks tap a complex, shallow reservoir that swells and contracts with each episode of activity. Most eruptive episodes are preceded by inflation of the summit region, but subsequent eruptions lower on the flanks deflate the reservoir and usually bring activity to an end until a new batch of magma initiates another cycle. Most of the compositional variations of lavas of this stage can be related to differing abundances of olivine phenocrysts, but few of the olivine phenocrysts could have crystallized from the liquid in which they now reside because the magnesian–iron ratio of the groundmass has little relationship to the amount of olivine. Moreover, the bulk-rock magnesium contents are too high for the picrites to have been melts in equilibrium with mantle peridotites. The crystals, many of which shows signs of having been deformed, must be torn from segregations that accumulate during the decades or centuries that batches of magma reside in the high-level reservoir. Geophysical evidence and the geometry of inflation of the summit region indicate that this reservoir is a steep, narrow system of dikes and sills, mainly in the upper half of the volcano. These may not be large, discrete bodies of magma but rather masses of crystals with differing amounts of interstitial liquid. No seismic evidence has been found for largely molten magma under volcanoes of this type, but the cores of a few deeply eroded volcanoes have sizable bodies of coarse-grained plutonic rocks. The Tahitian volcanoes described in the previous chapter are a good example.

As the Hawaiian volcanoes drift away from the hotspot, activity gradually declines, the lavas become increasingly olivine and pyroxene rich, and their compositions become more alkaline and differentiated. Eruptions from fissures on the flanks become less common while scattered vents near the summit begin to fill and eventually obliterate the caldera. Haleakala on the island of Maui is now in this "caldera-filling" stage, and Mauna Kea on the island of Hawaii has reached the point at which its caldera is no longer discernable. With the change from tholeiitic to alkaline basalts, *ankaramites* (very mafic basalts containing abundant phenocrysts of olivine and augite) become common. Some of the marginally alkaline basalts have differentiated to lavas that resemble basalts petrographically but have a more sodic plagioclase. Those with modal or normative andesine are called *hawaiites*, and those with oligoclase, *mugearites*. In rare instances, more felsic rocks, such as *benmoreites* and *trachytes*, form stubby lava flows and small viscous domes. The general trend of these differentiates straddles the division between nepheline-normative and hypersthene-normative compositions (**Fig. 8-14a**).

Eruptions become weaker and gradually come to an end, but in a few cases, the volcanoes have a brief episode of renewed activity before finally

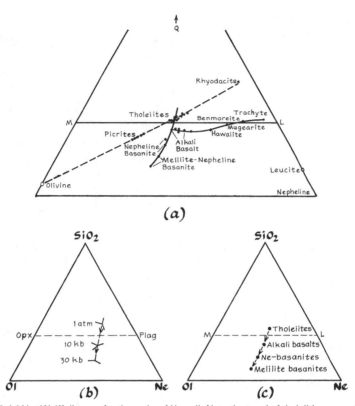

Figure 8-14 (a) Von Wolff diagram for the rocks of Hawaii. Note the trend of tholeiitic compositions directly away from olivine (along the dashed "control line") and the course of differentiation from alkaline basalts to trachyte close to the line of silica saturation, ML. The change of compositions of basalts during the main evolution of the islands is shown by the steep solid line from tholeiites to melilite nepheline basanites. Note that it does not include the alkaline lavas erupted from the very young volcano, Loihi, at the southern base of Kilauea. Construction of the diagram is explained in Figure 7-8. (Adapted from G. A. Macdonald, 1968, *Geol. Soc. Amer. Mem.* 116:477–522.) (b) The system olivine–nepheline–silica projected from anorthite is a close analog of the diagram in (a), even though it does not include as many components. When the Hawaiian basalts in (a) are plotted in the same way (c), they define a trend close to that of the invariant point in the synthetic system.

becoming extinct. The period of dormancy may last as long as 2.5 million years. The lavas discharged in this stage are not voluminous, but they are notable for their highly alkaline compositions. The rocks around Honolulu belong to this late, "rejuvenated" stage. They include nepheline basanites rich in melilite, a mineral found only in very silica-deficient mafic rocks. The vents break out mainly on the flanks, often close to the shoreline where they form palagonitic tuff cones, such as the familiar landmark Diamond Head.

The progressive increase of alkalinity of the lavas is consistent with the mineral assemblages of their mafic and ultramafic xenoliths, which, as we saw in the preceding chapter, correspond to progressively deeper origins of the basalts. When the normative components of the lavas are plotted on a von Wolff diagram (**Fig. 8-14a**) of the kind used to trace the Tahitian series (Fig. 7-8), they follow a pressure-related trend analogous to that of the system olivine-plagioclase-silica (**Figs. 8-14b** and **8-14c**). It is not clear, however, why these highly alkaline lavas appear after such a long hiatus when the island has drifted far from the hotspot. They seem to be related to other, largely independent processes that are still poorly understood.

Magma Generation in the Deep Mantle

The compositional features of intraplate magmas, together with seismic evidence that enables us to trace their deep sources, indicate that melting occurs in plumes of rising mantle initiated at or below the base of the transition zone at 660 kilometers (**Fig. 8-15a**). In some instances, the roots may be as deep as the core–mantle boundary.

One of the lines of evidence cited in support of this interpretation is the isotopic character of helium. The proportions of the two isotopes of helium, ^3He and ^4He, are considered a good indicator of the depth of origin of mantle-derived magmas. ^3He is a stable isotope, whereas ^4He is produced by radioactive decay of elements that are concentrated mainly in the lithosphere and upper mantle. As the abundance of ^4He decreases with depth in the mantle, $^3He/^4He$ increases accordingly. The fact that the ratios for ocean

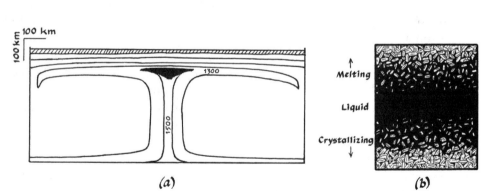

(a) *(b)*

Figure 8-15 (a) Temperature structure of the inferred mantle plume under Hawaii. The zone of melting (solid black) differs from the melting zone under spreading ridges (Fig. 8-10) in that the potential temperature and depth of the solidus decreases outward from the center. (From White and McKenzie, 1995, *Jour. Geoph. Res.* 100:17,543–17,585.) (b) When enough melt is formed, it may migrate upward more rapidly than the crystalline residue by melting its roof while crystallizing on the floor. If the rocks are already at or near their solidus temperature, the system can maintain a thermal balance because the heat given up by crystallization at the base provides the heat required for melting at the top.

island basalt are as much as three to four times larger than those of ocean–ridge basalts is taken as evidence that the former have deeper origins.

Regardless of the depth from which the magmas rise, the seemingly random distribution and varied compositions of hotspots suggest that they are caused by localized thermal or compositional anomalies that reduce the density and viscosity of the mantle and causes it to ascend. These inhomogeneities could be thermal, compositional, or some combination of the two. Compositional anomalies could come from remnants of ancient subducted crust that have carried water and lithophile elements deep into the mantle where they have the effect of lowering melting temperatures. Even if they do not induce melting directly, water, alkalies, and most lithophile elements lower the viscosity and density of the mantle and facilitate upwelling, which in turn leads to melting by decompression as the rocks rise to shallower levels and lower pressures.

As noted earlier, ultramafic nodules brought up from deep sources may contain small amounts of phlogopite or carbonate minerals that could contribute these solidus-lowering components. The water need not be in hydrous minerals, however. It can also be held in minerals such as olivine, pyroxene, and garnet that we normally think of as anhydrous. These minerals can take in several hundred parts per million of water as structurally bound OH. Clinopyroxene takes in the most—about 10 times as much as olivine or garnet—and orthopyroxene can hold five times as much. Hydrous minerals, of course, contain much more water, but they are rarely present in large amounts.

The amount of water in magmas coming from the mantle can be estimated from the water content of melt inclusions in olivine or in magnesium-rich submarine basalts that have been quenched to glass at pressures sufficient to inhibit vesiculation. The average is found to be of the order of 0.39 weight percent. If we assume that water is strongly partitioned into the first melt and is diluted by further melting, this amount of water could come from 10- or 11-percent melting of a garnet lherzolite with a water content of 450 ± 190 ppm.

We have no way of knowing how typical this is of the entire mantle. The water contents of mantle rocks are certainly not uniform; they probably increase with depth and are lower under the continents than under oceanic crust. Mantle xenoliths, such as those found in diamond pipes, indicate that the water content of the upper mantle under continents is about 270 ± 20 ppm, considerably less than that of the mantle under places like Hawaii. The sources of Hawaiian alkaline basalts, for instance, contain 525 ± 75 ppm water, while the source rocks of the basalts of normal oceanic ridges contain only 140 ± 40 ppm. Enriched MORB, as we would expect from their hybrid

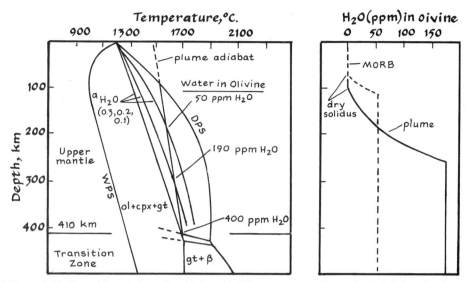

Figure 8-16 The melting relations of peridotite with different amounts of water (in the left figure) and the rate of melt production in a rising column of mantle peridotite. See the text for discussion. (After P. Wallace, 1998, *Geoph Res. Ltrs.* 25:3639–3642.)

origins, contain about 350 ± 100 ppm water, an amount intermediate between that of the source rocks of normal MORB and OIB.

The importance of the pressure dependence of the solubility of water can be seen in its effect on melting (**Fig. 8-16**). As mantle rocks rise, the amount of water they can contain decreases so that by the time an upwelling column reaches a depth of about 180 kilometers, the water content would decline to less than a third its initial value. The proportion of melt produced by the fluxing effects of such small amounts of water is correspondingly small, and substantial amounts of melt do not appear until the level of the dry solidus is reached at a much shallower depth.

It is significant that basalts of very similar major-element composition can differ widely in their trace elements and isotopic ratios. The difference illustrates the basic principle that the trace-element composition of magmas is largely independent of the strict controls that phase equilibria impose on the major elements. Even though their mantle source may differ widely from place to place, the major-element character of magmas is governed by the conditions under which they form and make their way to the surface. The crude correlation between the character of basalts and the age and thickness of the lithosphere indicates that a major controlling factor is the depths at which magma equilibrate as they rise from their primary source toward the surface.

The manner in which Hawaiian basalts vary during the life span of a volcano provides a valuable clue to the nature of the mechanisms that govern their compositions. If we include the basalts of Loihi, the small, newly formed volcano at the leading edge of the chain, the overall sequence is first from alkaline to more tholeiitic basalts during the period of active growth then a return to increasingly alkaline compositions as activity declines. This pattern is even more pronounced in the volcanoes on the island of Reunion in the Indian Ocean. The two large volcanoes, Piton de la Fournaise and Piton des Neiges, are similar in many ways to Kilauea and Mauna Loa on Hawaii, but we have a better record of their compositional evolution. Hundreds of lava flows exposed in deep canyons cut into their flanks contain a record of their activity over a period of half a million years or more.

The transition from alkaline to more tholeiitic compositions on Reunion continued during much of the shield-building stage—much longer than in Hawaii. Experimental studies indicate that the proportion of residual olivine increased at the expense of pyroxene as the depth of equilibration of the magmas rose slowly from about 30 to 25 kilometers, and as in Hawaii, the expansion of the olivine field with declining pressure yielded liquids with greater silica contents. The basalts then remained remarkably constant for a hundred thousand years or more, until activity slowly declined and the basalts returned to more alkaline compositions.

It is unlikely that such a sequence could be due to systematic changes of the depths of melting deep in the mantle, for as we noted earlier, the processes responsible for magma generation at hotspots have no apparent relationship to plate motion. Moreover, the nearly constant isotopic ratios of intraplate lavas over long periods of time indicate that their mantle sources have remained fairly uniform. Even though the overall strength of the hot spot does not change, however, the depth and amount of melting could vary radially, and the compositions of magma might differ with distance from the center. Radial variations of this kind could be responsible for the temporal changes in the character of basalts as a new volcano approaches a hotspot, reaches a peak of activity over its center, and then declines as it recedes from its source.

■ Ridge–Hotspot Interaction

Some of the most instructive clues to the nature of hotspots are found in places where they interact closely with an oceanic ridge. Just as nearby hotspots affect the composition of ocean–ridge basalts, the reverse is also true; seamounts close to ridges produce lavas that are very similar to the basalts of oceanic ridges. Iceland is especially interesting in this regard

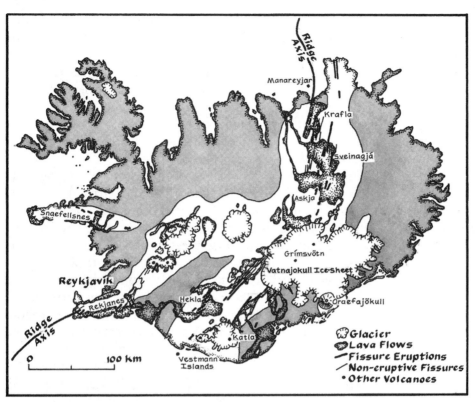

Figure 8-17 The island of Iceland stands directly astride the Mid-Atlantic Ridge. The axis of the ridge emerges along the Reykjanes Peninsula and is displaced eastward before running into the Arctic Ocean. The exposed Tertiary rocks (shaded) reach Eocene ages in the far northwestern part of the island.

because the hotspot is directly under the ridge and the structures and volcanic activity can be observed on dry land (**Fig. 8-17**).

Iceland

The Mid-Atlantic Ridge emerges from the sea on the Reykjanes Peninsula and is offset along a series of disconnected segments toward south central Iceland, where it joins a second rift zone that has been the focus of most recent activity. It then follows a nearly straight course toward the Arctic Ocean. Basalts erupted from the rift zone are typical E-type MORB, whereas those erupted at some distance from the rift have many of the characteristics of ocean island basalts. Tensional fractures and fissure eruptions have resulted in a dilation of about 30 meters during the last 3,000 to 5,000 years, but the present rate of spreading is highly sporadic. Sudden extensions of as much as several centimeters are associated with intrusions of shallow dikes, which appear to be injected horizontally from sources spaced at intervals along the active rift. Many of the dikes have no surface manifestation other than dilation of the crust.

The record of volcanism, which extends well back into Tertiary time, shows an intensity at least 10 times greater than that of the main Mid-Atlantic Ridge. The fact that the long-term rate of spreading in Iceland is not measurably different from the rate on less productive sections of the ridge demonstrates that there is no simple relationship between the rates of spreading and magma production.

The greatest recorded outpouring of lava of historic time was the 1783 eruption of Laki volcano, which discharged more than twelve cubic kilometers of basalt over an area of 565 square kilometers. Magma poured from a fissure 25 kilometers in length, at a rate of 5,000 cubic meters per second. Where the fissure passed under a broad ice sheet, floods of melt water were released, causing wide devastation along the southern coast. When basalt erupts under ice or water, its surface is quenched to a type of clear glass called *sideromelane* (or *tachylite* if it is the more slowly cooled opaque variety). The glass quickly absorbs water and becomes a yellowish brown, amorphous material called *palagonite*. Palagonite of the kind that covers much of Iceland is a major component of marine sediments near large submarine volcanoes.

Most lavas of recent Icelandic fissure eruptions are olivine-poor tholeiitic basalts similar to rocks dredged from the Mid-Atlantic Ridge. In the recent past, however, eruptions west and east of the main rift zone have produced lavas that tend to be increasingly alkaline with distance from the main axis. A few large volcanic complexes of Tertiary age are scattered over a broad region, mainly east of the central rift. The tholeiitic centers have some of the most extensively differentiated rocks on the island, including ferrobasalts, icelandites, and rhyolites.

Extensive sheets of rhyolite are interlayered with Tertiary flood basalts on the flanks of the rift, and smaller volumes of rhyolitic pumice and obsidian are closely associated with basalt in some of the large volcanic complexes, such as Hekla and Askja. The intimate association of these basalts and rhyolites at Hekla led a German geologist, Robert Bunsen, to propose that these were two primary magmas of global extent and that all intermediate rocks were simply mixtures of two end-members. This concept seemed eminently logical when it was first proposed in 1851 and dominated much of the thinking throughout the later half of the 19th century.

A few small bodies of sodic granophyre, locally referred to as "granite," are probably subvolcanic intrusions. The volumes of these felsic rocks, variously estimated at 3 to 9 percent of the Cenozoic section, are much greater on Iceland than on other oceanic islands, and their abundance led early geologists to speculate that Iceland might be a small continental fragment covered by a thin veneer of basalt. Geophysical and petrologic evidence now rules out any possibility of sialic crust under the island. The siliceous differentiates have little or no potassium feldspar and differ from typical continental rhyolites and granites in their trace-element abundances and isotopic ratios. Most are now thought to be products of partial melting of amphibolites in the deep lithosphere.

The Galapagos Islands and Adjacent Ridge

The contrasting character of magmatism at oceanic ridges and hotspots is especially well displayed where the Galapagos Spreading Center passes close to the cluster of islands forming the Galapagos Archipelago (**Fig. 8-18**). For 20 million years or more, the Galapagos hotspot was directly under the spreading axis, and the augmented production on that part of the ridge produced the Carnegie and Cocos ridges as the Nazca and Cocos plates drifted toward the southeast and northeast, respectively. Even after migrating north of the hotspot, this section of the ridge has continued to produce greater volumes of basalt than more distant sections to the east and west.

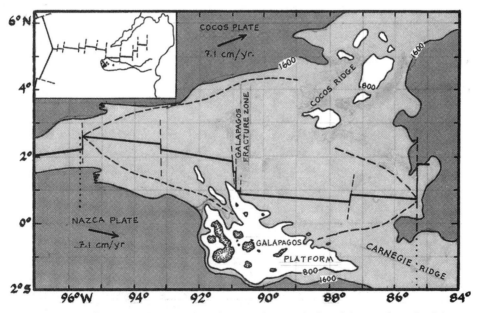

Figure 8-18 The Galapagos Islands stand on a broad platform south of the Galapagos Spreading Axis in the eastern equatorial Pacific. The Carnegie and Cocos ridges extend from the Galapagos hotspot in the directions of spreading indicated by arrows on the Nazca and Cocos plates on opposite sides of the spreading axis.

The basalts erupted where the ridge is closest to the leading edge of the hotspot have all the characteristics of enriched MORB. They have greater concentrations of incompatible elements and more radiogenic strontium and neodymium, and in this respect, their compositions are intermediate between those of normal ridges and intraplate volcanoes. Similar relationships are found in the Atlantic, most notably at ridges near the hotspots underlying the Azores, Jan Mayen, and Tristan da Cunha. The compositions of enriched MORB in these regions differ not only from normal MORB but also from one another (Table 8-1, nos. 3, 4, and 5).

Most of the compositional variations in the anomalous section of the Galapagos spreading ridge are symmetrical about a point opposite the leading edge of the hotspot (**Fig. 8-19**), but the eastern segment of the ridge, which

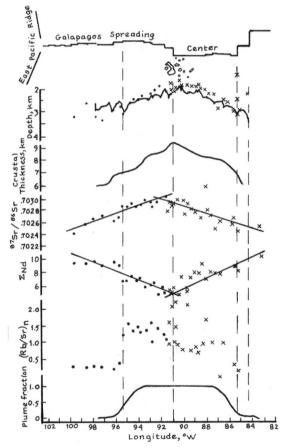

Figure 8-19 Variations of depth elevation of ridge, crustal thickness, Sr and Nd isotopic ratios, Rb/Sr, and a calculated contribution of the adjacent hotspot to the Galapagos Spreading Center. (Adapted from Schilling et al., 2003, *Geoch. Geoph. Geosystems* (G^3), vol. 4: no. 4.)

is offset to the south and 110 kilometers closer to the hotspot, has been more productive and is about 500 meters higher than the western limb. Note also that the Rb–Sr ratios have a sharp break at the transform fault near 95.5° East, but the $^{87}Sr/^{86}Sr$ ratios show no corresponding offset. The absolute abundances of trace elements are affected by a variety of conditions such as the depth of melting and amount of differentiation, but these processes have not affected the isotopic ratios inherited from deeper levels of the mantle.

The form of the hotspot has been defined as a zone of low seismic velocities directly under the Galapagos Platform, but instead of being a simple vertical column, it has been deflected toward the southeast, the same direction as the drift of the Nazca Plate. This may account for the fact that, unlike Hawaii and other intraplate chains, the Galapagos volcanoes are scattered over a broad platform rather than in a narrow chain of islands that vary in composition as they drift downstream. Most historic eruptions have been concentrated in the large, youthful shield volcanoes close to the leading edge of the system, and although the islands are generally older toward the east, several have had renewed activity after drifting as far as 250 km downstream. These young lavas differ little from those erupted closer to the hotspot, presumably because their shallow source has been carried along with the drifting Nazca Plate.

While the lavas follow no regular sequence with age or distance from the hotspot, they have a remarkable spatial zoning that seems to have persisted for at least 5 million years (**Fig. 8-20**). With increasing distances north and south of the central axis, the basalts become richer in olivine, alkalies, MgO, and radiogenic Sr, suggesting they come from melting of a deeper source. All the differentiated rocks are confined to a narrow belt down the center of the archipelago. Part of the explanation for this zoning is suggested by Figure 8-20d, which shows the minimum Mg numbers for individual eruptive centers. The low values near the center reflect the differentiated character of the ferro-basalts, icelandites, and rhyolites described in the previous chapter. Higher values near the margins correspond to more restricted ranges of differentiation. The fact that the east–west symmetry of strontium isotopes along the ridge does not extend through the Galapagos platform indicates that the melting regimes are essentially independent.

Similar evidence is found in the trace-element patterns of the lavas (**Fig. 8-21**). When the abundances of two incompatible elements, such as Zr and Th, are plotted against one another, they define linear trends at constant ratios for lavas that are related by crystal fractionation under conditions where both elements are about equally excluded. Because their

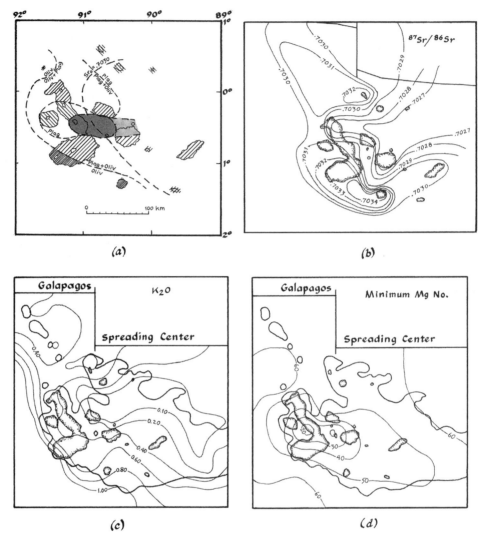

Figure 8-20 Compositional zoning of the Galapagos. (a) The most important phenocrysts of basalts are plagioclase in the central part of the archipelago, but olivine becomes more important toward the margins at the same lime that the basalts become more alkaline. Differentiated lavas and plutonic xenoliths are found in the central zone with dark shading. (b) The variation of strontium isotopes resemble the mineralogical zoing. (c) The K_2O contents of basalts (at a uniform value of 8-percent MgO) increase outward from the central axis. Basalts of the central islands closely resemble those of midocean ridges, whereas those toward the outer margins are more like basalts of intraplate settings. The edge of the Galapagos Platform is shown by a heavy line corresponding to the 1,000-fathom contour. (d) When the minimum MG numbers ($MqO/MgO + FeO^*$) for individual eruptive centers are contoured for the archipelago as a whole they reflect the strong differentiation just downstream from the hotspot and the gradational change to more primitive compositions at the margins. (After White et al., 1993, *Jour. Geoph. Res.* 98:19,533–19,563.)

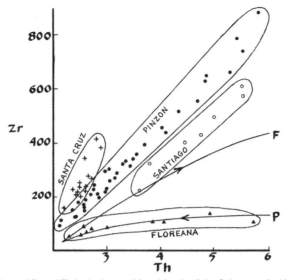

Figure 8-21 Variations of Zr and Th in the lavas of four islands of the Galapagos Archipelago. Curves labeled F and P are calculated variations of Zr and Th for crystal fractionation and partial melting, respectively. The arrows indicate the directions of increasing differentiation and melting. See the text for an explanation. (After C. S. Bow, 1979, PhD Thesis, University of Oregon.)

distribution coefficients for the crystallizing minerals are equally small, the two elements are concentrated to the same relative degree in declining fractions of remaining liquids, and slightly different initial Zr–Th ratios of the parental magmas are preserved throughout the differentiated series. Note that the trend for these elements in the lavas of Floreana Island on the southern edge of the platform differs from that of more central islands in having a greater enrichment of Th relative to Zr. Using the equations of trace-element distributions for crystal fractionation and partial melting (Equations 5-4 and 5-11), the observed distribution can be compared with that predicted for crystallization or melting of a mineral assemblage that includes different possible mantle minerals, namely spinel and plagioclase. Because the distribution coefficient of Zr for spinel is slightly larger than for plagioclase, a closer fit to the observed pattern of relatively slow Zr enrichment is obtained if one assumes that spinel was in the melting residue. The calculated curve for partial melting reproduces the observed variation much more closely than the one for crystal fractionation. One can conclude, therefore, that the variations near the margins of the hotspot are due mainly to differing amounts of melting at a level below the plagioclase-spinel transition. This is consistent with the more alkaline character and higher strontium isotope ratios of basalts in the same region. The general pattern, therefore, is consistent with that which was deduced for other hotspot–related systems, such as Hawaii and Reunion.

■ Flood and Plateau Basalts

Somewhat akin to MORBs are the great flood basalts that from time to time overwhelm vast areas of the continents and seafloor (**Fig. 8-22, Table 8-2**). Both are discharged in great volume from fissures rather than central-vent volcanoes. Wherever feeder systems of flood basalts have been exposed by erosion, they are seen to be mazes of dikes, typically of great number, extent, and complexity (**Fig. 8-23**). Where they differ from MORBs is in the concentration of vastly more intense magmatism in shorter eruptive periods. While ocean ridges produce greater total volumes of basalt, this is only because their activity is more prolonged.

The most familiar example in America is the group of basalts forming the Columbia River Plateau of Oregon and Washington. Hundreds of lava flows, some with volumes exceeding 2,000 km^3, originally covered about 163,000 km^2 and had a total volume of about 174,300 km^3. Yet the Columbia River province ranks only 10th in size among continental flood

Figure 8-22 View of the basaltic flood lavas of the Columbia River plateau of Washington and Oregon in the northwestern United States.

Table 8-2 Notable features of some of the more important flood-basalt provinces. Types of basalt are listed according to their normative classifications and their estimated relative volumetric proportions. The original dimensions of pre-Tertiary provinces are very conjectural. The Brito-Arctic Province encompasses parts of Eastern Greenland, Ireland, Scotland, and the Faeroe Islands.

		Volume	Types of Basalt		
Province	Age	$\times 10^3\ km^3$	Qtz Thol.	Oliv. Thol.	Alk. Bas.
Lake Superior	Precambrian	1,300	42	51	7
Siberia	Permo-Triassic	> 2,000	28	69	3
Karoo, S. Africa	Jurassic	2,000	57	37	6
Paraná, Brazil	Cretaceous	> 1,500	72	28	0
Ontong-Java	Cretaceous	40,000	20	80	0
Brito-Arctic	Eocene	1,800	5	50	45
Deccan, India	Eocene	1000–2500	55	35	10
Columbia River	Mid-Miocene	174	30	70	0

Figure 8-23 Dikes exposed in cliffs along the eastern coast of Greenland fed basaltic flood lavas along a rift formed at the time of the opening of the North Atlantic. As Greenland separated from northern Europe, the activity became focused in what is now the Mid-Atlantic Ridge.

basalts. The Mesozoic Karoo basalts of South Africa are roughly twelve times larger, both in areal extent and volume.

By far the greatest volume of basalt in the Columbia River group was erupted within a period of less than two million years during the mid-Miocene (**Fig. 8-24**), and most of the Eocene flood basalts of the Deccan province in India, one of the largest on Earth, were erupted in less than a million years. Individual eruptions were separated by periods of repose lasting thousands of years, but the rates of discharge during a single eruption were of the order of tens or hundreds of cubic kilometers per day. Not all the magma was erupted as lava flows; in some provinces, a large part was intruded as sills. This is particularly true where the magma rose into deep sedimentary basins. Because of their greater densities, basaltic magmas are unable to rise through the light sediments and intrude laterally along planes of equal gravitational potential.

The total section of Columbia River Basalts reaches a thickness of about 10 kilometers near their center, and yet the surface onto which the lavas flowed seems never to have been high or to have developed much relief. As the basalts poured out, the underlying rocks subsided to form a basin with almost the same depth and areal extent as the lavas that filled it. The process seems to have been one of transferring mass from the mantle to the surface while displacing the lower crust downward by an equivalent

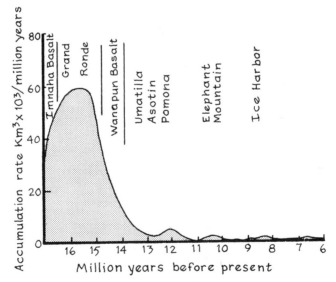

Figure 8-24 Volumes of Columbia River basalt erupted as a function of time. (After P. R. Hooper, 1988, in *Continental Flood Basalts,* J. D, Macdougall, ed., pp. 1–33.)

amount. If so, the immediate source of a large part of the magma must have been within the lithospheric mantle well above the convecting asthenosphere.

The terms "plateau basalts" or "cone fields" is often applied to extensive lavas that were not the products of voluminous fissure eruptions but of many scattered sources, each of which produced relatively small individual flows. Where these flows merge and overlap they form seemingly continuous sheets of great lateral extent, but their volumes are generally smaller than those of true flood basalts. Despite these differences, however, the lavas have much in common with flood basalts, both in tectonic setting and composition, and one may simply be a weaker counterpart of the other.

Oceanic flood basalts of Cretaceous age cover a large region of the southwestern Pacific Ocean between New Guinea and the Marshall Islands. The largest group forms the Ontong-Java Plateau, covering an area of a million and a half square kilometers with a relief of about 3 kilometers above the normal seafloor. Their vent systems cannot be seen, but it seems likely that they were discharged from fissures. Coming from sources near but slightly west of the Cretaceous spreading axis, they resemble mid-ocean ridge tholeiites, but their incompatible elements are somewhat less depleted (Table 8-1).

Flood basalts differ widely in composition, even within an individual province. While most are tholeiitic, some have substantial proportions of alkaline basalts. With few exceptions, the basalts are more evolved than the basalts of oceanic spreading axes. Their MgO contents are normally less than 8 weight percent, and many are less than 5. The chief exceptions are picrites that are found among the earliest lavas and sills of several major provinces, most notably the Karoo basalts of South Africa. The resemblance of these picrites to the magmas of certain hotspots is consistent with the notion that the two are simply different expressions of the same basic phenomenon.

Although the Columbia River basalts differ from one source area to another, the series as a whole shows no consistent compositional changes with time. The small differences that have been noted are difficult to explain by any simple scheme of crystal fractionation or mixing. The same is true to an even greater degree in other provinces. An extreme example is the suite of rhyolites erupted intermittently during the outpouring of Deccan basalts. These felsic magmas have no direct genetic relationship to the basalts; most likely, they resulted from melting of the continental crust.

It is difficult to imagine masses of basaltic magma of such dimensions being products of crystal fractionation of even larger bodies. The size of the magma body required to produce batches of well-mixed magmas with volumes measured in hundreds or thousands of cubic kilometers would be comparable to that of the Bushveld Intrusion, one of the largest bodies of igneous rock on Earth. This is especially surprising when one considers the variety of tectonic settings in which they are found. Some were associated with continental rifting and opening of oceanic basins. The basalts along the east coast of Greenland, for example, correspond to similar lavas in northern Ireland and Scotland and date from the opening of the North Atlantic in Eocene time. Others, such as those of the Lake Superior region and Siberia, seem to have been associated with rifting that was less extensive or aborted at an early stage. The origin of the Columbia River basalts is even less clear. They were erupted behind an active belt of subduction-related volcanism that was then at its peak intensity, but most geologists consider this a coincidence and favor instead an origin related to the beginning of the hotspot now under Yellowstone. The hotspot relationships are not evident, however, because the dikes that fed most of the flows are not on the track of igneous centers leading back to Yellowstone. The plume, if it exists, seems to have been deflected northward by the inhomogeneous structure of the lithosphere in this region.

Seismic studies of the source region of the Columbia River basalts indicate that the magmas were stored within the crust beneath a cluster of granitic plutons in what are now the Willowa Mountains of eastern Oregon. It has been postulated that a dense residuum left when the granitic magmas were extracted from the lower crust became delaminated and sank into the mantle to make space for a large reservoir in which the magma collected before erupting through swarms of fissures in what is now eastern Oregon. Anomalously high seismic velocities in the crust just below some flood basalt provinces indicate that differentiation or partial melting left dense masses of mafic crystals, principally olivine and pyroxene.

Large concentrations of excluded trace elements are typical of many flood basalts, especially the earliest, most primitive lavas. In some provinces, most notably the Karoo, the most magnesian lavas are those with the greatest concentrations of incompatible elements. The explanation offered for this apparent anomaly is that the incompatible elements were more readily scavenged from the lithosphere by the early, exceptionally hot, picritic magmas and that subsequent magmas were less affected because they rose through rocks that had already been depleted. An alternative explanation is that the picrites are products of a process of zone melting that buffers the concentrations of included elements while permitting accumulations of excluded elements as the first waves of melting pass through an undepleted lithosphere.

When eruptions of flood basalts finally come to an end, the regions rarely have further igneous activity. This has been true of most if not all sites of great basaltic eruptions. They appear to be triggered by a sudden intense surge of energy, and having expended that energy, the mantle seems inca-

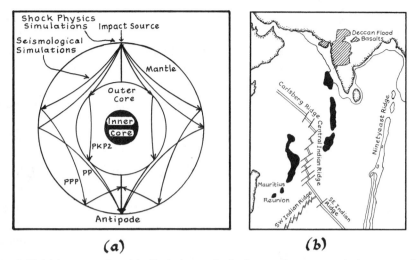

Figure 8-25 (a) A cross-section of the Earth shows why shock waves from a meteorite impact near the Yucatan Peninsula would be focused at the antipode and could possibly have triggered the Deccan flood basalts. (After M. B. Boslough, et al., 1996, *Geol. Soc. Amer. Spec. Paper 307*:541–550.) (b) The Deccan flood basalts of India mark the initiation of a hotspot about 65 million years ago, essentially the same time as the meteorite impact. As the Indian plate drifted northward over the stationary hotspot at the rate of 1.7 centimeters per year, a ridge and chain of seamounts were created along the floor of the Indian Ocean. The hotspot is presently under the active volcano Piton de la Fournaise on the island of Reunion. (Adapted from R. A. Duncan, et al., 1989, *Jour. Volc. Geoth. Res.* 36:193–198.)

pable of further melting in the same region. Recent studies have indicated that at least some of these huge outpourings were triggered by the impact of meteorites (**Fig. 8-25a**). They do not necessarily occur at the point of impact, as was the case with the Sudbury event described in the opening chapter, but at the antipode where the energy is focused by shock waves converging in the asthenosphere on the opposite side of the globe. Thus, the meteorite that struck the Earth near the Yucatan Peninsular at the end of the Cretaceous is said to have triggered the Deccan flood basalts in what is now India. It is unlikely that every meteorite impact triggered flood basalts or that all eruptions of flood basalt were due to meteorites, but if conditions were suitable, the shock may have been enough to initiate eruptions. The fact that many mass extinctions of species seem to have coincided with these two types of catastrophes is consistent with an interaction of this kind.

■ Lunar Flood Basalts

The basaltic lavas of the Moon offer a number of instructive contrasts to those of the earth. The Moon's surface is divided into two distinct types of topography, the highlands, consisting of plagioclase-rich anorthosites, and the flat lowlands and mare basins filled with flood basalts. Although the basalts cover about 17 percent of the surface, they are so thin that they account for less than a percent of the total mass of the moon. Because their viscosities were exceptionally low—even lower than those of terrestrial flood basalts (Fig. 2-10a)—many flows spread for distances of hundreds or even thousands of kilometers on nearly flat surfaces.

The greatest volumes were erupted between about 3 and 4 billion years ago, with only very minor activity for a few hundred million years thereafter. The period during which most lavas were discharged followed formation of the lunar highlands and came shortly after bombardment by meteorites reached its peak intensity. The distribution of volcanism is extraordinarily asymmetrical, being confined largely to the earth-facing side of the moon. The far side has an equivalent density of impact craters, but few are filled with lava. The explanation for the scarcity of lavas on the far side is based on the distribution of mass in the moon. Because of the gravitational attraction of the earth, the center of mass is not at the geometric center but slightly closer to the earth, and the radius from the center of mass is greater on the far side. It is thought that magmas generated at depths of a few hundred kilometers were unable to erupt as lavas and intruded sills into lighter rocks not far beneath the surface.

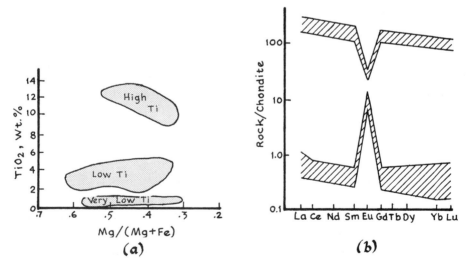

Figure 8-26 (a) Lunar basalts can be separated into three distinct groups on the basis of then TiO_2 contents. Note that their Mg ratios overlap, showing that they cannot be related to each other by any normal mechanism of differentiation. (b) The rare-earth concentrations in lunar anorthosites (below) have a reciprocal relation to those of basalts (above). Note the marked spike for Eu, which is preferentially concentrated in plagioclase, and its mirror image in basalts that result from fractionation of the plagioclase-rich anorthosites (adapted from compilations by S. R. Taylor, 1992, *Planetary Science' A Lunar Respective*, Lunar Planet Inst, 481 p.).

Compositions of flood basalts of the continents, oceans, and the moon. FeO* is total iron as FeO. All norms except 4 and 5 are calculated assuming a uniform ratio of Fe_2O_3 to FeO of 0.15.

Table 8-3					
	1.	2.	3.	4.	5.
SiO_2	46.87	49.60	48.35	42.16	45.03
TiO_2	1.64	1.08	1.57	9.43	2.90
Al_2O_3	9.31	14.20	15.49	9.89	8.59
FeO*	12.17	10.74	10.98	19.11	21.03
MnO	0.17	0.20	0.17	0.28	0.28
MgO	15.91	8.02	7.03	5.67	11.55
CaO	9.06	12.44	9.92	12.15	9.42
Na_2O	1.32	2.01	2.76	0.45	0.23
K_2O	0.56	0.06	0.51	0.11	0.06
P_2O_5	0.20	0.10	0.24	0.04	–
Molecular Norms					
Ap	0.43	0.21	0.52	0.09	–
Ilm	2.32	1.54	2.26	14.23	4.23
Mt	2.14	1.52	1.96	–	–
Or	3.36	0.36	3.12	0.70	0.37
Ab	12.04	18.45	25.63	4.38	2.16
An	18.12	30.20	29.34	26.70	23.31
Di	20.99	25.77	15.95	30.67	20.55
Hy	21.56	20.30	13.66	19.95	36.87
Ol	19.05	1.65	7.58	–	12.50
Q	–	–	–	3.27	–
Trace elements (ppm)					
Ba	272	2	280	88	69
Zr	129	52	8	334	107
Rb	23	1	6	0.9	1.1
Sr	445	88	274	194	96
Cr	848	342	188	1230	3780
Ni	675	125	77	< 2	52
La	13.4	2.7	7.4	11.3	6.1
Sm	5.3	2.3	3.5	14.4	4.5
Lu	0.2	0.4	0.4	1.0	0.6
$^{87}Sr/^{86}Sr$	.7056	.7037	.7037	.6991	.6991

1. Picritic Karoo basalt (R. M. Ellam and C. K. Cox, *Earth Planet. Sci. Ltrs*. 92:207–218 and 105:330–342).
2. Typical ocean plateau basalt, Nauru Basin (P. A. Floyd, 1989, *Magmatism in the Ocean Basins*, A. D. Saunders and M. J. Norry, eds., pp. 215–230).
3. Columbia River Basalt, Picture Gorge type.
4. High-Ti lunar basalt, Apollo 11.
5. Low-Ti lunar basalt, Apollo 12.

The most conspicuous compositional feature that sets most lunar basalts apart from their terrestrial counterparts is their greater concentrations of titania, which in some samples reach 16 weight percent. This is particularly true of the earliest basalts. With time, TiO_2 shows a general decrease to values closer to those of terrestrial basalts. The decline was not a smooth one, however. The compositions of basalts fall into three groups, each with a distinctive TiO_2 content (**Fig. 8-26a and Table 8-3**). Because of the more reducing conditions of the moon, there is no ferric iron, and a few rocks even contain small amounts of metallic iron. Water is also absent, and Na_2O, K_2O, and other components of feldspar are much less abundant than they are in terrestrial basalts.

The basaltic lavas are in some ways complementary to the anorthositic rocks of the highlands (**Fig. 8-26b**). The latter appear to be formed by a plagioclase-rich fraction that, being lighter, accumulated at the surface while a more mafic residue accumulated at a depth of 200 to 400 kilometers. The origins of the anorthosites and their relationship to the basalts are still being debated.

Selected References

Foulger, G. R., et al., 2005, Plates, Plumes and Paradigns, *Geol. Soc. Amer. Spec. Paper* 388, 881 p. A large collection of papers dealing with various aspects of oceanic magmatism, including evidence bearing on the existence of plumes.

Batiza, R., 1989, Seamounts and seamount chains of the Eastern Pacific, in Winterer, E. L., Hussong, D. M., and Decker, R. W., eds., *The Geology of North America, Vol. N, The Eastern Pacific Ocean and Hawaii,* Boulder, CO: Geological Society of America, pp. 289–306. An excellent summary of the structure, morphology, and tectonic relations of seamounts near the East Pacific Rise.

Keszthlyi, L., S. Self, and T. Thordarson, 2006, Flood lavas on Earth, Io and Mars, *Jour. Geol. Soc.,* 163: 253–264. New interpretations of the eruptive mecanisms of flood basalts.

Morgan, J. P., D. K. Blackman, and J. M. Sinton, eds., 1992, Mantle Flow and Melt Generation at Mid-Ocean Ridges, *Geophys. Monogr.* 71. Amer. Geophys. Union. An important collection of papers dealing with the magma generation under oceanic ridges.

Macdougall, J. D., ed., 1988, *Continental Flood Basalts.* Kluwer Acad. Publ., 341 p. A collection of outstanding papers on the major Phanerozoic flood basalts and their origins.

Nicolas, A., 1989, *Ophiolites and Dynamics of Oceanic Lithosphere,* Kluwer Acad. Publ., 367 p. A detailed account of ophiolites with special emphasis on Oman. The treatment of dynamic processes is especially thorough.

9 | Magmatism at Convergent Plate Boundaries

The rocks considered thus far have been mainly those of relatively simple geological settings where the composition and structure of the crust are less important than conditions in the mantle. The settings at convergent plate boundaries are more complex, and the igneous rocks far more varied. In considering these subduction-related rocks, we first examine their characteristic geological settings and compositional features and then attempt to find how the two may be linked.

■ Tectonic and Structural Settings

Modern orogenic rocks occupy a distinctive place in global tectonics (**Fig. 9-1**). With few exceptions, they erupt only in island arcs or along active continental margins, mainly around the Pacific Ocean, where oceanic lithosphere is subducted into the mantle beneath chains of calc-alkaline volcanoes. Although the descending lithosphere is in all cases oceanic, the overlying plate on which volcanism is focused may be either oceanic or continental. Where it is oceanic, the volcanoes tend to form arcuate chains of islands; where continental, they form lines of majestic cones near the leading edge of the continent.

In most cases, a deep oceanic trench marks the surface expression of subduction and parallels the volcanic chain at a distance of about 150 to 250 kilometers (**Fig. 9-2**). At depth, the subducted slab is delineated by earthquakes concentrated in a dipping plane, known as a *Benioff zone*, which descends at angles of 45 degrees or so beneath the volcanic axis.

The most distinctive lavas erupted in these settings belong to the calc-alkaline series of which andesite is the most typical, although not necessarily the most abundant member. An "andesite line" drawn around the margins of the Pacific Ocean separates these andesitic volcanoes from the interior of the oceans where such rocks are totally absent. This is not to say that all subduction-related rocks are calc-alkaline; many, especially those

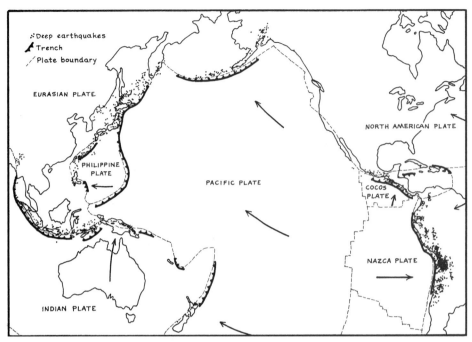

Figure 9-1 The principal setting of orogenic volcanism today is around the margins of the Pacific Ocean, where oceanic lithosphere spreading outward from the East Pacific Rise converges on the surrounding plates and plunges into the mantle. The nearly continuous line of trenches marking the convergent plate boundaries corresponds to an "andesite line," separating the chains of andesitic volcanoes from the interior of the oceanic basin where no calc-alkaline andesites are to be found. The distribution of volcanoes in island arcs and continental margins corresponds closely to the distribution of Benioff zones where earthquakes extend to depths as great as 700 km directly beneath the active volcanic belts.

produced in the early stages of island arcs, have tholeiitic affinities. In some systems, individual volcanoes evolve from tholeiitic to more calc-alkaline compositions with time, and in a few modern chains, such as those of Central America and Japan, tholeiitic and calc-alkaline volcanoes stand side by side in the same chain. Nor are all andesites restricted to the margins of convergent plates above subduction zones. Some are found in the interiors of continents far from the oceans. Nevertheless, calc-alkaline andesites are so characteristic of island arcs and active continental margins that subduction must be the main process responsible for their distinctive compositions and occurrence.

The principal volcanoes are aligned along a well-defined "volcanic front." Few if any volcanoes are found on the trench side of this line, but in

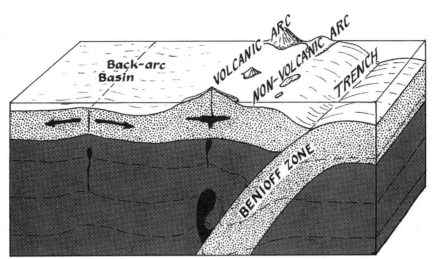

(a)

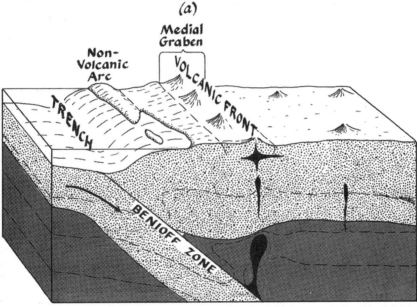

(b)

Figure 9-2 The volcanic chains of island arcs (a) and continental margins (b) share many structural elements. A trench, which in some places may be filled with sediments, marks the surface expression of subduction. Immediately inward and parallel to the trench a nonvolcanic arc or coast range consists of uplifted and often highly deformed older sedimentary and igneous rocks. On continental margins, a second, less conspicuous structural uplift rises along the axis of the volcanic front, and the major volcanoes are normally concentrated within or along the margins of a medial graben. These same features, although less apparent, are also present in some island arcs. In both settings, volcanoes may be distributed behind the volcanic front, but they are smaller and fewer in number. Some island arcs have a back-arc basin in which volcanism is concentrated along a spreading axis similar to but much weaker than normal oceanic ridges.

the opposite direction, smaller cones, many of alkaline compositions, may be scattered over wide regions behind the main volcanic axis. A low ridge forms an outer "nonvolcanic arc" or coast range between the trench and volcanic front. Even where this structural block has no surface expression, geophysical profiles show that it lies below a cover of youthful shelf sediments. At continental margins, the volcanic axis follows the crest of a second uplifted block that may have a shallow, axial graben, especially where the volcanoes are closely spaced and very productive. The eruptive products of these volcanoes may be accumulated in the depression so that only a subdued topographic expression marks the much greater structural relief of the basement.

The back-arc basins behind islands arcs seem to be regions of crustal extension. In the case of New Zealand and Japan, they were formed when fragments of continental lithosphere became detached and drifted seaward. Some have spreading axes similar to oceanic ridges but on a much smaller scale. The volcanoes in these marginal seas, as well as those behind the volcanic front on continental margins, are small, widely scattered, and generally of alkaline composition.

Subduction and zones of earthquakes are not invariably accompanied by volcanism. This is seen most clearly in the Andes (**Fig. 9-3**). Although the Peru–Chile trench is continuous offshore and is everywhere associated with a dipping Benioff zone, volcanoes are active only in three segments where the angle of descent is steep and a wedge of asthenosphere lies between the Benioff zone and overlying lithosphere. Where the inclination of the zone of earthquakes is shallow and remains in the lithosphere to distances well behind the axis of the Andes, volcanoes are scarce or totally absent. This relationship indicates that the source of magmas must lie in a wedge of asthenosphere between the Benioff zone and lithosphere, and only in that region are thermal and structural conditions appropriate for generation and rise of magmas.

Eruptive Behavior and Compositional Trends

Orogenic volcanoes differ so widely in composition and eruptive behavior that it is difficult to identify any single pattern they have in common. Even within a single volcanic chain, adjacent volcanoes may develop in totally different ways. Most large andesitic volcanoes have probably grown from one or more cinder cones, commonly on dilational fissures, which, with time, became centralized in a symmetrical cone over a central vent complex. Growth is often rapid, especially in the early stages of cone building. Some volcanoes have grown to heights of 500 or 600 meters in periods of only a

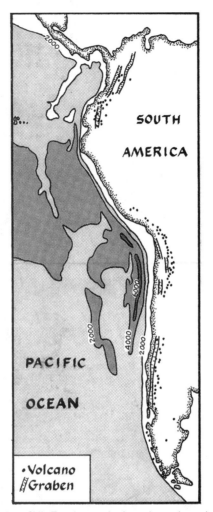

Figure 9-3 Even though the Peru-Chile Trench extends along almost the entire length of South America, volcanism is confined to three segments in which the dip of the Benioff zone is steep enough to descend below the base of the lithosphere so that a wedge of asthenosphere lies directly below the main axis of the Andes. Medial grabens follow the volcanic axis and are absent from those parts of the chain that have no large active volcanoes.

few centuries (**Fig. 9-4**). The early stages commonly produce proportionately more basaltic lava and scoria, but as volcanoes continue to grow, their magmas become more varied and their activity more erratic. Some mature volcanoes produce domes of viscous lava, whereas others may have voluminous eruptions of siliceous pumice that trigger the collapse of a caldera. Still

Figure 9-4 The volcano Izalco in El Salvador is an example of the rapid growth of andesitic cones. Since it first erupted in 1770, it has reached a height of about 500 meters. Eruptions of ash and lava were very frequent during its first century but have gradually declined. As can be seen in the foreground, much of the lava has accumulated around the base, while most of the volume of the cone is made of pyroclastic material.

others have alternating eruptions of basalt and dacite or rhyolite, usually in small cones and domes around their base.

One of the most characteristic features of andesitic volcanism is its high proportion of explosive eruptions. Unlike the shield volcanoes of oceanic islands, the large andesitic cones of the circum-Pacific system consist largely of pyroclastic ejects with only subordinate amounts of lava. Not all of this material is preserved in the cones. Some of the most voluminous discharges from explosive eruptions, particularly those of the "Plinian" type, are spread over vast regions and tend to be carried away by erosion, leaving little evidence in the stratigraphic record. This presents a problem when one attempts to measure the volumes and average compositions of the eruptive products of large, mature volcanoes. Deep-sea sediments may preserve a record of their activity in the form of ash layers carried out to sea by winds or ocean currents, but the record is erratic and unreliable. A similar problem arises is dealing with the basal units on which many large volcanoes stand. Flat-lying lavas are topographically inconspicuous, but they may have a great areal extent and large total volume—often more than that of the large cones that tend to capture our attention.

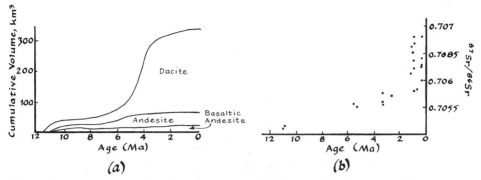

Figure 9-5 (a) As shown by this example from the central Andes, the proportion of more differentiated lava in a long-live volcanic center tends to increase with time. Note that the volumes are cumulative and were produced mainly in two relatively brief episodes. There are few if any basalts. Over the same period the strontium isotopic ratio became more radiogenic (b) (A. Grunder, personal communication).

While bearing these sampling problems in mind, it is probably safe to say that the compositional trends in island arcs are somewhat different from those on continental margins. The former tend to have proportionately larger volumes of andesitic basalt that, together with lesser amounts of andesite and dacite, form a continuous, differentiated series in which volumes decline with increasing silica content. On continental margins, the average silica contents of the rocks are somewhat greater. The volcanoes of the Andes, for example, produce proportionately large amounts of andesite, dacite, and rhyolite, but little if any basalt. As a general rule, volcanoes on thick continental crust have larger proportions of silica-rich magmas than those on thin crust of more mafic composition, and this tendency seems to increase with time (**Fig. 9-5**).

The products of many continental volcanoes are distinctly bimodal, with large proportions of basic and siliceous rocks but few of intermediate compositions. This discontinuity is often referred to as a *Daly Gap*, after the geologist, Reginald Daly, who first drew attention to its importance. The bimodality may be expressed in a group of volcanoes, each producing a narrow range of compositions, either basic or siliceous, or it may characterize the products of a single volcanic center that erupts magmas of contrasting composition from one or more closely associated vents. The most extreme examples of bimodal assemblages are found in block-faulted continental regions where contrasting lavas erupt in seemingly random order from separate but nearby vents. Basalt–rhyolite complexes of this kind are found in

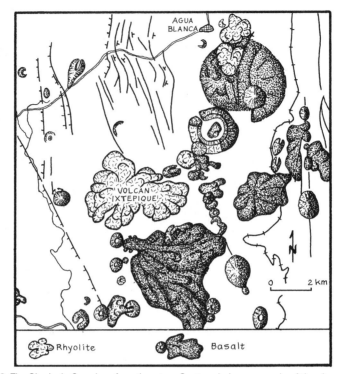

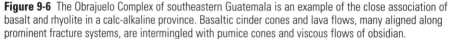

Figure 9-6 The Obrajuelo Complex of southeastern Guatemala is an example of the close association of basalt and rhyolite in a calc-alkaline province. Basaltic cinder cones and lava flows, many aligned along prominent fracture systems, are intermingled with pumice cones and viscous flows of obsidian.

most volcanic provinces, not only at the margins of continents but in the interiors of continents as well. An example from southeastern Guatemala is illustrated in **Figure 9-6**. One of the most distinctive compositional features of calc-alkaline volcanism is this close association in space and time of magmas of very different degrees of differentiation.

Although individual volcanoes rarely remain active for more than a million years or so, the magmatism of the province as a whole can continue much longer. The San Juan volcanic complex of Colorado, for example, has been active more or less continuously for 12 million years, and the Aucanquilcha complex of the central Andes has persisted almost as long. Activity shifts to new volcanoes or plutonic intrusions within the same general region, and although the intensity of magmatism may vary, there is often a continuity in the character of the evolving magmas. The example shown in Figure 9-5 illustrates a long-term trend toward more silica-rich compositions and increased strontium isotopic ratios, while the volumes of erupted material declines. These changes are a clear sign that the locus of the magmatic system is becoming shallower with time, and the amount of interaction with the crust is increasing.

Despite the importance of andesite in many provinces, especially those with thick continental crust, basalt is the most primitive member of the orogenic magma series and, as such, must be the parental magma from which andesites have differentiated. The once-popular notion that andesites are primary magmas derived directly from melting of subducted oceanic crust would greatly simplify the problem of explaining their origin, but this interpretation is untenable in the light of what is now known about the rocks. Andesite is a derivative magma, and its parent is basalt.

The calc-alkaline and tholeiitic series that are often closely associated in these settings stem from basalts that are so similar in their major-element compositions that one could argue that they are essentially one and the same (**Table 9-1**). Calc-alkaline basalts grade into a distinct type of high-alumina basalt that tends to be somewhat richer in alkalies and excluded trace elements than common subalkaline basalts. The differences between the calc-alkaline and tholeiitic series are most pronounced in their intermediate members. Calc-alkaline rocks lack the strong enrichment of iron so characteristic of ferrobasalts and icelandites but instead contain greater amounts of alumina and plagioclase. Few intermediate calc-alkaline rocks contain more than 10-percent total iron oxide, but almost all have at least 16-percent alumina. These differences are shown most clearly when the proportions of iron oxides are plotted against alkalies or silica (**Fig. 9-7**). As a consequence of their greater iron enrichment, many intermediate members of the tholeiitic series have pigeonite rather than hypersthene in their groundmass, and few contain hornblende or biotite. They are less porphyritic, and as a group, they tend to have larger proportions of basalt and basaltic andesite and proportionately fewer rocks of more siliceous compositions.

Calc-alkaline rocks, being richer in alumina, contain large amounts of plagioclase, particularly as phenocrysts (**Fig. 9-8**). The phenocrysts can have normal, reversed, or oscillatory zoning, and the compositions and types of zoning may differ from one crystal to another, even in a single thin section. Most phenocrysts are more anorthitic than either the plagioclase of the groundmass or the normative plagioclase of the rock as a whole. It is not uncommon, especially among the rocks of island arcs, to find phenocrysts of bytownite or anorthitic in rocks in which the normative plagioclase is sodic labradorite or even andesine.

Table 9-1	Typical compositions of the tholeiitic and calc-alkaline rocks of the Cascade Range of central Oregon. The tholeiitic rocks are of Oligocene age and come from the Western Cascades, whereas the calc-alkaline compositions are from the Quaternary High Cascades.

Tholeiitic Series

	Basalt	Basaltic Andesite	Tholeiitic Andesite	Dacite	Rhyolite
SiO_2	52.7	55.9	58.9	65.4	70.1
TiO_2	1.3	1.5	2.2	1.1	0.5
Al_2O_3	16.8	16.1	15.5	14.6	15.1
FeO^*	10.4	9.1	7.9	5.8	3.3
MnO	0.3	0.2	0.2	0.2	0.2
MgO	5.2	3.3	2.7	1.8	0.8
CaO	10.1	9.5	6.7	4.8	2.7
Na_2O	2.7	3.6	4.1	4.5	5.6
K_2O	0.3	0.5	1.1	1.2	1.6
P_2O_5	0.2	0.3	0.4	0.1	0.1
Ba	101	158	240	225	635
Zr	100	168	175	240	262
Rb	10	11	21	30	88
Sr	275	365	220	150	190
Co	47	34	28	24	< 5
Ni	27	22	5	1	–

Calc-Alkaline Series

	Basalt	Basaltic Andesite	Andesite	Dacite	Rhyolite
SiO_2	51.7	55.9	60.4	65.6	73.0
TiO_2	1.2	1.0	0.9	0.7	0.3
Al_2O_3	17.5	18.1	17.5	16.4	14.2
FeO^*	9.5	7.7	6.4	4.7	2.4
MnO	0.2	0.1	0.1	0.1	0.1
MgO	6.2	4.6	2.8	1.7	0.5
CaO	9.0	7.6	6.2	4.4	1.7
Na_2O	3.7	3.9	4.3	4.6	4.6
K_2O	0.8	0.9	1.2	1.6	3.1
P_2O_5	0.3	0.2	0.2	0.2	0.1
Ba	377	469	614	660	780
Zr	82	140	152	166	189
Rb	17	31	35	48	77
Sr	585	618	719	429	226
Co	36	26	20	7	–
Ni	92	48	20	9	2

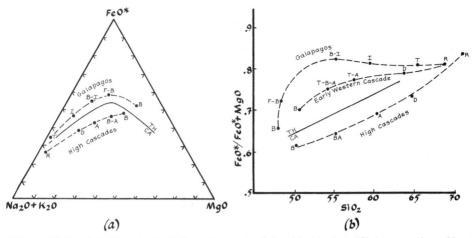

Figure 9-7 The tholeiitic and calc-alkaline series can be distinguished by their differing proportions of iron, alkalies, and silica. The AMF diagram (a) is a plot of $Na_2O + K_2O$ versus MgO and FeO^* (total iron as FeO). (b) A Harker diagram showing the rate of iron enrichment relative to silica. In both diagrams, the solid line is an empirical division between tholeiitic and calc-alkaline trends.

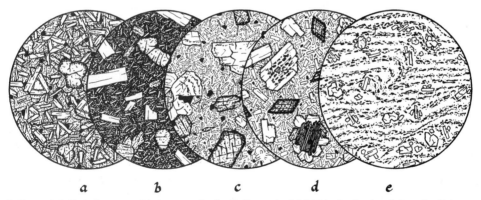

Figure 9-8 Typical petrographic features of calc-alkaline rocks. (a) A high-alumina basalt from the High Lava Plain of Central Oregon. It consists of abundant laths of plagioclase and rounded phenocrysts of olivine in a subophitic groundmass. Angular spaces between the plagioclase crystals give the rock a distinctive *diktytaxitic* texture. (b) A basaltic andesite from Paricutin Volcano in Mexico. Phenocrysts of labradorite and resorbed olivine are set in a matrix of finer plagioclase, augite, and opaque glass. (c) An andesite from the summit of Mt. Jefferson in the central Oregon Cascades shows the abundant phenocrysts and aggregated *glomerocrysts* of augite, hypersthene, and strongly zoned plagioclase that typify calc-alkaline andesites. (d) A dacite from the active dome of Mt. St. Helens, Washington, consists of *seriate* plagioclase zoned from calcic labradorite to medium oligoclase. Smaller phenocrysts of hypersthene, augite, and oxyhornblende are set in a hyalopilitic groundmass containing microlites of plagioclase, pyroxene, magnetite, and colorless glass. (e) A rhyolitic lava from the Platoro Caldera in the San Juan Mountains of Colorado has sanidine, oligoclase, and resorbed quartz in a flow-banded glassy matrix.

Subcalcic augite is almost ubiquitous in calc-alkaline rocks. Hypersthene is somewhat less common but is found in many andesites, dacites, and occasionally in rhyolites. Olivine phenocrysts, deeply embayed and encased in pyroxene, may occur in rocks having as much as 60-percent SiO_2, and it is not unusual to find as many as four or five mafic silicate minerals—olivine, augite, hypersthene, hornblende, and biotite—in the same rock. These complex assemblages are especially common in andesites and dacites of viscous domes and dikes. The hydrous minerals, hornblende and biotite, are unstable at surface conditions and may be rimmed with or totally altered to iron oxides, pyroxenes, and other anhydrous phases. Hydrous mafic minerals, as well as disequilibrium assemblages in general, tend to be more common in large, mature volcanoes built on thick continental crust.

Many of these disequilibrium relationships can be attributed to the sluggish equilibration of high-viscosity magmas at relatively low temperatures. Other factors, such as contamination with older rocks or other magmas, may be equally important. Whatever the reason, it is remarkable that, despite their great textural and mineralogical diversity, the chemical compositions of these rocks can be monotonously uniform. The fact that variations in the proportions and compositions of minerals are only weakly reflected in the bulk composition of the rocks suggests that if mixing or contamination takes place, it does not produce wide deviations of the major-element trends of differentiation.

Equally distinctive is the tendency for calc-alkaline suites to have linear variations between their mafic and felsic end members so that their components plot close to straight lines on almost all variation diagrams. A few calc-alkaline suites show minor nonlinear variations, especially in elements such as Mg, Ni, and Cr in the most mafic members of the series, but these are uncommon and, in most instances, only weakly developed.

The ratios of CaO to $Na_2O + K_2O$ are notably greater in the calc-alkaline rocks of island arcs than in those of continental margins, whereas the rocks of volcanoes in the interiors of continents tend to be relatively richer in alkalies, especially K_2O. Rocks of this latter type are often assigned to a separate potassic or "shoshonitic" series (**Fig. 9-9**).

■ Examples of Orogenic Volcanic Provinces

Calc-alkaline volcanic systems differ widely in structural settings, intensity of activity, and evolutionary history, but they have so much in common that a few well-studied examples can illustrate most of the distinguishing elements of the group as a whole. We shall look first at the Cascade province of western North America, a familiar example of a continental margin

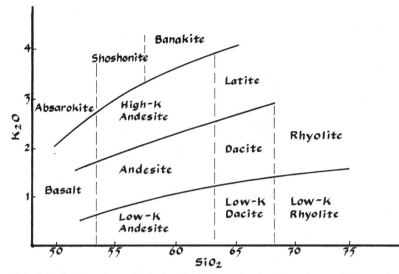

Figure 9-9 Calc-alkaline rocks can be further divided into subseries, depending on their potassium contents. Low-potassium rocks are more common in island arcs, whereas high-potassium, or "shoshonitic," rocks are found almost exclusively in the interior regions of continents.

system, then at the Aleutian Islands, a well-developed island arc extending from the Alaskan Peninsula across the northern Pacific.

The Cascade Range

The chain of andesitic volcanoes of the Cascade Range extends a thousand kilometers along the margin of the North American continent from northern California to southern British Columbia (**Fig. 9-10**). Although the system has neither a trench nor a Benioff zone, it has most of the basic elements of a zone of plate convergence. (The trench has been flooded with sediments, and earthquakes, although rare in historic times, have occurred at intervals in the recent past.) A well-preserved sequence of older rocks affords a rare perspective of the development of such a system through time.

Perhaps the single most striking feature of the record of earlier volcanism is that the linear chain of volcanoes we instinctively associate with subduction today has seldom been a recognizable feature of the system in the past. Before the High Cascade Range developed a million years or so ago, volcanism was scattered, both in space and time, and although subduction continued at a nearly constant rate, long periods passed in which volcanism was weak, if not totally absent.

The earliest activity broke out toward the end of the Mesozoic Era. Plutonic and weakly metamorphosed sedimentary and volcanic rocks of this

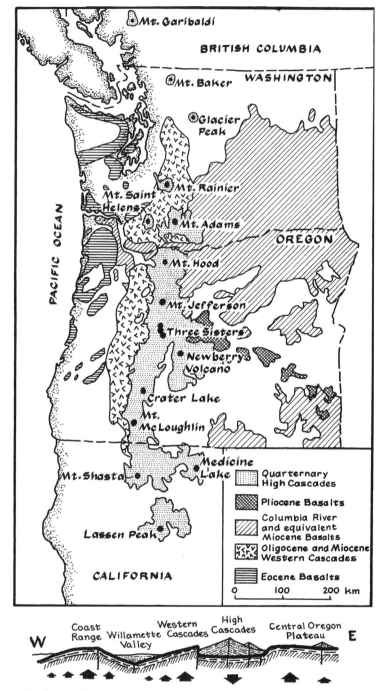

Figure 9-10 The Cascade Range extending along much of the northwestern coast of North America is a prime example of a chain of large andesitic volcanoes built on an active continental margin. Pre-Cenozoic metamorphic and plutonic rocks crop out along a broad arc extending southeastward from the northern part of Washington to eastern Oregon and then in scattered exposures to the southwestern corner of Oregon and adjacent parts of California. These continental crustal rocks have not been found in the central part of the chain; instead, the crust in that region consists mainly of basic volcanic rocks. A schematic east–west section through western Oregon (below) shows the principal structural elements of the Cascades and Coast Range.

age are exposed in a great sigmoidal belt extending from British Columbia to northwestern California. Unfortunately, erosion and an extensive cover of younger rocks limit any detailed reconstruction of the Eocene system as a whole. Intense igneous activity marked the opening stages of the Cenozoic era in what is now the Coast Range of Oregon and Washington, but most if not all of the rocks in that region seem to be oceanic in character. Calc-alkaline rocks appeared at the end of the Early Eocene, following a strong tectonic event, and by the end of the Eocene, a swarm of small volcanoes formed a zone about 150 kilometers wide extending from British Columbia through Washington and into central Oregon. Volcanism increased in intensity and spread westward and southward until, by the early Oligocene, it covered a broad zone across most of western Oregon and Washington. Some of the lavas erupted along the western margin of this zone were tholeiitic, but those farther inland were distinctly calc-alkaline. With time, differentiated rocks became more important as basaltic lavas gave way to larger proportions of siliceous pyroclastic tephra discharged from several large, low-rimmed calderas in what is now central Oregon. Evidence of large composite volcanoes is hard to find, but in the more deeply eroded northern part of the system, stocks and plutons up to 10 kilometers or so in diameter probably represent the roots of important volcanic centers.

After an early Miocene episode of faulting, uplift, and erosion, renewed volcanism broke out in the middle Miocene. Although brief, this episode was extraordinarily intense. By any measure, it was the most important igneous event of the Cenozoic era, not only in the Cascades but throughout much of the Circum-Pacific. Most mid-Miocene volcanic centers in Oregon and Washington were located along a line trending slightly east of north and close to the axis of the present chain. Other large andesitic centers have been recognized far behind the main axis. Apart from the Columbia River basalts, which were discharged during this same period but have no apparent relationship to the orogenic system of the Cascades, the products of this mid-Miocene episode are strongly calc-alkaline with large proportions of andesite. Today, the principal centers are marked by stocks, mainly of quartz diorite, and by broad aureoles of hydrothermal alteration and mineralization. The volcanoes seem to have been the first to have the form and alignment we associate with andesitic belts today, but they were restricted to a shorter, broader belt, mainly in central and northern Oregon.

Again, igneous activity was terminated by faulting, gentle folding, and broad tilting. Two minor episodes of widely scattered volcanism followed, one about 9 to 10 million years ago and another a few million years before the modern High Cascades became active. Most of the growth of the large Quaternary cones has taken place during a remarkably brief period. The fact that very few of their lavas have reversed magnetic polarities, even in the lowest levels of deeply glaciated cones, indicates that by far the greatest volumes must have been discharged since the present period of normal magnetic polarity began about 670,000 years ago.

Block faulting of the Central Cascade Range resulted in uplift and westward tilting of the Western Cascades, while subsidence of the basement below the active volcanoes began to form a steadily deepening axial graben. The graben is not topographically conspicuous because it has been partly filled by the products of Quaternary volcanoes that rise within it. It is most pronounced in the central Cascades where volcanism has been strongest and dies out toward the north and south where individual volcanoes are more widely spaced and the total volume of Quaternary volcanic rocks is small.

Andesite is the dominant rock of all the major volcanoes of the High Cascades. Taking the total volume of post-Pliocene rocks as a whole, however, andesitic cones account for a subordinate fraction of the total mass of erupted magma. The proportion of andesite is large only in those parts of the chain where the total volume of Quaternary rocks is small, namely in Washington and California (Fig. 9-11).

The rocks of the High Cascades have all the petrographic and chemical characteristics of calc-alkaline rocks of modern continental margins (Table 9-1). Apart from the most basic basalts and the most siliceous rhyolitic obsidians, almost all the rocks are strongly porphyritic and rich in plagioclase. Olivine phenocrysts are not uncommon in the basaltic andesites and andesites, but they are normally corroded and rimmed with pyroxene. Titaniferous magnetite, although abundant in the groundmass, is rarely an

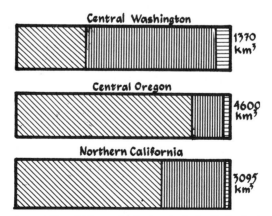

Figure 9-11 Relative volumes of basalt (diagonal ruling), andesite (vertical ruling), and rhyolite and dacite (horizontal ruling) in three sections of the Cascade Range.

important phenocryst. Most andesites contain two pyroxenes, augite and hypersthene, which vary little in composition, even in rocks of very different silica contents. Hornblende is restricted almost solely to dacites and rhyolites, and even in these rocks, it readily breaks down to oxidized clots of anhydrous minerals.

Plagioclase, normally in the range of sodic labradorite to calcic andesine, is by far the most conspicuous phenocryst throughout the full range of rocks. The pattern of zoning tends to be increasingly complex as the rocks become more differentiated. In many andesites and dacites, it even differs from grain to grain in the same rock. The composition of groundmass plagioclase is somewhat more sensitive to the bulk-rock alkali and silica contents but is rarely more sodic than oligoclase.

Some large volcanoes, such as Mount Rainier and Mount Baker, are of a "coherent" type, consisting mainly of monotonously similar andesite. Others, such as Mount Mazama (the volcano containing the caldera of Crater Lake) and the Three Sisters (**Fig. 9-12**), are strongly "divergent" in the sense that a main cone of andesite and dacite has evolved in later eruptions to contemporaneous basalts and rhyolites or rhyodacites. Newberry volcano in Oregon and the complex of vents in and around the flooded caldera of Medicine Lake, California, are extreme examples in which andesites are distinctly subordinate, and the basaltic and rhyolitic rocks form a strongly bimodal assemblage.

The Aleutian Arc

The volcanic arc of the Aleutian Islands (**Fig. 9-13**) is of special interest because it is a good example of an igneous belt that crosses from a continental margin to an island arc built on oceanic crust. The arc stretches some 2,500 km from the continental mainland of Alaska into the Bering Sea. Like most island arcs, it has a line of nonvolcanic islands and a structural ridge offshore from the volcanic chain of the Alaskan mainland. A well-defined Benioff zone dips below the arc at an angle of about 45 degrees. A trench extends along the entire length of the Aleutian Islands, and because of the curvature of the arc with respect to the motion of the northern Pacific plate, the angle of incidence ranges from nearly perpendicular in the eastern part to parallel near the western extremity. The corresponding rates of convergence range from about 5.5 cm per year along the Alaskan Peninsula to zero at the northwestern end of the arc. Subduction began about 70 million years ago, and magmatism has continued in an episodic fashion for most of the Cenozoic era. Thus, something of the order of 500 to 1,000 km^3 of oceanic lithosphere has been subducted for each cubic kilometer erupted during this period.

(a)

Figure 9-12 (a) The volcano South Sister in the central Oregon Cascades exemplifies large, composite cones of the "divergent" type in which late eruptions of basaltic lavas and rhyolite have broken out on the flanks of a cone consisting mainly of andesite and dacite. The viscous flow of obsidian in the foreground is slightly younger than the basaltic cone immediately in front of it, but the reverse relations are also found. An extensive series of nearly flat-lying basaltic lavas underlies the entire region.

The detailed geologic history of the arc is poorly known, particularly in the more remote islands, but the broad elements of the stratigraphic sequence have been determined from studies of the major eastern islands. The section consists of three major volcanic units. The oldest of these, the Finger Bay Series, is a thick section of highly altered lavas and pyroclastic sediments, probably of early Eocene age. The overlying Andrews Bay strata consist of marine volcanoclastic beds of middle to late Eocene age. No important Tertiary units of younger age have been recognized on land. The Recent volcanic series, most of which has been laid down in the last million years, is separated from the Eocene units by an unconformity representing all of middle and late Tertiary time.

A more complete record has been obtained from marine sediments off-shore. It shows that volcanism was renewed at the beginning of the Miocene

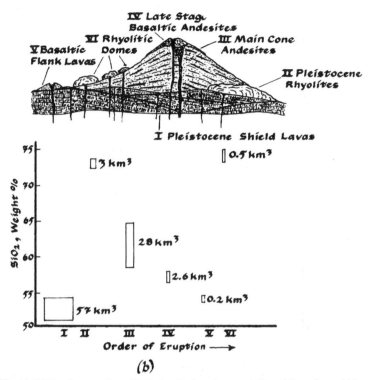

Figure 9-12 (Cont.) (b) The diagram above illustrates the long-term evolution of the volcano. Volumes of rocks in each stage are shown by the relative areas of the rectangles. The midpoint on the vertical dimension of each rectangle is placed at the mean value of SiO_2 for the rocks of that stage, and the vertical height of the edge indicates one standard deviation for the silica value.

and continued, with especially strong pulses about 15 and 10 million years ago. Thus, the present volcanoes stand on islands that were probably Tertiary volcanic centers and were subjected to repeated uplift and erosion. In the eastern part of the chain, some of these centers are underlain by sub-volcanic plutons that have been correlated in age with an early Miocene episode of volcanism.

This pattern of multiple episodes of similar activity at the same locations through much of Cenozoic time is in contrast to the record in the Cascades, where volcanism shifted from place to place and seldom had a strong linear distribution. The episodic character of the two regions was similar, however. In fact, periods of intense volcanism seem to have developed simultaneously, not only here but around much of the Pacific rim.

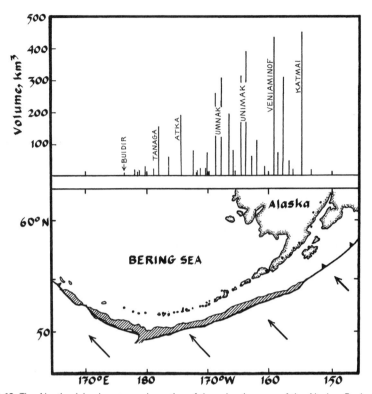

Figure 9-13 The Aleutian Islands are a prolongation of the volcanic range of the Alaskan Peninsula and are an excellent example of a belt that passes from a continental margin to an island arc. When the volumes erupted from the volcanoes are plotted according to their position along the length of the Aleutian Arc, they tend to decrease from east to west as the rate of subduction become less because of the decreasing angle of convergence. This relationship is strong evidence that the amount of magma generated in the arc is a direct function of the amount of subducted oceanic crust. (After B. D. Marsh, 1982, The Aleutians, *Andesites*, R. S. Thorpe, ed., pp. 99–114.)

The modern volcanoes form an imposing chain parallel to the trench and about 110 km above the Benioff zone. Only two small volcanic centers stand behind the arc. The distance between the trench and volcanic front increases systematically from west to east, as one would expect from the change of strike relative to the direction of subduction and the greater amount of sediments scraped from the descending slab and accreted to the leading edge of the arc. Large volcanic centers are regularly spaced at intervals of 60 to 70 kilometers along the entire arc, but the chain is segmented by offsets and minor changes of strike corresponding to a similar segmentation of the underlying Benioff zone. Volcanoes situated near these offsets tend to be more tholeiitic than those in other parts of the chain.

Although the most recent episode of activity has lasted less than 3 million years, the total volume erupted during that period is estimated at 4,700 km^3. The remarkable distribution of this volume (Fig. 9-13) relative to the position of volcanoes along the arc indicates that the intensity of volcanism is a direct function of the angle of plate convergence.

The Recent volcanic centers have evolved from broad shields composed mainly of thin basaltic lavas to high composite cones of andesite with large proportions of pyroclastic material. Greater volumes of silica-rich pyroclastic rocks have been erupted from volcanoes on the Alaskan Peninsula, but as one proceeds along the Aleutian chain from east to west, basalts give way to greater proportions of andesite so that the compositions of the most primitive magmas become richer in silica and alkalies and poorer in magnesium, iron, and calcium. At the same time, the isotopic ratios of strontium, neodymium, and lead become less radiogenic. The significance of these variations will become apparent when we consider the origin of such magmas in a later section.

■ Shallow Differentiation

Compared with the relatively homogeneous magmas erupted from most oceanic volcanoes, the products of subduction-related volcanoes have smaller volumes and may differ in composition from one eruption to another, often in irregular ways. During the main stages of growth, it is rare to see an andesitic volcano produce a series of progressively more differentiated lavas that can be related to steady differentiation. Instead, they often erupt highly differentiated magmas then revert to more primitive compositions, even in the course of a single eruption. These irregularities would indicate that the magma is stored and differentiated in small, shallow reservoirs, possibly a network of dikes and sills, rather than a single large chamber, but this may not always be the case. The dimensions of calderas resulting from sudden discharges of voluminous magma indicate that sizable bodies of magma can develop directly beneath volcanoes that are still potentially active.

Plutonic intrusions several kilometers across can be seen under many deeply eroded volcanoes. They seem to have stoped their way upward near the central conduit and even into the base of the volcano itself. Although they send out dikes and sills that may have reached the surface, it is unclear whether all of these bodies were reservoirs in which magma differentiated during quiet intervals between eruptions.

One of the notable peculiarities of calc-alkaline volcanism is the relatively large volumes of highly differentiated magmas that can evolve in short periods of time. Another feature, possibly related to the first, is the close association of magmas of contrasting compositions erupted from the same or closely related vents or in single outpourings of compositionally graded magmas. The degrees of differentiation are too great, the volumes too large, and the time intervals too brief for these differentiates to be produced by progressive fractional crystallization of the kind seen in layered intrusions. Moreover, the bimodality of many calc-alkaline suites shows that they differentiate in ways that enable the undifferentiated magma to erupt together with the products of its own crystallization.

Volcanologists have often observed that, during their long intervals of repose, mature volcanoes tend to evolve felsic differentiates in the uppermost levels of some sort of reservoir, and the longer the interval between eruptions, the greater the volume and explosivity of the first liquids discharged when an eruption finally occurs. This behavior may be the result of a process of differentiation based on the mechanism of liquid fractionation outlined in Chapter 5. When crystallization at a steep wall produces a liquid that is less dense than the main magma, the light layer may rise along the walls and accumulate as a gravitationally stable unit overlying the main body of denser, more basic magma (**Fig. 9-14**). Such a process offers an attractive explanation for sharply zoned magmas, such as that discharged during the great caldera-forming eruption responsible for Crater Lake (**Fig. 9-15**).

Similar effects would be expected if dense mafic magma melts more felsic wall rocks, as seems to have happened in the case of Paricutin Volcano in Mexico. In this case, however, the light, differentiated magma was not the first to be erupted but followed the denser mafic lavas. This seeming anomaly has been explained as a result of the mechanics of withdrawal from a

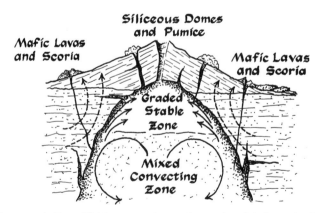

Figure 9-14 The concentrations of light components near the margins of shallow calc-alkaline intrusions may produce a buoyant boundary layer of differentiated liquid that rises and accumulates in a compositionally zoned body under the roof. Eruptions from different parts of the intrusion could account for the association of mafic and felsic rocks erupted from the summits and flanks of large andesitic volcanoes. The diagram is schematic and not to scale.

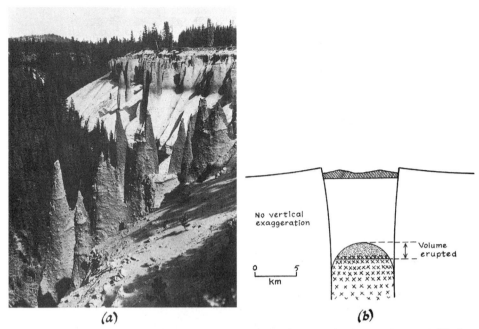

Figure 9-15 (a) The siliceous pumice discharged during the great caldera-forming eruption responsible for Crater Lake, Oregon, is compositionally graded through most of its thickness but changes abruptly to a hornblende-rich scoria near the top of the section. The magma appears to have been erupted continuously from a zoned magma chamber in which the siliceous magma at the top was the first to be discharged and the dark scoria was part of a denser, more mafic magma below. (Photograph by the Oregon State Highway Department.) (b) The size, depth, and zonation of the magma chamber have been reconstructed from chemical and experimental studies. (Constructed from data of J. L. Ritchey, 1980, *Jour. Volc. Geoth. Res.* 7:373–386, and R. L. Smith, 1979, *Geol. Soc. Amer. Spec. Publ.* 180:5–27.)

chamber in which a light, viscous magma overlies a heavier, more fluid one (**Fig. 9-16**). When magmas of differing properties rise through a channel tapping a compositionally zoned reservoir of this kind, the lower liquid tends to be drawn up and enters the central part of the conduit. If the lower liquid is less viscous, as it is in this case, it flows more rapidly, and even though it is denser and starts at a lower level, it can reach the surface before the more viscous differentiated liquid. As the rate of discharge declines, the draw-up diminishes, and the lighter, more viscous liquid is able to escape without entraining the lower part of the reservoir.

■ Experimental Studies of Calc-Alkaline Rocks

Most of the compositional features of orogenic magmas of the tholeiitic type can be explained in terms of shallow fractionation of the same minerals—plagioclase, olivine, augite, and magnetite—that account for the compositions

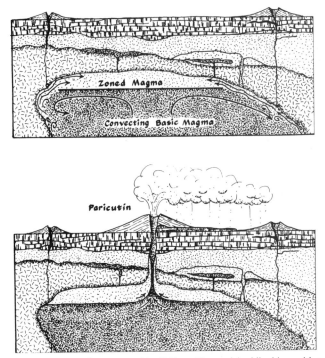

Figure 9-16 Before Paricutin volcano erupted in 1943, a silica-enriched liquid, resulting from melting and assimilation of crystal rocks, is thought to have accumulated above the denser parental magma. Tapping of these two zones led to more rapid rise of the denser but less viscous lower magma and later eruption of a more silica-rich fraction when the rate of discharge declined.

of common tholeiitic magmas. It is more difficult to explain the differentiation of calc-alkaline magmas in this way. The problem is best illustrated by comparing the temperatures at which these minerals crystallize in magmas of differing compositions.

Table 9-2 shows the temperatures of crystallization of plagioclase, olivine, and augite from three typical calc-alkaline rocks at atmospheric pressure. Note that, unlike the examples of tholeiitic rocks in Chapter 8, the minerals begin to crystallize at extraordinarily high temperatures and over a wide range. Moreover, the liquidus temperature of plagioclase *increases* going from basalt to dacite. It is obvious that if the three rocks represent cogenetic liquids they could not have been produced by crystal fractionation of these same minerals under the conditions at which their liquidus temperatures were determined in the experiments. Some basic aspect of the melting experiments must have differed from the conditions under which the natural system evolved.

Table 9-2 Liquidus temperatures of calc-alkaline rocks determined by melting natural rocks at atmospheric pressure

Sample	SiO$_2$	Liquidus Temperatures Plagioclase	Olivine	Augite
Olivine basalt	50.5	1215°	1185°	1175°
Andesite	60.7	1255°	–	1180°
Dacite	69.9	1275°	–	1180°

(From G. M. Brown and J. F., Schairer, 1971, *Geol. Soc. Amer. Mem.* 130:139–157).

The minerals could have anomalously high liquidus temperatures because they were not precipitates of their host liquid but crystals picked up from some other source. This seems unlikely, however, because the bulk compositions of most calc-alkaline rocks, especially andesites, are too regular to be random mixtures of liquids and xenocrysts. A more likely explanation is that the liquidus temperatures were lowered by some component that is no longer in the rock. An obvious candidate for such a "lost" component would be a volatile, such as water.

The Role of Water

As is normally the case when another component is added to a system, water has the effect of lowering the temperatures of crystallization of all anhydrous minerals. Moreover, the amount by which liquidus temperatures are lowered differs from one mineral to another. As the simple system diopside-anorthite illustrates (**Fig. 9-17**), the liquidus temperature of plagioclase is depressed much more than that of a mafic mineral, such as pyroxene. Note that the liquidus curves of both minerals are lower at 10 kilobars water pressure than at atmospheric pressure, and because of the greater depression of the liquidus of anorthite, the eutectic composition is shifted to higher anorthite contents. A magma precipitating these minerals at an elevated water pressure would be more plagioclase rich, but if raised to a shallower level where much of the water is exsolved, the same composition would be far below its low-pressure liquidus and would crystallize large amounts of plagioclase before pyroxene. This is exactly what was observed in the low-pressure melting carried out on the natural rocks cited in Table 9-2.

The amount of water required to change the liquidus temperatures of plagioclase at elevated pressures is not large, probably less than one weight percent. A small initial concentration must increase as anhydrous minerals crystallize and the water is concentrated in smaller and smaller amounts of

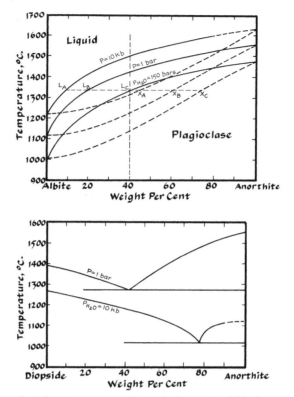

Figure 9-17 The effect of water pressure on the plagioclase system (a) is the opposite of that of dry load pressure—it lowers the temperatures of both the liquidus and solidus. The effect of water pressure on plagioclase is greater than on pyroxene (b). Note that the liquidus of anorthite is lowered more than that of diopside, and the eutectic is shifted to more anorthite-rich compositions.

liquid. Thus, even though water contents of basaltic magmas are not great, they could affect the stability relations of the major phases, and this effect would increase with differentiation, as the data in Table 9-2 seem to require.

Water affects not only the relative amounts of plagioclase but its composition as well. Note in Figure 9-17a that at a constant temperature and elevated pressure, a dry melt, L_A, would be in equilibrium with crystals of composition X_A, and on rising to the surface (P = 1 bar), the compositions would shift to L_B and X_B, respectively, and crystals would be reversely zoned to more anorthitic rims. A wet magma, on the other hand, would have the composition L_C and An-rich crystals, X_C, and if it exsolves water as it rises, this liquid would produce crystals zoned to more albitic compositions. It must be remembered, of course, that the binary system portrayed here is highly simplified, and the form of the liquidus and solidus curves can be very different in more complex systems (Fig. 4-5).

Lack of Iron Enrichment

A simple way to explain a lack of iron enrichment during differentiation of any series of magmas is to postulate early crystallization of large amounts of an iron oxide mineral, such as magnetite. Experiments with iron-bearing systems (Fig. 4-17) show that the field of magnetite expands greatly with increasing oxidation. It overruns the stability field of olivine, preventing liquids from evolving toward higher iron contents and diverting their course toward a greater enrichment of silica. This effect seems particularly appropriate for calc-alkaline rocks formed in the oxidizing environment of the crust.

Magnetite is a common mineral in many basaltic lavas, particularly those of island arcs. It is less common, however, in intermediate calc-alkaline lavas or their plutonic equivalents. Melting experiments carried out on natural andesites at elevated water pressures show that magnetite becomes less important as calc-alkaline magmas differentiate and are enriched in water. **Figure 9-18** summarizes results of a study of the phase relations of an andesite from Mount Hood, Oregon, over a range of water pressures up to ten kilobars. Again, the effect of water is to lower the liquidus temperatures of anhydrous minerals, particularly plagioclase, while raising that of the hydrous

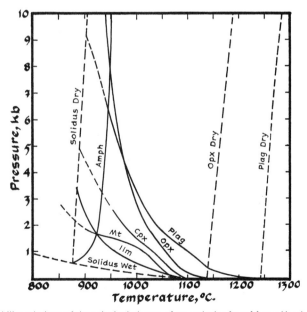

Figure 9-18 Stability relations of the principal phases of an andesite from Mount Hood, Oregon, as functions of temperature and water pressure. (After Eqgler & Burnham, 1973, *Geol. Soc. Amer. Bull.* 84:2517–2532.)

mineral, amphibole. At pressures of about seven kilobars, plagioclase, hyper-sthene, and amphibole all have nearly the same liquidus temperature of about 950°C. Note that the liquidus temperatures of magnetite and ilmenite are also lowered by water pressure, but these iron-oxide minerals are not liquidus phases at any water pressure or geologically reasonable oxygen fugacity. Above about two kilobars, they are replaced by amphibole.

The amount of water required to stabilize amphibole in place of mag-netite is not great, especially at the relatively low liquidus temperatures of andesites. Common hornblendes contain large proportions of both ferric and ferrous iron, and they normally have a lower silica content than the calc-alkaline liquids in which they crystallize. It seems reasonable, therefore, that amphibole, rather than magnetite, could limit iron enrichment in the middle stages of differentiation of calc-alkaline magmas.

Enrichment of Silica

The effect of elevated water pressure on the silica contents of liquids can be seen in the relatively simple three-component system forsterite–pyroxene–silica illustrated in **Figure 9-19**. Comparison of the stability fields at

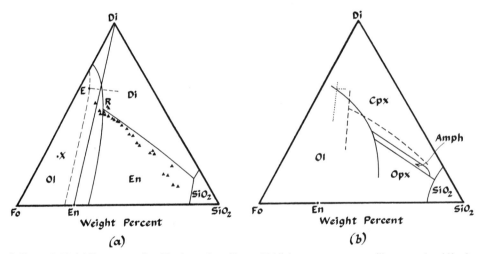

Figure 9-19 (a) The system diopside–forsterite–silica at 20 kilobars water pressure illustrates the shift of compositions to greater silica contents with increasing water pressure. Phase boundaries under dry conditions are indicated by broken lines. The compositions of calc-alkaline rocks from the Oregon Cascades (triangles) fall close to the two-pyroxene cotectic for 20 kilobars and extend from basaltic andesites near the invariant point at R to rhyolites at the eutectic near the SiO_2 corner of the diagram. (b) When anorthite is added to the system, a field of amphibole appears in the middle range of compositions between the pyroxenes. At the same time, the boundaries shift toward SiO_2, and compositions of liquids come closer to the calc-alkaline rocks. Fields are projected from anorthite, which is a stable phase throughout the system. Twenty kilobar field boundaries for dry and wet conditions in (a) are shown by dotted and dashed lines, respectively.

high pressure under dry conditions with those at an equivalent pressure wet shows that all the main phase boundaries are shifted toward higher silica contents. Note in particular that the invariant point, E, is shifted to R across the pyroxene join toward the silica corner of the triangle. Unlike the effect of pressure under dry conditions, which increases the stability of enstatite relative to olivine, high pressures of water have the opposite effect; the stability range of olivine increases relative to pyroxene.

The consequences of this difference can be appreciated by comparing the compositions of liquids that would evolve under the two conditions, low-pressure dry and high-pressure wet. The compositions of basaltic liquids produced by melting a mantle consisting of olivine and two pyroxenes, corresponding to points E and R, do not differ greatly, but subsequent cooling and crystallization would cause the liquid evolving at higher water pressures to follow a much longer course of differentiation ending at a final composition greatly enriched in silica. The analogy to calc-alkaline rocks becomes quite clear if the compositions of natural calc-alkaline rocks are plotted on the same diagram (Fig. 9-19a). Apart from a slight divergence due, no doubt, to the effects of the additional components in natural melts, the course of the calc-alkaline magmas is close to the simplified system at elevated water pressure. An even closer approach to natural rocks is achieved if plagioclase is added to the same system (Fig. 9-19b). Amphibole appears as a stable liquidus mineral in the region of the cotectic between enstatite and diopside.

These effects offer a possible explanation for some of the differences between tholeiitic and calc-alkaline series and for the close link between the latter and subduction. Although the tholeiitic and calc-alkaline series have very similar parental basalts and are commonly found in close association, different conditions of differentiation cause them to evolve along very different trends. The same or very similar basalts can evolve along paths leading to a tholeiitic or calc-alkaline trend depending on the depth and amount of water present where they differentiate.

■ Origins of Calc-Alkaline Magmas

There is scarcely any petrogenetic scheme that has not at some time been invoked to explain the compositional features and geological occurrence of andesites. Despite decades of intense research, much still remains to be learned about their origins.

It was long thought that calc-alkaline rocks were simply products of basaltic magmas that had been contaminated by continental crust. Because andesite, the most distinctive member of the calc-alkaline series, is typical of regions where continental crust can be assimilated, it seemed obvious that basalts rising through tens of kilometers of sialic rocks could scarcely reach the surface without being highly contaminated. The only question was how this came about.

This question came to the forefront in the 1920s, when a lively debate broke out between Norman Bowen and Clarence Fenner over the origin of the calc-alkaline rocks in and around Katmai Volcano on the Alaskan Peninsula. Fenner visited the region shortly after the great eruption of 1912 had poured 7 cubic kilometers of rhyolitic ignimbrites into the Valley of Ten Thousand Smokes. Near the base of Mount Katmai, he observed a spectacular group of hybrid rocks, similar to the sample illustrated in Figure 5-12, and deduced that very hot rhyolitic magma of the main eruptive unit had melted and mingled with older mafic lavas in the interior of Mount Katmai. He went on to propose that mixing of two liquids, one mafic and the other felsic, could have produced the full range of intermediate compositions found elsewhere in the same region.

As support for this interpretation, Fenner pointed to the linear compositional variations of the rocks and argued that crystal fractionation would not produce such a trend (**Fig. 9-20a**). As noted in Chapter 5, crystal fractionation normally produces nonlinear variations because the compositions and proportions of minerals being removed are constantly changing as crystallization advances. Fenner also drew attention to the lack of strong iron enrichment normally seen in intermediate differentiates, such as those that were then being studied in western Scotland.

Bowen responded by pointing out that melting of mafic rocks by a more evolved magma, such as rhyolite, was thermodynamically absurd. The enormous amount of heat required to melt the mafic rock would not be available unless the rhyolite were superheated to an unrealistic degree. Even in the unlikely event that the rhyolitic magmas were originally superheated, that is, at temperatures far above that at which they would begin to crystallize, the difference between the heat capacities of silicates (about 0.3 calories per gram) and their heats of fusion (about 80 calories per gram) is so great that a mafic rock caught up in superheated rhyolite would absorb so much heat that it could yield only trivial amounts of liquid.

Bowen went on to argue that Fenner's linear variations resulted from plotting compositions that had been altered by accumulation of phenocrysts and that if only aphyric rocks were used (Fig. 9-20b), the trends were slightly curved and could be explained by crystal fractionation. This argument seemed valid at the time, but it is now thought that most porphyritic calc-alkaline rocks do in fact approach liquid compositions.

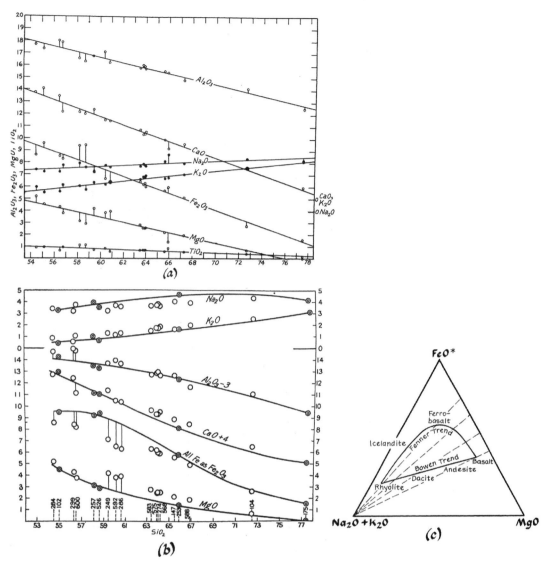

Figure 9-20 The Harker diagram shown in (a) is Fenner's original plot of the compositions of rocks from the Katmai region. Fenner inferred that their variations were linear and that they resulted from mixing of two end-members, one rich and the other poor in silica. The diagram in (b) is Bowen's interpretation of the same data. He disregarded all porphyritic rocks (open circles) and argued that those without phenocrysts (double circles) defined curved lines of the kind that would be expected for a series related by crystal fractionation. Here we have an example of how two petrologists, if they are determined to do so, can use the same data to support directly opposing conclusions. The basic differences in their interpretations can best be illustrated by an AFM diagram (c) showing the iron enrichment, or "Fenner trend," common in tholeiitic series, which Fenner considered typical of liquids related by crystal fractionation. He attributed the straighter "Bowen trend" of calc-alkaline rocks to mixing. Note that the trends associated with the two names are the opposite of those the two individuals saw in the rocks of Katmai. (The dashed lines indicating constant Fe–Mg ratios show that the differences are not in the relative proportions of the mafic components but in the proportions of felsic components represented by alkalies.)

Even if the trends of major-element variations were indeed curved, as Bowen maintained, they did not rule out other processes, such as combined crystal fractionation and assimilation. As noted in Chapter 3, assimilation need not alter the trends of a liquid line of descent; it only changes the proportions of end products. So many of the geological relationships of andesites seem to point to interaction between basic magma and crustal rocks that many geologists concluded that Fenner's scheme was simply backward. It was more logical that mafic, high-temperature magmas assimilate crustal rocks to produce intermediate magmas of andesitic compositions. A number of observations lend support to this reasoning.

■ The Role of Crustal Assimilation

The fact that calc-alkaline andesites are not erupted from volcanoes in the deep oceanic basins but are found almost exclusively along continental margins or in island arcs where subduction carries crustal material below the volcanic chains is impressive evidence that the distinctive character of the rocks is in some way a result of interaction between mantle-derived basaltic magma and the crust.

A number of detailed studies of calc-alkaline suites have shown how this might come about. One of the most influential of these was a study by Ray Wilcox of the lavas discharged by the Mexican volcano Paricutin between 1943 and 1952. During that period, 1.4 cubic kilometers of compositionally graded magma were discharged under conditions that permitted close observations and continual sampling. Silica contents of the lava increased from 54 percent in the first flows discharged in 1943 to over 60 percent at the close of the eruption 9 years later (**Fig. 9-21**). At the same time, olivine decreased in abundance and developed reaction rims of pyroxene, while hypersthene became more abundant. Wilcox explained these progressive changes as the result of a fractionation of the observed phenocrysts and concurrent assimilation of xenoliths of granitic rocks that were found in all stages of melting and incorporation into their host. Using a mass-balance calculation to relate the compositions of analyzed rocks and the minerals they contain, it is possible to show that the progressive variations in the major and trace elements are perfectly consistent with simultaneous fractional crystallization and assimilation of granitic basement rocks. The heat liberated by crystallizing minerals, by itself, would be insufficient to melt the necessary quantity of xenoliths, but it is reasonable to postulate that additional heat could be contributed by a larger body of convecting magma at greater depth.

At first glance, it might seem a simple matter to determine the degree to which such a process of assimilation might account for most of the

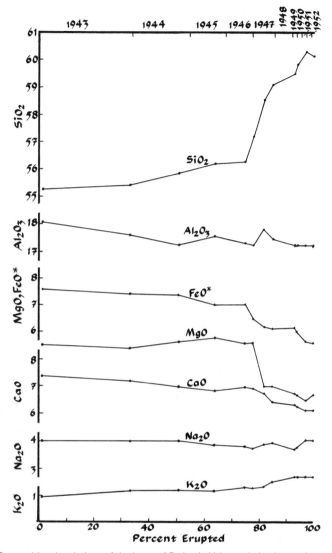

Figure 9-21 Compositional variations of the lavas of Paricutin Volcano during its continuous activity between 1943 and 1952. Note that the rate of eruption, as shown by the upper scale, declined with time and that the most silica-rich magma was the last to appear.

compositional features of calc-alkaline rocks elsewhere. For example, by examining long volcanic chains like the Alaskan-Aleutian Arc that cross from one type of crust to another, one can compare lavas from volcanoes on thick sialic crust with those erupted where the underlying section consists entirely of basic igneous rocks or volcanic sediments that differ little from the magmas passing through them. This test has been applied to several volcanic arcs, and in every case, the lavas of volcanoes on thick continental crust are found to have major-element trends of differentiation indistinguishable from those on oceanic crust. Assimilation, if it has played a role, has produced no obvious compositional difference in the rocks. On closer scrutiny, however, this observation proves to be less conclusive than it at first appears.

Consider first the compositions of crustal-rocks and how they compare with the magmas in which they are being assimilated. If the crustal material has an unusual composition rich in components that are alien to common magmas, even small amounts of contamination may have very conspicuous effects. For example, when the Mexican volcano El Chichon erupted with such violence in 1981, the ejecta contained small crystals of the sulfate mineral anhydrite, a phase that is very rare in igneous rocks. This anhydrite is easily traced to the thick section of evaporites underlying the volcano. Another volcano, Concepcion, near the southern end of Lake Nicaragua, is underlain by Cretaceous sediments so rich in phosphates that they are a potential source of commercial fertilizer. The lavas of Concepcion contain extraordinarily large crystals of the phosphorus-bearing mineral, apatite, that must be a product of contamination by the sediments through which the magma rises.

Minerals like anhydrite or large crystals of apatite are conspicuous anomalies in calc-alkaline rocks, but if the assimilated material consists of felsic metamorphic or plutonic rocks of more common compositions, its contribution is much less obvious. Elements, such as silicon and the alkalies, that are normally enriched during igneous differentiation are also concentrated in the continental crust, and it is difficult to say whether these components are gained from assimilation or from differentiation; both processes have the same effect.

Another consideration is inherent in the basic phase relations governing assimilation. Even if a magma were contaminated with rocks, such as limestone or pelitic shales, with bulk compositions far from those of common igneous rocks, the added components, such as CaO, Al_2O_3, and SiO_2, are the same as those entering minerals that normally crystallize at some stage of cooling of the magma. In such circumstances, the effects of assimilation may be concealed by the way in which the magma responds to an addition of these components. It simply crystallizes greater amounts of minerals containing large proportions of the added elements. As long as crystal–liquid relations control the course of differentiation, the major-element compositions of evolving liquids are constrained to follow a liquid line of descent that cannot deviate far from that defined by crystal fractionation. The principal effect of assimilation of crustal material is to change the proportions of end products while leaving the major-element compositions of liquids essentially unchanged. Thus, if the effects of assimilation are to be detected, the evidence is not likely to be found in trends of differentiation. Instead, one should expect to find it reflected in the relative volumes of differentiated rocks. Magmas that assimilate large amounts of felsic continental crust should produce larger proportions of rock types, such as andesite, dacite, or rhyolite, that are richer in the components of continental lithosphere than those rising through mafic crustal rocks of oceanic character.

A test of this kind has been applied to the volcanic chain of the Cascade Range. It will be recalled that the northern and southern ends of the range are underlain by thick sections of metamorphic and plutonic rocks, whereas the crust in the central part of the belt seems to be largely oceanic in character. Although the trends of differentiation are found to be nearly the same from one end of the range to the other, the relative proportions of rock types differ widely. As we saw earlier (Fig. 9-11), andesites and more differentiated rocks are relatively more abundant where magmas have risen through a thick sialic section.

Even this evidence, impressive as it is, may not be conclusive proof of wholesale assimilation, for the differences in proportions of rock types could have a much simpler explanation. Magmas passing through thick sections of cool continental crust must lose heat and crystallize on the way to the surface. Moreover, they rise more slowly through light continental rocks, and basic magmas, being denser than many shallow crustal rocks, may never reach the surface. Thus, the continental crust may form a thermal barrier and density trap that inhibits dense mafic magmas of high temperatures from rising unaltered through cooler crustal rocks of low density. If so, differences in the proportions of rock types may be due not so much to assimilation as to the physical limitations on the rise of magma through continental crust.

Isotopic Evidence

What is needed to resolve this dilemma is a geochemical tracer that can serve as a measure of crustal contamination while being unaffected by any form of differentiation. A distinctive component that has different concentrations in the mantle or continental crust but is insensitive to subsequent crystal fractionation would afford a way of detecting the relative contributions of the crust and mantle, no matter how the magma may have evolved. The isotopes of certain elements, such as strontium and lead, can provide just such a clue because they have different concentrations in rocks of the crust and mantle but enter igneous minerals in amounts that are independent of their isotopic character.

As we saw in the opening chapter, strontium has several isotopes, one of which, ^{87}Sr, is produced from radioactive decay of ^{87}Rb. The other isotopes of strontium, including ^{86}Sr, are not radiogenic and do not change with time. Rubidium, like other alkalies, is a lithophile element and is strongly enriched in continental crust. With time, the radioactive isotope, ^{87}Rb, decays and the ratio of ^{87}Sr to nonradiogenic ^{86}Sr in the continental crust has increased (**Fig. 9-22**). As a result of this gradual evolution, most continental rocks now have ratios of the order of .713 to .725, whereas mantle-derived rocks have lower ratios, normally between .701 and .704. Hence, any contribution of continental crustal rocks to melts of the mantle is easily detected.

The isotopic ratio of a contaminated magma depends not only on the ratios of the initial magma and contaminant but also on the total concentrations of Sr, as well as any loss to fractionated crystals. The amount of Sr in a contaminated magma (CM) can be calculated from a simple mass balance

$$(M + C - F)(Sr)_{CM} = M(Sr)_M + C(Sr)_C - F(Sr)_F \tag{9-1}$$

where M and C arc the weight fractions of original magma and contaminating material, and F is the amount of any crystals removed by fractionation.

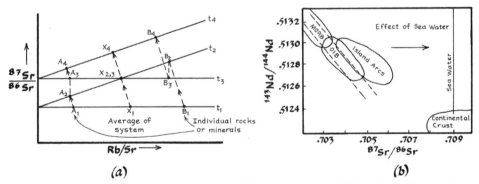

(a)   *(b)*

Figure 9-22 A plot of Sr isotopes against the ratio of Rb to Sr shows the individual effects of age and composition on isotopic ratios. If left to evolve, rocks or minerals of differing Rb/Sr ratios will increase in $^{87}Sr/^{86}Sr$ by amounts that are directly proportional to time and their Rb contents. At time t_2, they will define a line or "isochron" with a slope that is a direct function of age. If raised to high temperatures, the minerals in a rock may re-equilibrate and take on a more uniform ratio, even though their Rb/Sr still differ. After a rock is "reset" in this way at time t_3, ^{87}Rb continues to decay, and the Sr ratio will again evolve from its new baseline to give another isochron at time t_4 that is a measure of time elapsed since the last "setting of the clock." If the rock is then melted and the liquid differentiates to a series of compositions, only the Rb/Sr ratio is affected. More differentiated compositions will be richer in Rb and poorer in Sr, but their Sr ratios will be the same until more time elapses and more ^{87}Rb decays to ^{87}Sr. (b) The distribution of Nd and Sr isotopes in lavas of island arcs does not follow the "mantle array" defined by oceanic basalts but diverges toward larger Sr ratios. Although the trend of values extends in the direction of continental crust, the absence of such crust under island arcs indicates that seawater is a more plausible explanation.

Sr_{CM}, Sr_M, etc., are the Sr contents in parts per million of Sr in the contaminated magma and the various added or subtracted components. The isotopic ratio of the contaminated magma results from the combined weighted contributions of strontium from both the original magma and its contaminant in the proportions of their strontium contents and isotopic ratios. The equation expressing this relationship is

$$(M + C - F)\, Sr_{CM}\left[\frac{^{87}Sr}{^{86}Sr}\right]_{CM} = M Sr_M\left[\frac{^{87}Sr}{^{86}Sr}\right]_M + C Sr_C\left[\frac{^{87}Sr}{^{86}Sr}\right]_C - F Sr_F\left[\frac{^{87}Sr}{^{86}Sr}\right]_F \tag{9-2}$$

where $^{87}Sr/^{86}Sr$ is the isotopic ratio of each of the components CM, M, C, and F.

In the case of Paricutin, a mass-balance calculation indicated that the amount of assimilated crustal material required to explain the major elements of the andesites was about 10 percent. The calculated amount of concurrent crystallization was negligible, probably because the crystals remained suspended in the lava as phenocrysts. The basic lavas contain about 556 ppm Sr. Felsic rocks, on the other hand, have much less, the average for the xenoliths and basement rocks at Paricutin being about 227 ppm Sr. The amount of Sr in the andesite should then be

$$Sr_{CM} = (0.9)(556) + (.1)(227) = 523 \text{ ppm}$$

This is consistent with the measured Sr contents of the most differentiated andesites, which range between 511 and 556 ppm.

The isotopic ratios of the most basic and most differentiated lavas are 0.7037 and 0.7043, respectively. Although this difference seems small, it requires assimilation of crustal material with a much larger ratio:

$$(523)(0.7043) = .9(556)(0.7037) + .1(227)(^{87}Sr/^{86}Sr)_C$$
$$(^{87}Sr/^{86}Sr)_C = 0.7144$$

Because the crustal rocks have much less Sr than the initial magma and contribute a proportionately small part to the andesite, they must have much larger isotopic ratios to have any significant effect.

This example illustrates the factors that must be taken into account if Sr isotopes are to be used as a test of crustal assimilation. Because the Sr contents of mafic magmas are so much greater than those of felsic rocks, assimilation will be apparent only if the crustal rocks have large isotopic ratios or contribute very large proportions of Sr to the contaminated magma. These limitations explain why isotopic evidence is so often ambiguous. In the Cascade Range, for example, the Sr ratios of andesites vary as much in a single volcano as they do throughout the chain as a whole, and no systematic difference is observed between the Sr in volcanoes standing on thick continental crust and those in regions where such crust is thin or totally absent. On average, however, the Sr ratios of the andesites of continental margins are statistically greater than those of island arcs, and the difference seems to increase with time. In the Andes, for example, it is found that the strontium has become increasingly radiogenic over a period of nearly 11 million years (Fig. 9-6b), suggesting that the magmas are reacting with progressively shallower levels of the continental lithosphere. Alternatively, it could mean that the isotopic character of subducted material is changing or greater amounts are being incorporated into the magmas.

■ The Role of Subducted Crust

Just as we see that "all the rivers run into the sea, yet the sea is not full," we also find that sediments have been eroded from the continents for billions of years, and yet the oceans are still deep and continents stand high above them. Continental material entering the oceans, either as solid erosional debris or as soluble products of weathering, must be returned in some way to the continents. This balance is maintained in two ways. As oceanic lithosphere converges on zones of subduction, part of its blanket of sediments is scraped from the underlying basalt and added to the leading edge of the continent. At the same time, another part is carried down with the descending oceanic plate to be incorporated into the magmas that add volcanic and plutonic rocks to the continental crust. The sediments following these two routes have distinctly different characters. Those accreted to the continental margins are mainly clastic debris eroded from the adjacent continent or island arc, whereas those that are more likely to be subducted are chiefly pelagic sediments laid down in the interior of the ocean basins.

An obvious place to test whether any of these sediments are recycled in calc-alkaline magmas is in an island arc with no continental rocks directly under the volcanoes. If the character of sediments entering the trench differs along the axis of the arc, it should then be possible to see corresponding variations in magmas of the adjacent chain of volcanoes. The arc where this test was first applied was the Lesser Antilles, which is entirely on oceanic crust but has a very asymmetrical trench. The southern part receives sediments from the continent of South America, while the northern part has only pelagic sediments from the floor of the Atlantic Ocean and volcanic debris from nearby islands. The volcanoes do not differ markedly from one end of the chain to another, but those nearer the continent tend to be richer in potassium and other lithophile components as would be expected if greater amounts of continental material were being subducted at that end of the chain.

Further evidence comes from the isotopic character of subduction-related magmas. The lavas of island arcs differ from oceanic basalts in that their strontium and neodymium isotope ratios deviate from the mantle array toward higher $^{87}Sr/^{86}Sr$ at nearly constant Nd ratios (Fig. 9-22b). This divergent trend has a very simple explanation. The Sr ratio of modern seawater, about 0.709, is substantially greater than that of modern mid-ocean-ridge basalts because radiogenic ^{87}Sr is carried into the oceans by rivers draining the continents, but seawater contains only trace amounts of neodymium. Hence, when shallow oceanic crust is hydrothermally altered by seawater, it takes in more radiogenic Sr while retaining its original Nd ratios. If this material is subducted and contributes to the magmas of island arcs, the magmas should have higher Sr ratios but unchanged Nd. This is exactly what is observed.

Several methods have been devised to compare the rates at which individual components of the sediments are subducted with those at which these same components are returned in subduction-related magmas. Knowing the rate of convergence of a plate, the thickness and composition of the sediments, and the proportions of the sediments that are subducted, one can calculate the flux of a given element into the mantle. Estimating how much of this sedimentary flux is incorporated into magmas is more difficult. One must first allow for the effects of differentiation and crustal assimilation by adjusting the average compositions of the most basic aphyric lavas to a reference level, usually taken as 6 weight percent MgO. The calculated abundance of the selected element in this adjusted composition is then multiplied by the rate of eruption of magma to obtain an amount per unit time that can be compared with the corresponding flux of the element into the adjacent trench.

In addition to potassium, barium has proved especially useful for such a comparison. It is relatively easy to estimate the amount of barium being subducted from analyses of drill cores through the blanket of marine sediments. To determine how much of the barium in magmas comes from this source, one must allow for the amount that comes from melting of the mantle. This can be done by comparing the abundance of barium in the lava to that of an element that serves as a measure of the amount of melting. We saw earlier (Fig. 7-23) that sodium is one such element that varies in a regular fashion with the amount of melting. Normalizing the amount of barium to the amount of sodium in a primitive basalt provides a measure of the abundance of Ba in an undifferentiated magma independent of the degree of melting.

When this is done (**Fig. 9-23**), the abundances of lithophile elements in primitive calc-alkaline basalts are found to have a remarkably linear correlation

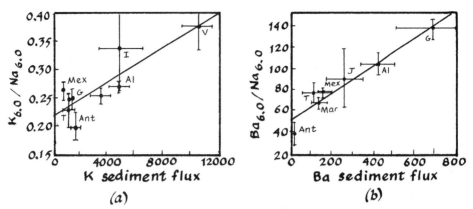

Figure 9-23 The flux rates of certain lithophile elements in subducted sediments have a remarkable correlation with the enrichment of these elements in subduction-related basalts. Sediment fluxes are in grams per year per centimeter of arc length, and the abundances of K and Ba are normalized to that of Na at an MgO content of 6 weight percent. Letters beside each data point refer to individual volcanic chains. (After Plank and Langmuir, 1993, *Nature* 362:739–743.)

with the flux of subducted sediments. The pattern is not always regular, however. Individual volcanoes may deviate from the averages for the chain as a whole, probably because the elements are not uniformly incorporated into the magma. When one makes a mass balance of the input and output of an individual element, only part of the total amounts that are subducted ends up in the magma. In the case of barium, for example, only about 20 percent is returned to the crust. By far the greater part must be going into the deep mantle. In the preceding chapter, we noted that the anomalies responsible for oceanic hotspots may originate from residues of subducted material that descended into the deep mantle.

Geochemical Tracers

Another radioactive isotope, beryllium-10, gives a quantitative measure of the rates at which subducted material is transferred through the system. [10]Be is produced in the atmosphere by the effect of cosmic rays on oxygen and nitrogen, and when absorbed in seawater, it adheres to particles of clay deposited on the seafloor. Having a half-life of only 1.5 million years, only a small fraction is left after the 10 million years or so required to reach depths where it can enter magma forming below a volcanic axis. Nevertheless, the small amount that remains must be returning to the surface because the amounts of [10]Be found in the lavas of subduction-related volcanoes are too great to have come from the mantle or older crustal rocks, where this isotope has long since been depleted by radioactive decay.

In practice, the absolute amounts of ^{10}Be in rocks (a few million atoms per gram) are too small to be measured accurately, but a mass spectrometer can measure the relative amounts of two beryllium isotopes with great precision. Thus, by measuring the ratio of ^{10}Be to the more abundant stable isotope, ^{9}Be, the proportions of the former can be accurately determined. Because the amount of ^{10}Be declines with time, the difference between this ratio in sediments entering a trench and in lava coming from nearby volcanoes gives us a measure of the time that has elapsed, but it tells us nothing about the amounts of sediment going through the system during the same time interval. For that, we need an element that does not decay with time. Boron is a good choice for this because it is relatively plentiful in sediments but very rare in the mantle. If its abundance is measured in both the sediments and lavas, the amount exchanged in a given time can be estimated by plotting the ratio B/Be against ^{10}Be/Be.

The data for individual volcanic arcs (**Fig. 9-24**) tend to fall on straight lines between two major components. Because mantle rocks have very little ^{10}Be or B, the main source for these components in the lavas must be subducted sediments. They cannot come from direct assimilation, however, because the bulk compositions of oceanic sediments do not fall on the trajectory of the observed variations in calc-alkaline rocks. A more likely explanation is

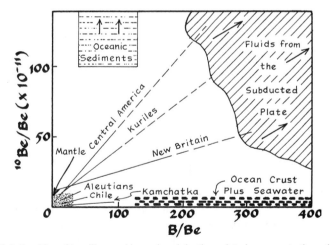

Figure 9-24 Relationships of beryllium and boron in subduction-related magmas to the ratios of the same components in subducted oceanic sediments, crust, and mantle. The measured values for lavas in individual chains of volcanoes define mixing lines between mantle compositions and those inferred for fluids distilled from subducted sediments. They cannot be the result of direct melting of oceanic crust or of sediments alone but must have a component that is the result of fractional distillation of boron and beryllium from the subducted slab. (The diagram is based on work by Louis Brown and others as summarized in *Yearbook 88* of the Carnegie Institution of Washington, 111–118).

that differing amounts of boron and beryllium are transferred into the zone of melting in fluids driven out of the subducted crust. The strong linear trends for individual arcs suggest that the proportions of the two elements are the result of differing degrees of fractional distillation, and the conditions governing the fractionation are essentially constant in a given subduction zone.

Further evidence of volatile transfer is found in the proportions of incompatible elements in calc-alkaline magmas. If these elements are distilled from subducted sediments, their abundances should reflect their relative solubilities in high-temperature hydrous solutions. The solubilities of lithophile elements at elevated temperatures and pressures differ widely according to their ionic potential (i.e., the ratio of charge to ionic radius). Small ions with large charges, such as Ti, Zr, U, Th, Nb, and Ta, commonly referred to as high-field-strength elements (HFSE), have much lower solubilities than large-ion lithophile elements (LILE) with smaller charges. The latter include elements like Cs, Rb, K, Ba, Sr, and Pb, all of which have relatively large ionic radii and small distribution coefficients with respect to a mantle peridotite. The rare-earth elements are intermediate between these two extremes.

If melting is triggered by an influx of hydrous fluids that carry these trace elements in solution, the ratio of LILE to HFSE of the melt should be elevated relative to that of other mantle-derived magmas, such as ocean ridge basalts. The ratio Ba/La in calc-alkaline rocks, for example, normally exceeds 15, whereas it averages only 4 in normal MORB. Similar contrasts are seen in ratios of other LILE and HFSE, such as Ba/Th or Rb/Nb. All of these geochemical characteristics are consistent with a volatile flux that contributes proportionately greater amounts of the incompatible elements that are most soluble in hydrous solutions. This has an important bearing on the question of how subduction-related magmas are generated.

■ Generation and Rise of Subduction-Related Magmas

Magmas produced at convergent plate boundaries pose an obvious paradox in that they come from regions in which cold, subducted lithosphere should have the effect of drastically depressing the temperature of the mantle. Of the various mechanisms for generating magmas outlined in Chapter 1, only two could be effective under such conditions. One derives heat from the mechanical energy of subduction; the other lowers melting temperatures by introducing subducted material into the mantle.

The strong seismicity concentrated in Benioff zones testifies to the large amounts of mechanical energy released along the upper part of the descending plate. If part of this energy were converted to frictional heating of the subducted lithosphere and adjacent mantle, it might raise the temperature enough to cause melting. At first glance, the distribution of earthquakes seems to support this interpretation. There is a relative deficiency of earthquakes directly beneath the main volcanic axis, and the amount of "missing" seismic energy is of the same order as that required to melt the quantities of magma reaching the surface. It has been suggested that mechanical energy could be converted to heat through a positive feedback mechanism in which heat produced by shear lowers the viscosity of the rocks which in turn channels the shear and heat production into a progressively narrower zone until it causes melting. Seismic studies confirm that the zone of fewer earthquakes is indeed one of decreased viscosity, but this is attributed to the effects of fluids coming from the downgoing slab. The amount of heat generated in rocks of such low viscosity is relatively minor, and it seems doubtful that shearing could cause much melting. If it could, we should see volcanoes along large strike-slip faults, such as the San Andreas fault in California.

The effects of fluids expelled from the subducted crust could be responsible for another, more likely form of melting. We saw ample evidence in the preceding section that volatile components driven out of sediments and hydrothermally altered basalts could play an important role in melting. The lithospheric plate descending to regions of higher pressures and temperatures beneath the volcanic arc must re-equilibrate with its new surroundings, and as it does so, pore water that has not already been expelled by compaction would be incorporated into hydrous minerals that, in turn, become unstable at increased temperatures and pressures. Depending on their individual stability limits, these hydrous minerals can persist to differing levels. Clays, chlorite, and serpentine break down at relatively low temperatures and reach only shallow depths (**Fig. 9-25**), but amphibole can survive somewhat longer, and the magnesium-rich mica, phlogopite, can reach the greatest depth of all. Thus, as the plate descends, a series of reactions can take place as each assemblage becomes unstable and is replaced by a new one of higher density. As the ultimate stability limits of hydrous minerals are reached, water must be driven out along with other lithophile components that are soluble in water at elevated temperatures and pressures. In this way, the descending slab would be progressively dehydrated and partly depleted of its mobile elements. If these components rise into hotter levels above the slab, they could lower the melting temperature of the mantle and generate magmas with no addition of heat. According to this hypothesis, the volcanic front stands above a level in the mantle wedge where the temperature is high enough for an influx of fluids to trigger melting. The depth and rate of melting will depend in large part on the rate at

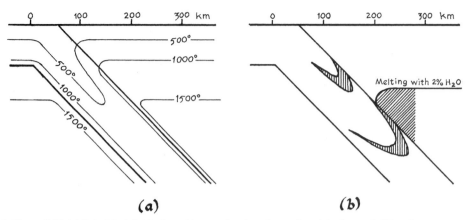

Figure 9-25 (a) Calculated distribution of temperature in a descending slab of oceanic lithosphere. (b) Stability limits of the principal hydrous minerals in a slab of subducted oceanic crust. The upper zone of vertical ruling indicates the region in which the temperature and pressure would cause serpentine, clay, and hornblende to break down; the lower zone indicates the stability limits of the Mg-rich mica, phlogopite. The upper zone lies approximately below the positions of coast ranges or nonvolcanic outer arcs, whereas the lower zone is below the inner uplifted blocks and volcanic fronts. A possible melting curve is shown for mantle rocks with about 2-percent water. The diagonally ruled area indicates the region in which breakdown of phlogopite might produce enough magma to mobilize diapirs under the volcanic front. At shallower depths where temperatures are lower, rising fluids would hydrate the mantle rocks and cause a volumetric expansion and elevation of the surface. (After Fyfe, W. S., and A. R. McBirney, 1975, *Amer. Jour. Sci.* 275-A:285–297.)

which volatile components are supplied by subduction and percolate into the hotter overlying rocks.

The regular spacing of large cones along the axis of a volcanic front can be related to the gravitational instability of light, partly molten mantle under a denser asthenosphere (**Fig. 9-26**). Evenly spaced mantle diapirs are thought to be produced along a long, narrow zone somewhere near the base of the mantle wedge. These are not masses of magma or even partly molten peridotite but mantle rocks that because of the influx of water-rich fluids have a relatively low density and viscosity. Experimental studies indicate that melting does not begin until these diapirs reach depths of about 60 kilometers and temperatures of 1000 degrees or more. Depending on the amount and composition of the fluid, the initial flux melting accounts for only about 10 to 30 percent of the final magma. The remainder comes from decompression melting as the partially melted mantle continues to rise.

Melting of this kind draws on the stored heat and low-melting components of the overlying mantle and would inevitably decline with time unless the source region is constantly renewed. In those regions where we know something about the rates of volcanism over extended periods, we see episodes of differing intensities but little evidence of a long-term decline. The

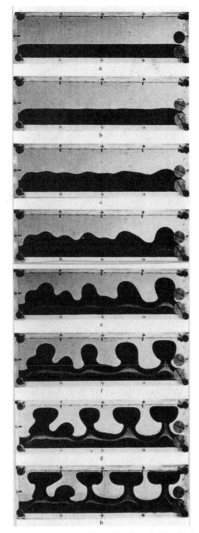

Figure 9-26 Diapirs of low-density mantle are believed to form with nearly equal spacing along a line directly below the volcanic front. This sequence of photographs by H. Ramberg shows a model in which a lighter liquid that has an initially even distribution (top) forms a wave-like surface and rises as separate plumes (bottom).

zone above the subducted plate seems to be an inexhaustible source of magma, and the best explanation is that a system of forced convection is bringing new material into the mantle wedge. It is postulated that mechanical coupling of the descending plate and adjacent mantle is dragging the lower part of the mantle with it and that this generates flow in the opposite sense at shallower depths (**Fig. 9-27**). In this way, the mantle wedge continues

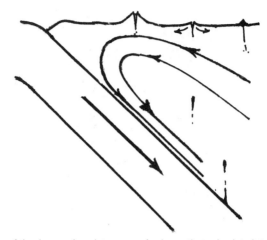

Figure 9-27 The drag of the down-going plate causes fresh mantle to circulate into the mantle wedge and in that way renew the heat and low-melting components of the mantle. The upper limb of this current tends to pull part of the overlying lithosphere toward the trench and can cause an extensional rift to form behind the main volcanic axis. In addition, small amounts of magma may reach the surface throughout the back-arc basin. The diagram is schematic and not to scale.

to provide the necessary heat and chemical components to sustain long-term magmatism.

This pattern of mantle flow could also explain the unusual nature of volcanism behind the main volcanic front. Although the onset of melting comes in abruptly along a sharp line close to the leading edge of the mantle wedge, it dies out less rapidly with greater distance inward. The amount and character of "back-arc volcanism" differs widely from place to place and has a highly diverse chemical character. In a few island arcs it is strong enough to produce small-scale spreading centers similar to oceanic ridges. Many of the basalts erupted from these rifts resemble those of normal ocean-ridge basalts and may have a similar origin.

It would certainly be a mistake to assume that all subduction-related magmas come from flux melting of the overlying mantle. This is shown most clearly in the earlier example of the Aleutian Arc, where we saw that the character of magmas changes systematically as the angle of convergence decreases along the length of the arc (Fig. 9-13). The volumes of erupted magma decrease, and the compositions of the most primitive melts are more felsic. The isotopic ratios become less radiogenic, and there is little if any correlation between factors like the Ba/La ratio and the flux of subducted sediments. These observations suggest that volatiles distilled from these sediments are not the only cause of melting. The melting process can also produce a magnesium-rich variety of andesite, sometimes referred to as *adakites*, after the island where they were first noted. Their REE have the distinctive heavy-depleted signature of melts produced from a garnet-rich source. The conclusion one must draw is that the subducted crust reaches levels where the basaltic layer changes to eclogite and eventually begins to melt. The Aleutians are the only province in which andesites of this unusual type make up a significant proportion of the lavas, but the gradual transition along the length of the arc suggests that the two types of melting—flux melting of mantle peridotites in the east and melting of basaltic crust converted to eclogite in the west—are two end-members of a spectrum.

The most important lesson in this may be that there is more than one way to produce an andesite. We instinctively associate lavas of this type with convergent plate boundaries, but large andesitic volcanoes are also found within the continental interior far beyond the possible reach of subducted oceanic crust. Mount Taylor in New Mexico and San Francisco Peak in Arizona are conspicuous examples. In the final chapter we consider these other tectonic settings in greater detail.

Selected References

Eiler, J., 2003, Inside the Subduction Factory, *Amer. Geoph. Union Monograph* 138:311. A compilation of papers on all aspects of subduction-related magmatism.

Gill, J., 1981, *Orogenic Andesites and Plate Tectonics*, Springer-Verlag, 390 p. A very detailed survey of orogenic rocks, including both tholeiitic and calc-alkaline types with strong emphasis on trace-element geochemistry.

Grove, T. M., and R. J. Kinzler, 1986, Petrogenesis of andesites, *Ann. Rev. Earth & Planet. Sci.* 14:417–454. A review of experimental work on differentiation of calc-alkaline and tholeiitic andesites.

Kelemen, P. B., K. Hanghoj, and A. R. Green, 2004, One view of the geochemistry of subduction-related magmatic arcs, with an emphasis on primitive andesite and lower crust. *Treatise on Geochemistry Vol. 3, The Crust*, R. L. Rudnick, ed., pp. 593–659.

Thorpe, R. S., ed. 1982, *Andesites*. John Wiley, 724 p. An especially broad and comprehensive compilation of papers dealing with the geology and petrology of andesitic volcanism.

10 Granitic Plutons and Siliceous Ignimbrites

The large, intrusive bodies found in deeply eroded orogenic terranes are often portrayed as the deep roots of the volcanic belts described in the preceding chapter. Most share many of the chemical characteristics of calc-alkaline volcanic rocks, particularly in regions of thick continental crust, but plutonic rocks of intermediate to high silica contents are relatively more abundant than in their volcanic counterparts. It is mainly this last feature that distinguishes these plutons from the large gabbroic intrusions discussed in Chapter 6.

We refer to such rocks loosely as "granites" or "granitoids," even though they usually encompass a broader range of compositions than granite in the strict compositional sense. The dual use of the term "granite" as both a general term for any leucocratic, coarse-grained rock and as a name for a specific composition can be confusing but is probably too deeply ingrained to be changed. In most instances, the intended meaning is evident from its context.

Many of the attributes of plutonic rocks, particularly their large proportions of felsic components, also characterize the great sheets of rhyolitic and dacitic pyroclastic rocks erupted as ignimbrites in interior regions of continents. In fact, many siliceous ignimbrites have been traced to sources in shallow intrusions of similar compositions. This is not to say that all plutons had a volcanic expression at the surface, nor that all large, eruptive sheets come from plutonic intrusions, but the two are so often closely related and gradational one into the other that one can scarcely doubt their common origin.

Several broad categories of felsic plutonic rocks can be distinguished on the basis of their chemical and mineralogical compositions, tectonic settings, and ages.

■ Petrographic and Mineralogical Classifications

Of the various types of plutonic rocks listed in Chapter 2, we are concerned here with only those in the silica-saturated upper part, shown in Figure 2-1. The series quartz diorite-tonalite-granodiorite-granite encompasses over 90 percent of the volume of most batholithic rocks, just as the equivalent volcanic rocks, dacite-rhyodacite-rhyolite, make up all but a minor fraction of the large ignimbrite sheets.

Mineral Assemblages and Textures

Felsic plutonic rocks owe the large sizes of their crystals to slow cooling and the effects of volatiles, mainly water, that facilitate growth. Their microscopic textures (**Fig. 10-1**) differ in several ways from those of more mafic rocks in layered intrusions. Very few granitic plutons have fine-grained chilled margins. Layering is much less common and limited almost entirely to thin, wispy streaks of dark minerals near the margins. Apart from mafic xenoliths, large modal contrasts are rare.

Although less diverse than their volcanic counterparts, plutonic rocks have a variety of textures reflecting differing conditions of crystallization. Porphyritic textures, which are rare in gabbros and peridotites, are quite common in granites. Subhedral megacrysts of potassium feldspar 5 or 10 centimeters in length are not uncommon. Hypidiomorphic-granular textures (Figs. 4-24a and 10-1b, 10-1c, 10-1d, and 10-1e) record the mutual spatial accommodations of crystals in their most stable forms as they interfere with one another during growth. How much of their form is due to recrystallization at subsolidus conditions is difficult to say, but it is doubtful whether

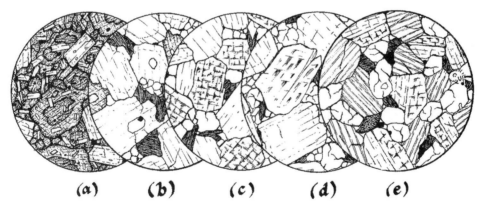

(a)   (b)   (c)   (d)   (e)

Figure 10-1 Textures of some common plutonic rocks. (a) Pyroxene hornblende gabbro from the Peninsula Batholith of Baja California. Hornblende appears to have replaced much of the earlier pyroxene. (b) Tonalite with hornblende, biotite, andesine, quartz, and minor alkali feldspar, Italian Alps. (c) Granodiorite from northern Nicaragua with quartz, oligoclase, and microcline. Biotite and hornblende are the main mafic minerals. (d) Subsolvus granite from the Precambrian shield of South Africa. About equal parts of quartz, perthitic orthoclase, oligoclase zoned to albite, biotite, and accessory apatite. (e) Hypersolvus granite from the Isle of Skye, Scotland. A single feldspar has exsolved into two sub-equal parts of K-feldspar and sodic plagioclase. The most abundant dark mineral is the sodic amphibole, riebeckite.

many of the textures of slowly cooled plutonic rocks are those of crystals grown directly from a liquid.

The granophyric textures so common in siliceous rocks of differentiated basic intrusions are rare in granitic plutons. Instead, coarser intergrowths of quartz and alkali feldspar may give the rock a *graphic* texture in which the crystallographic planes of the crystals have controlled the orientations of co-precipitated quartz and feldspar. Equally common are *myrmekitic* textures in which worm-like blebs of quartz or plagioclase have formed within larger crystals, usually orthoclase. All three of these textures—granophyric, graphic, and myrmekitic—result from crystallization of a eutectic composition consisting of quartz and alkali feldspars.

Feldspars and quartz account for well over half of the mode of felsic plutonic rocks. Pyroxenes and other anhydrous minerals are less common than amphibole and biotite. They are confined chiefly to rocks that crystallized at low water pressures. Although most common in shallow intrusions, pyroxenes are also typical of dry, high-pressure conditions. Muscovite rarely crystallizes from granitic melts at water pressures less than about 5 kilobars. It is found only in rocks with normative corundum, that is, with an excess of alumina over the amount needed for feldspars.

Subsolvus and Hypersolvus Granites

The character of the alkali feldspars is one of the most useful and widely applicable guides to the conditions of crystallization. We noted in Chapter 4 that at low water pressures the system albite-orthoclase has a temperature minimum between the two solid-solution series, and at subsolidus temperatures, a solvus limits the amount of solid solution between the two end-members (Fig. 4-6). It will be recalled that crystallization at low water pressures produces a single feldspar of intermediate composition, which, on cooling to the two-feldspar solvus, may exsolve into two forms, one rich in albite and the other in potassium feldspar. The result is a patchy intergrowth of two feldspars in a single grain. At increased water pressures, the solidus curves are lowered until they intersect the solvus, and instead of precipitating a single feldspar that exsolves only after an interval of subsolidus cooling, separate sodic and potassic feldspars crystallize directly from the melt and quickly start to exsolve their counterpart to form perthite or antiperthite, respectively.

These relationships can be used to divide granitic rocks into two broad groups. *Subsolvus granites* crystallized at elevated water pressures have two separate alkali feldspars (Fig. 10-1d); *hypersolvus granites* crystallized at a lower water pressure have a single feldspar (Fig. 10-1e). In either case, the feldspars may show exsolution.

A distinctive type of hypersolvus granites with *rapakivi* textures is distinguished by large, rounded phenocrysts of alkali feldspar overgrown with plagioclase, normally oligoclase or andesine. Most have two generations of alkali feldspar and quartz. These unusual rocks are largely confined to Precambrian shield areas. They are thought to owe their textures to low-temperature recrystallization under shallow, possibly subvolcanic, conditions.

Sanidine and other high-temperature alkali feldspars are confined to the very shallowest intrusions and to contact zones that have been heated to high temperatures by basic magma. The low-temperature potassium feldspar, microcline, is found in granites that have recrystallized under metamorphic conditions. It is not stable at magmatic temperatures.

Accessory Minerals

Other minerals, such as tourmaline, zircon, topaz, or allanite, are common accessories and may be particularly distinctive of the granites and rhyolites of certain regions where the magmas contain anomalously large amounts of boron, zirconium, fluorine, or cerium. Carbonates, sulfides, and oxides of iron, copper, zinc, lead, tin, molybdenum, and other metals may either be disseminated or concentrated in veins or pods in and around an intrusion. Some of these components are introduced along with water, carbon dioxide, sulfur, and chlorine from surrounding rocks by hydrothermal solutions circulating through the plutons after they have solidified; others are inherent in the original magma and concentrated by late-stage processes of differentiation.

Pegmatites

Special mention must be made of the extraordinary group of very coarse-grained rocks, pegmatites, that form dikes and pod-like segregations in many granitic plutons and metamorphic rocks. Although they differ widely in form and composition, most pegmatites form at late stages, probably after the main mass of magma has crystallized and a hydrous fluid has formed a separate phase. Individual crystals can reach enormous sizes, with single feldspars as large as 2 by 3.5 by 9 meters and micas 4 meters thick and 10 meters across.

Although most pegmatites are simply very coarse-grained rocks with compositions similar to those of normal granites, certain rare types are extraordinarily enriched in the elements concentrated in late-crystallizing granitic liquids. The high solubility of silica, alkalies, and other elements in this water-rich fluid enhances transfer and concentration of residual components, including rare and economically important elements, such as lithium, cesium, beryllium, tin, niobium, zirconium, uranium, thorium, boron, phosphorus, and fluorine. Many rare minerals, including beryl

($Be_3Al_2Si_6O_{18}$), the lithium pyroxene spodumene ($LiAlSi_2O_6$), and certain gemstones, such as aquamarine, are found mainly, if not exclusively, in pegmatites.

Siliceous Volcanic Rocks

At the opposite extreme of grain size from pegmatites are the glassy, volcanic counterparts of granitic rocks. All but a small fraction of these are laid down either as pumiceous pyroclastic falls or flows. Viscous crystal-rich domes of dacite or rhyolite may be extruded in relatively small volumes, but voluminous lavas of these compositions are all but unknown.

Although sometimes referred to as "ash-flow tuffs," ignimbrites are composed of fragmental debris with a much greater size range than ash, and they do not necessarily form indurated rocks as the word "tuff" implies. They are laid down by hot pyroclastic flows, typically of large volume, that move across the surface with great mobility and speed. Most are thought to be erupted from fissures along the boundary faults of large calderas, but, no eruptions on this scale have been observed in historic times, and much of what is written about them is necessarily conjectural. Although infrequent on a historical scale, they were common in the recent geological past, especially during the middle Tertiary, when vast areas of the western United States, Mexico, Honduras, and the western slopes of the Andes were blanketed with sheets covering thousands of square kilometers with thicknesses of hundreds of meters.

One of the closest historic analogs, the 1912 eruption from Novarupta near the base of Mount Katmai, Alaska, flooded the Valley of Ten Thousand Smokes with about 7 km^3 of rhyolitic pumice. Although large by historic standards, the volume is small beside that of many prehistoric eruptions, in which individual sheets exceed a thousand cubic kilometers. The cumulative volumes discharged from the Taupo calderas of New Zealand and from the San Juan calderas of Colorado exceed 10,000 km^3. The total volumes of these vast sheets are comparable to those of flood basalts.

Most large-scale pyroclastic rocks are rhyolites, quartz latites, or dacites; only a few are andesites, and hardly any are basalts. Phenocryst contents range from nil to nearly half the volume of the rock. The matrix is made up of myriads of glass shards formed by disruption of highly vesiculated pumice. Depending on temperature, thickness, and volatile contents, these fragments of glass may be loose and porous or strongly welded and compact (**Fig. 10-2**). Under normal conditions, temperatures on the order of 500°C to 600°C are required for welding, but in the presence of abundant water vapor, less heat and pressure is required to weld the pumice fragments into layers or lenses of dense black glass.

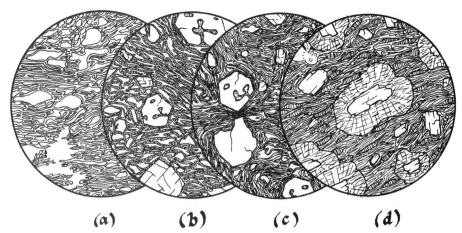

(a)   (b)   (c)   (d)

Figure 10-2 The textures of a series of specimens from a single ignimbrite sheet in central Honduras show progressive compaction with depth. (a) is from near the top, while (b) and (c) are from about 5 and 2 meters above the base, respectively. In (c), the shards have been partly remelted under the effects of high temperature and trapped water vapor to form lenticular "fiamme" of dense black glass. In (d), devitrification has produced radiating spherulites that have grown outward from centers around phenocrysts and across the original shards of glass.

Many ignimbrites, perhaps the majority, are compositionally zoned with a basal part more differentiated than the overlying section. These compositional variations appear to be inherited from stratified magma chambers tapped from the top downward in essentially uninterrupted outpourings.

■ Chemical and Tectonic Classifications

A confusing number of classification schemes have been proposed for granites. Some are based on tectonic settings or timing with respect to orogenic events. Others are based on the inferred source rocks or the mechanism of magma generation, and still others are a combination of two or more of these elements. For our present purposes, it will suffice to consider only the most generally accepted classifications.

Chemical Classifications

It can be seen from Tables 2-1 and 2-2 that the chemical compositions of the common felsic plutonic rocks, quartz diorite through granite, are close to those of calc-alkaline volcanic rocks. Plutonic rocks of orogenic batholiths and continental interiors differ from the siliceous differentiates of other environments primarily in their greater contents of potassium and other large-ion lithophile components. Those with enough potassium to have subequal amounts of plagioclase and potassium feldspar can be assigned to a high-potassium series (Fig. 9-9) of which latites and quartz latites are the most common volcanic members, and monzonite and syenite are their plutonic equivalents.

In this respect, felsic continental rocks differ from the plagiogranites of oceanic environments and ophiolites. The latter lack the potassium feldspar that is so conspicuous in their calc-alkaline equivalents. Usually referred to broadly as *trondhjemites*, they are also common in ancient continental shields, island arcs, and other regions of thin crust. The distinction between trondhjemitic and calc-alkaline series is shown most clearly in their contrasting trends of enrichment of alkalies (**Fig. 10-3a**).

Although found in both continental and oceanic settings, trondhjemitic rocks of these two environments have subtle differences. Continental trondhjemites normally have in excess of 14.5- to 15-percent Al_2O_3 and, in many cases, normative corundum, C (Al_2O_3), whereas equivalent oceanic rocks have less than that amount of Al_2O_3 and are diopside normative. Another difference is in their proportions of rare-earth elements (REE). Continental varieties are relatively enriched in the light rare earths, whereas oceanic rocks have much flatter patterns (**Fig. 10-3b**).

Most granitic rocks can be assigned either to an "S-type" derived from sedimentary or metasedimentary source rocks or to an "I-type" produced by

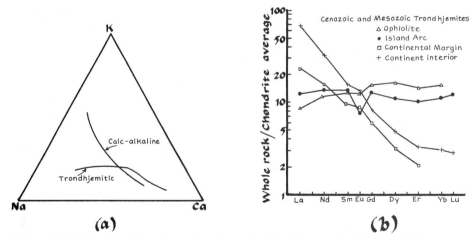

Figure 10-3 (a) Trondhjemitic series are best characterized by their normative feldspars, which become more albitic with differentiation than those of calc-alkaline granitic series. (After F. Barker and J. G. Arth, 1976, *Geology* 4:596–600.) (b) Continental trondhjemites have REE abundances that, when normalized to chondritic meteorites, are relatively enriched in light REE. Oceanic rocks of similar major-element compositions, such as plagiogranites of ophiolite complexes and island arcs, have flatter curves. (After J. G. Arth, 1979, in *Trondhjemites, dacites, and related rocks*, F. Barker, ed., Elsevier, p. 123–132.)

melting rocks that are dominantly igneous. The most conspicuous compositional differences are in their alumina contents and concentrations of certain trace elements inherited from their sources. Like continental trondhjemites, granites produced by melting of sedimentary rocks have more alumina than can be accommodated in feldspars. The excess is expressed as normative corundum. The I-type has normative diopside. The former is also richer in transition metals and in several lithophile elements, such as Ba, Rb, Th, and light REE.

Tectonic Classifications

Granites can be grouped into two broad categories, *orogenic* and *anorogenic*, based on whether or not the rocks are associated with a major mountain-building event. The former has at least two subtypes, *syntectonic* and *pos-orogenic*, but a *late-orogenic* type has also been proposed.

Syntectonic granites are uncommon, especially in orogenic systems undergoing strong deformation. Most appear only when compressional forces are weak, but a few have been generated by thrusting of hot, deep-seated metamorphic rocks over cooler, water-rich horizons. They have strong fabrics produced by deformation in the plastic state. Late-orogenic granites include most of the large batholithic belts, such as the cordilleran system of North and South America. They resemble calc-alkaline volcanic rocks with which they undoubtedly share a common origin. Postorogenic granites are said to result from relief of pressure on crustal rocks that have been deeply buried during a preceding orogenic event. They are very heterogeneous, as would be expected if they come from a combination of older crustal rocks and subduction-related igneous material produced during earlier stages of plate convergence.

Anorogenic plutonic intrusions have an even wider diversity of rock types, including anorthosites, rapakivi granites, and both alkaline and peralkaline rocks. We shall reserve this group for special attention in the chapter that follows. **Table 10-1** compiles the salient features of the felsic plutonic and volcanic rocks discussed in this chapter.

■ Crustal Environment and Internal Structure

Siliceous magmas are intruded at a wide range of crustal levels. Their forms and cooling histories are determined mainly by the physical and chemical contrasts between the magmas and the environment into which they are emplaced. Shallow bodies have textures and contact relationships reflecting a strong temperature contrast with their surroundings; deep-seated granites have more gradational boundaries, especially where their host rocks are regionally metamorphosed crustal rocks of felsic compositions. These different settings or

Table 10-1						
	1. S-type Granite	2. I-type Granite	3. Archean Trondhjemite	4. Massif Anorthosite	Bishop Tuff 5. Early	6. Late
SiO_2	73.48	64.89	68.38	53.29	77.4	75.5
TiO_2	0.22	0.48	0.23	0.34	0.07	0.21
Al_2O_3	13.41	14.85	16.39	26.42	12.3	13.0
FeO*	1.87	5.24	1.97	2.44	0.7	1.1
MnO	0.04	0.10	0.03	0.03	0.03	0.02
MgO	0.71	2.73	1.00	1.09	0.01	0.25
CaO	1.70	5.51	1.75	10.26	0.45	0.95
Na_2O	2.63	1.83	7.09	4.32	3.9	3.35
K_2O	4.63	2.55	0.78	0.91	4.8	5.55
P_2O_5	0.08	0.09	0.08	0.10	0.01	0.06
Trace Elements (ppm)						
Rb	336	126	11	–	190	95
Zr	100	109	127	–	85	140
Sr	97	127	554	–	5	110
Ba	350	355	609	–	<10	465
Th	19	14	–	–	21	13
U	6	2	–	–	7	3
Nb	10	8	5	–	>25	<5
Y	33	11	–	–	25	12
La	23	23	–	–	19	61
Ce	49	49	–	–	45	98
Molecular Norms						
Ap	0.17	0.20	0.17	0.21	0.02	0.13
IL	0.32	0.69	0.32	0.47	0.10	0.30
Mt	0.30	0.86	0.31	0.38	0.11	0.18
Or	28.13	15.64	4.62	5.33	28.68	32.83
Ab	24.29	17.06	63.82	38.44	35.42	30.39
An	8.14	25.72	8.18	49.56	1.90	4.23
Di	–	1.64	–	0.19	0.23	0.11
Hy	4.10	13.06	4.88	5.39	0.73	1.72
C	1.32	–	0.98	v	–	–
Q	33.24	25.14	16.72	0.04	32.81	30.12

1. and 2. S- and I-type granites (B. W. Chappel, 1984, *Phil. Trans. Roy. Soc. London* A310:693–707).
3. Trondhjemite (J. Tarney, B. Weaver, and S. A. Drury, 1979, in *Trondhjemites, dacites, and related rocks*, F. Barker, ed., Amsterdam: Elsevier, p. 287).
4. Massif-type anorthosite (B. F. Windley, 1973, *Geol. Soc. South Africa, Spec. Pub.* 3:319–332).
5. and 6. Early- and late-stage rhyolites of the Bishop Tuff (W. Hildreth, 1979, *Geol. Soc. Amer. Spec. Paper* 180:43–75).

levels of emplacement can be divided into three broad zones. The deepest, the *catazone*, is exposed only in deeply eroded terranes of high-grade metamorphic rocks (**Fig. 10-4a**). They have margins with a concordant foliation that merges into surrounding gneisses with similar mineral assemblages. Some must have formed at or near their present level under *ultrametamorphic* conditions, that is, close to the solidus temperatures of siliceous crustal rocks. Others appear to have been intruded into a ductile crust with temperatures and physical properties approaching those of the magma.

In the *mesozone*, magmatic and metamorphic rocks are more easily distinguished, both in texture and mineralogy (**Fig. 10-4b**). Contacts may be concordant, as in the catazone, but most are sharply cross-cutting, often with abundant angular xenoliths in the margins. The plutons tend to have less internal structure, but many are compositionally zoned, both vertically and concentrically.

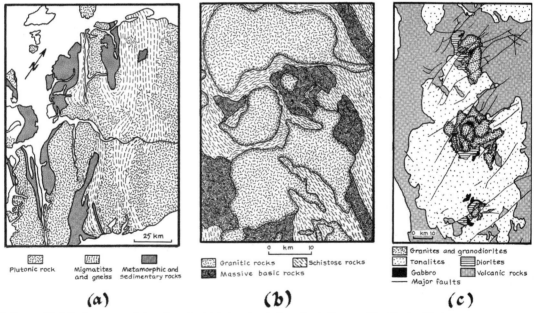

Figure 10-4 Environments of plutonic rocks. (a) A deep level of erosion is illustrated by this simplified map of the Mesozoic batholith of British Columbia. Note the concordance between the "tad pole" shapes of the granites and their surrounding metamorphic rocks. (After W. W. Hutchison, 1970, *Can. Jour. Earth Sci.* 7:376–405.) (b) Plutons exposed at intermediate depths are exemplified by this simplified map of part of the Sierra Nevada Batholith of California. Note the cross-cutting relations and more equidimensional forms of the intrusions. (After Bidwell Bar Quadrangle, *U. S. Geol. Surv.*) (c) Typical structures at the shallowest levels of intrusion are shown by this example from the coastal batholith of Peru. Note the fault-controlled structures and ring-like intrusions in close association with volcanic rocks. (Compiled from various maps, including those of J. S. Myers, 1975, *Bull. Geol. Soc. Amer.* 86:1209–1220, and M. A. Bussell, W. S. Pitcher, and P. A. Wilson, 1976, *Can. Jour. Earth Sci.* 13:1020–1030.)

Shallowest of all are the plutons that have risen to the *epizone* within a few kilometers of the surface (**Fig. 10-4c**). Typical of these are the stocks and ring-complexes emplaced in brittle sedimentary or volcanic country rocks. Many are associated with contemporary volcanism. They have narrow zones of contact metamorphism and may be hosts of economic mineral deposits associated with extensive alteration by hydrothermal solutions that circulated through the intrusions and their wall rocks. The contacts are sharply discordant. Dikes and sills penetrate the wall rocks, dislodging blocks of all sizes. The margins are often crowded with xenoliths, some of which are wallrocks and others the mafic residue of melting, usually referred to as *restites*.

The depths assigned to these three zones are imprecise. They depend less on pressure than on the thermal gradient and physical contrast between the magma and the environment into which it is intruded. Shallow epizonal bodies are thermally and rheologically disharmonious with their surroundings, whereas the contrast between catazonal plutons and the rocks they intrude may be detectable only in gradational differences of bulk composition, mineralogy, and texture.

We should add to this series a fourth category, the volcanic zone, in which intrusions breach the surface with voluminous pyroclastic eruptions. Bodies of this kind are exposed in eroded calderas and ring-dike complexes. In some instances, one can even see the coarse-grained plutonic rocks grade into dikes that were the feeders of huge volcanic eruptions.

The long-held view of plutons as steep-sided bodies emplaced by large, slowly rising diapirs has been largely discredited. Diapirs cannot reach shallow levels because they lose too much heat during their slow ascent from their source. They are possible only in the deep, ductile crust of the catazone and possibly part of the mesozone. Shallower intrusions are more likely to be sheet-like or funnel-shaped bodies fed by repeated injections from dikes. The magma does not force aside the surrounding rocks but rises passively through fissures formed by extensional stresses in the crust, and because it rises rapidly, it loses less heat and has less reaction with the country rocks than would a diapir. Magma intruded into thick sills requires no lateral displacement of the wall rocks; it simply elevates the roof rocks. The roof may be wedged upward by planar injections or elevated along steep, pre-existing fractures. In some cases, the floor is depressed by displacement into space left by the transfer of magma for a deeper level.

The rates at which granitic magmas are generated are comparatively slow relative to those at which they can rise and intrude the crust. For this reason, most plutons do not grow by a single surge of magma but by repeated injections. Studies in the Sierra Nevada Range have shown that emplacement of a typical tabular intrusion required times ranging from a few months for a body less than a kilometer across to more than a million years for one measuring 100 kilometers. Most are thought to be emplaced in less than 1000 years.

Although they are rarely layered, most granitic plutons have some sort of compositional zoning. Several types are illustrated in **Figure 10-5**. Sharply bounded zones may represent successive intrusions (Fig. 10-5a) of progressively more differentiated magma, each more or less in the core of a less differentiated predecessor. In other cases (Fig. 10-5b), new magma has intruded along steep ring dikes at the same time that a central block of the roof subsides. The compositional relations in these bodies tend to be more discordant. At the other end of the spectrum, the internal boundaries can be so

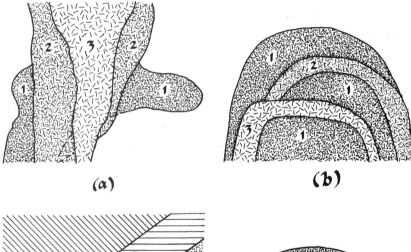

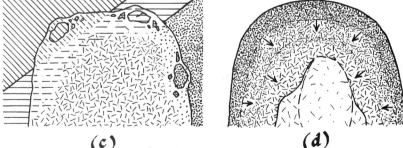

Figure 10-5 Zoning in granitic plutons may result from (a) successive cross-cutting intrusions, (b) subsidence of earlier rocks and injection of "bell-jar" intrusions or ring dikes, (c) contamination by different wall rocks, or (d) inward crystallization and differentiation. In the last case, the late-stage liquids may become lighter than the overlying rocks and rise into them.

diffuse and gradational that they are difficult to recognize in the field. Compositional differences in the wall rocks may be reflected in subtle differences in the character of the adjacent plutonic rocks (Fig. 10-5c). A body of magma crystallizing inward from its walls (Fig. 10-5d) tends to develop concentric zones, and so long as the magma is stable, the changing composition is seen mainly in modal and mineralogical changes with few if any marked discontinuities. More often than not, however, the orderly sequence of inward crystallization is disrupted by later intrusions.

Rates of solidification can vary within wide limits. The greatest effect comes from an exsolution of volatiles, which greatly accelerates crystallization by raising the liquidus temperatures of the minerals. Rates of conductive cooling differ according to the temperature contrast and the properties of the mineral assemblage in the walls. Just as the latent heat of crystallization tends to retard the rate of cooling during crystallization, endothermic reactions in contact-metamorphic zones have a similar effect; they buffer temperatures and retard the rise of temperatures in the wall rocks. The thermal gradient that prevailed around a solidified pluton can sometimes be determined from the isotherms recorded in the metamorphic zoning of the rocks.

Internal convection of the magma and external convection of a hydrothermal system both increase the rate of heat transfer but the latter has the greatest effect. Both forms of convective heat transfer are limited by the much slower rate at which heat is conducted through the contact zone separating the two systems. It is the rate of conductive heat transfer through the margins that governs the rate of convective heat transfer, not the reverse.

■ Tectonic Settings

Large bodies of felsic rocks, either plutonic or volcanic, are confined mainly to the continents. The most notable, if not the only, exception is the series of Tertiary rhyolites in Iceland. Although small by the standards of continental ignimbrites, the Icelandic bodies are considerably larger than the small felsic lavas and intrusions on other oceanic islands. For many years, this was taken as clear evidence that Iceland was underlain by continental crust, but a more likely explanation, as we shall see shortly, is that the rhyolites owe their origin to melting of an unusual type of oceanic lithosphere under the Icelandic Plateau.

Many of the older continental plutonic rocks, such as those of the Precambrian greenstone belts and early Paleozoic metamorphic terranes, are scattered over wide areas of the continental interior. Younger plutons, especially those of the Circum-Pacific region, are concentrated mainly in broad composite belts near the continental margins, but a few are associated with rifting or intraplate hot spots within the continental interior.

Convergent Continental Margins

The most extensive plutonic rocks on the American continents are those of the Cordilleran System close to the Pacific continental margins. The long episode of Circum-Pacific plutonism lasting through most of the Mesozoic and Cenozoic Eras was one of the great magmatic events of the Earth's history. Batholiths rose into rocks of all ages and lithologic character, with little regard for major differences in the local structure or degree of metamorphism. The structural differences seen in individual segments of the Cordilleran batholiths are not due to their crustal environment so much as their depths of erosion. The coastal batholith of British Columbia, for example, is set in high-grade metamorphic rocks of the catazone (Fig. 10-4a), whereas the batholiths of the western Sierra Nevada intruded shallower rocks, mostly within the greenschist facies of metamorphism (Fig. 10-4b), and the coastal batholith of Peru is exposed at an even shallower level of brittle, weakly metamorphosed, sedimentary, and volcanic rocks (Fig. 10-4c).

Most of this enormous mass of granitic rocks was intruded in pulses between 225 and 18 million years ago. Few of these episodes coincided with major orogenic events, but many occurred during subsequent periods of uplift when compressive forces were relaxed. In some instances, volcanism preceded plutonism; in others it followed, and in some, the two were essentially contemporaneous. No simple reason has been found to explain why some periods are dominated by plutonism and others by volcanism.

The transition from plutonism to volcanism is shown very clearly in the coastal batholith of Peru. A belt of plutons, up to 65 kilometers wide and over 2400 kilometers long, was intruded as more than 800 separate plutons over a period of about 70 or 80 million years (110 to 30 million years ago). Swarms of intrusions ranging in size from centimeter-wide dikes to plutons tens of kilometers across came in at intervals of 10 or 20 million years. Each intrusive pulse consists of a spatially restricted group of consanguinous magmas that seems to stem from a common parent. Starting with gabbro, they tend to become more felsic with time, and the average compositions of successive batches of magma have proportionately smaller volumes of gabbro and relatively more of granite. They show the same temporal trend as the volcanic rocks of the central Andes (Fig. 9-6a).

No continuity is seen from one segment of the belt to another. Instead, magmas of differing compositions were intruded simultaneously in widely separated parts of the system. The impression gained from detailed radiometric dating and cross-cutting field relations in a given swarm of intrusions is that batches of magma formed at intervals of 18 million years or so. Individual intrusions took as long as seven million years to crystallize at depths of 5 or 10 kilometers but much less in shallow subvolcanic bodies.

Taken as a whole, the dominant pattern through the period of plutonism is one of diminishing volumes and lateral extent of the intrusions with time. In this respect, the evolution of the plutonic belt resembles that of the volcanic belt of the Cascade system where widespread voluminous volcanism of the early Tertiary episodes became more restricted in distribution and less intense with time until it formed the narrow rectilinear belt of the modern High Cascades.

Within an individual sequence of plutonic intrusions, early gabbroic bodies are tabular or lens-shaped, whereas later, more felsic bodies are narrower and more steep sided (**Fig. 10-6**). Dikes can be traced from the plutons into the walls and roof where they wedged apart blocks that then sank and helped make room for the rising magma. In places, large blocks are seen to have settled through the magma to accumulate on the floor of the intrusion. Despite the abundance of various types of xenoliths, they seem to have had little effect on the bulk composition of the magma.

The shallowest plutons were emplaced along ring fractures around a central, down-faulted block of the roof. They are thought to have been expressed

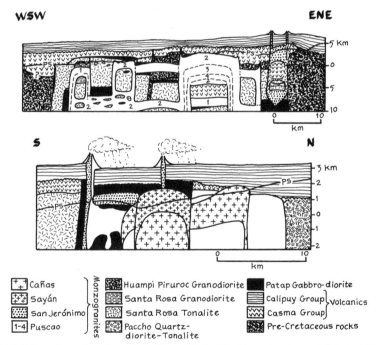

Figure 10-6 Schematic sections through the Coastal Batholith of Peru showing postulated relationship between intrusive and extrusive rocks. The present erosion surface is indicated by the lines labeled PS. (After J. S. Myers, 1975, *Bull. Geol. Soc. Amer.* 86:1209–1220, and M. A. Bussell, W. S. Pitcher, and P. A. Wilson, 1976, *Can. Jour. Earth Sci.* 13:1020–1030.)

at the surface by large calderas. Great volumes of dacitic and rhyolitic ig-nimbrites poured from fissures in and around these calderas, and in some instances, plutons continued to rise and intruded their own volcanic ejecta.

The magmas are mainly of the I-type derived from igneous sources with only a subordinate contribution from sedimentary sources. Their composi-tions range from gabbro to leucogranite. Except in the shallowest subvol-canic stocks, the felsic members are subsolvus types with two discrete alkali feldspars.

The gabbroic rocks initially crystallized plagioclase and pyroxene, but most were later amphibolitized by influxes of volatiles, partly from later intrusions but mainly from the surrounding rocks. The original igneous crys-tals are replaced by new assemblages with metamorphic textures (Fig. 10-1a). Subsequent intrusions consistently follow an order of increasing silica con-tent starting with quartz diorite and then evolving through tonalite and granodiorite to monzonite-granite and finally rocks intermediate between syenite and granite. These successive magmas tend to be nested in zoned bodies that continued to differentiate as they crystallized upward from the floor and inward from the walls. Crystal sizes decrease upward, possibly because of a loss of volatiles from the upper levels of the magma.

The compositional trends have much in common with those of orogenic volcanic rocks, such as the suites from continental margins discussed in Chapter 9. The early gabbroic rocks have a distinct trend (**Fig. 10-7**) not unlike the early tholeiitic rocks of volcanic series. They seem to have evolved by fractionation of pyroxene, plagioclase, and olivine. The later granitic rocks are distinctly calc-alkaline, with hornblende becoming more important than pyroxene, just as it does in comparable volcanic rocks.

As in most subduction-related magmatic systems, the felsic plutonic rocks became increasingly important with time. They are especially promi-nent in extensional settings with high surface heat flow. The Taupo volcanic zone of New Zealand, for example, was initially the focus of andesitic vol-canism that produced large composite volcanos that continued to grow over a period of about 300,000 years but then gave way to increasing amounts of dacite and rhyolite that culminated in enormous ignimbrite eruptions from the Taupo caldera. The geochemical properties of these later magmas show a strong gentic affinity to the older andesites, and it is thought that they are in large part the result of recycling of the latter by a rising zome of crustal anatexis. The influx of heat during the period of andesitic volcanism was augmented by increased heat flow that accompanied crustal extension and possibly upwelling of the asthenosphere. The large volumes of siliceous magma produced in this way collected in a shallow reservoir that ultimately erupted and formed a huge caldera. In some localities, such as the Eocene

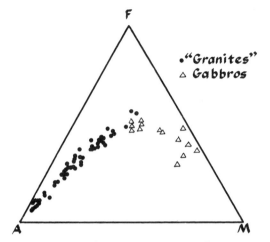

Figure 10-7 AMF diagram for rocks of the Coastal Batholith of Peru. The early gabbroic rocks of each batch of magma define a compositional trend that differs from that of the more mafic gabbroic rocks. Relationships such as these could arise from melting of a compositionally zoned lithosphere in the manner illustrated in Figure 10-21.

Glen Coe caldera of western Scotland, moderate amounts of erosion have exposed massive granitic rocks that rose to depths of a kilometer or less in the core of the caldera.

Granites in Zones of Continental Collision

When plate convergence culminates in a collision of two continents, the intense crustal deformation may be accompanied by plutonism, but usually on a much smaller scale than that associated with subduction. Plutons form only under unusual conditions and generally in relatively small volumes. The Alps and Himalayas are the main modern examples.

The two main intrusive complexes of the Alps, Ademello and Bregaglia-Iorio, were emplaced in the Oligocene during a brief period of relaxation immediately after the main period of compression. They rose into small, fault-bounded areas of tension close to the Insubric Line marking the boundary between the Eurasian and African plates. The activity ended abruptly when compression was renewed in the Miocene.

The plutons are composed chiefly of felsic calc-alkaline rocks, including ademellites, monzonites, and tonalites that take their names from localities in the Italian Alps. The earliest intrusions were very mafic hornblende gabbros, which soon gave way to larger volumes of increasingly differentiated rocks. Small differences in the proportions of hydrous and anhydrous minerals precipitated in the early stages of differentiation led to two distinct trends, much as they did in the Coastal Batholith of Peru.

In the case of the Himalayas, granitic rocks have been produced at two stages of the collisional event when India converged on Asia. The first, a syntectonic group generated during the period of strongest deformation, resulted from thrusting of hot metamorphic rocks over colder rocks of lower metamorphic grade (**Fig. 10-8**). Fluids driven out during heating of the lower unit rose into the overlying rocks, lowered their melting temperatures, and triggered melting. The magmas are typical of the S-type derived from sedimentary source rocks.

A younger group of granites that appeared after relaxation of the main compressional forces resembles the late-stage intrusions of the Alps. The magmas, which are of both the S and I types, come from sedimentary and igneous rocks at greater depths. Overthickening of the zone of maximum compression is thought to cause a condition of isostatic instability that leads to upwelling of the depressed crust. This brings hot crustal material to levels of lower pressure where it begins to melt and rise as dikes and stocks through high-grade metamorphic rocks of the catazone. Although commonly marked

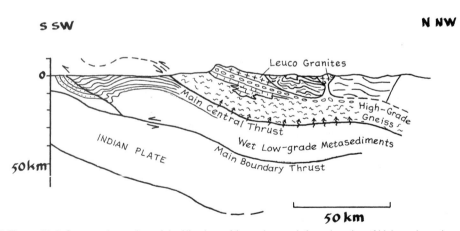

Figure 10-8 Syntectonic granites of the Himalayan Mountains result from thrusting of high-grade gneisses over wet low-grade meta-sediments. Water and other volatile components from the underlying unit infiltrated up into the hotter thrust sheet triggering partial melting at temperatures of about 700°C. Melting continued for more than 10 million years, while the heterogenous melts intruded as irregular sills and diapirs at depths of 10 to 15 km. (Adapted from C. France-Lanord and P. Le Fort, 1988, *Trans. Roy. Soc. Edinburgh, Earth Sci.* 79:183–195.)

by flow-related foliation, the granites have less tectonic fabric than those generated above thrust sheets.

Continental Rifts and Hotspots

Crustal extension is normally associated with fissure eruptions of basaltic lavas, but it has also produced equal or even greater volumes of felsic volcanic rocks. In places, such as the Great Basin of the western United States and large areas of Mexico and Central America, basalts are quite subordinate to great sheets of siliceous ignimbrites. Unlike flood basalts, however, most of the felsic magmas are not discharged from long fissures but from more centralized sources marked by calderas and dome complexes.

Mention has already been made of large igneous centers that have no apparent relationship to orogenic belts or continental rifting but have been attributed to hotspots over mantle plumes. Yellowstone (**Fig. 10-9**) is the best-

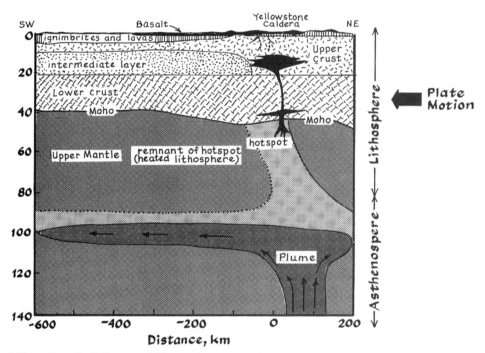

Figure 10-9 The Yellowstone Caldera marks the present location of a hotspot that began with the Columbia River flood basalts and has been migrating at a rate of 4.5 centimeters per year along the Snake River Plain. Basaltic magma rising from a mantle plume has melted the upper crust to produce huge volumes of rhyolitic ignimbrites. The residue of this melting forms an intermediate layer in the crust downstream from the hotspot. (After Smith, R. B., and L. W. Braile, 1994, *Jour. Volc. Geoth. Res.* 61:121–187.)

documented—some would say the only—hotspot beneath thick continental crust. As noted in our discussion of the Columbia River flood basalts in Chapter 8, it is thought that a mantle plume was initiated about 16 million years ago and has since formed a more or less linear succession of volcanic centers ending at its present location under the Yellowstone caldera. Evidence for the plume consists of anomalously low P-wave velocities extending to depths of about 200 kilometers. The overlying rocks are uplifted as much as 600 meters in a broad dome centered over the hotspot, while a body of basaltic magma ponded at shallow depths beneath the crust produces a surface heat flow about 30 times the continental average. Rhyolitic magma generated by partial melting of the silica-rich basement series has erupted periodically as enormous pyroclastic flows, and subsidence into the evacuated reservoir has produced a chain of large, eastward migrating calderas. As the lithosphere moves downstream, the influx of magma and heat has declined, but a few basalts continue to be erupted in the hotspot's wake along the Snake River Plain.

In North America, the most conspicuous manifestations of intracontinental siliceous magmatism are volcanic, but on other continents they are dominantly plutonic. At least part of the difference must be due to depths of erosion. Only rarely are both volcanic and plutonic rocks well exposed in the same region. A notable exception is the cluster of large calderas and related volcanic rocks situated over a complex of Tertiary granitic intrusions in the San Juan Mountains of Colorado.

The igneous history of the San Juan region has been divided into three stages extending through most of middle and late Tertiary time. Activity began about 35 million years ago with eruptions of lavas and volcanic breccias from large central-vent volcanoes. The products of this stage were mainly andesites with subordinate amounts of more siliceous rocks (**Fig. 10-10**). About 30 million years ago, the volume of eruptions increased dramatically. Enormous pyroclastic flows poured from fissure vents to cover an area of about 25,000 km^2. More than 16 major sheets of dacite and rhyolite are associated with a cluster of at least fourteen calderas that average about 20 km in diameter. Single sheets spread over as much as 15,000 km^2, reached local thicknesses of over a kilometer, and had volumes as great as 3000 km^3. Intermediate lavas were largely restricted to areas in and around calderas. They are petrographically similar to the rocks erupted from central-vent volcanoes during the preceding period but were distinctly subordinate to the siliceous pyroclastic rocks. Activity declined rapidly, and about the same time the eastern part of the field was faulted and tilted as a consequence of formation of the adjacent Rio Grande Rift.

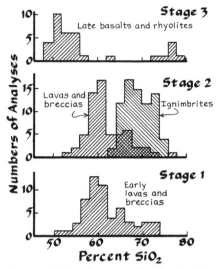

Figure 10-10 Volumes of rocks erupted during the three main episodes of igneous activity in the San Juan Complex. Note the progressive divergence of compositions with time. The third, most recent episode is thought to be related to more wide-spread regional tectonic events and has a wider distribution than the other two. (Adapted from P. W. Lipman et al., 1970, *Geol. Soc. Amer. Bull.* 81:2329–2352.)

This tectonic event ushered in a new and entirely different period of igneous activity marked by bimodal eruptions of basaltic lavas and rhyolitic domes and tephra. Compared with the preceding episode, the volumes of magma were relatively small. They came from widely scattered vents that seem to have little relationship, either structurally or genetically, to the earlier events. Most likely they were part of the widespread bimodal volcanism that accompanied crustal extension throughout much of the Great Basin in late Cenozoic time.

Despite its conspicuous discontinuities, both in time and composition, the magmatic suite as a whole forms a coherent calc-alkaline series. Differences are confined mainly to trace elements and isotopic ratios. In each eruptive sequence, large volumes of siliceous magma were discharged in what must have been very short intervals of time. Without historic examples to judge by, the duration of the huge eruptions responsible for each ignimbrite sheets can only be guessed, but they could not have lasted more than weeks or months. They have few if any erosional discontinuities or other signs of depositional breaks between eruptive units, and in this sense, they resemble flood basalts. Despite obvious differences in composition and eruptive behavior, they had comparable volumes and tectonic settings.

■ Plutonic Rocks of Precambrian Ages

We noted in the opening chapter that certain types of igneous rocks are very characteristic of ancient continental crust and that some are confined almost entirely to such settings. The ultramafic komatiites are a conspicuous example. Two others, trondhjemites and anorthosites, are at the opposite end of the compositional spectrum (**Fig. 10-11**).

Archean Trondhjemites

Although by no means confined to the Precambrian, the distinctive group of plutonic rocks known as trondhjemites is very common in Archean terranes, such as those of the Canadian Shield. As mentioned earlier, they are more sodic than most other granites (Fig. 10-3a) and have little or no K-feldspar.

Many trondhjemitic plutons form the cores of thick volcanic complexes composed originally of basalt but now altered to low- to medium-grade metamorphic rocks, mainly greenstones and amphibolites. The trondhjemitic magmas could be partial melts of this basaltic crust. Downwarping under the load of thick volcanic rocks may have depressed the crust to levels where it was first amphibolitized and then partially melted and mobilized. As we shall see shortly, partial melting of amphibole-rich mafic rocks can yield liquids rich in silica and with large ratios of sodium to potassium. Moreover, the abundance patterns of the REE in trondhjemites are symptomatic of liquids in equilibrium with amphibole (Fig. 10-3b).

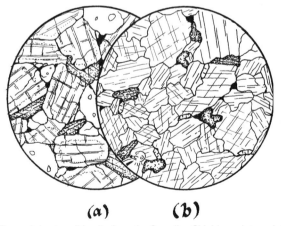

(a)   *(b)*

Figure 10-11 (a) A Precambrian trondhjemite from the Canadian Shield consists mainly of quartz and oligoclase but little if any K-feldspar. Hornblende is the principal mafic mineral, but sphene is an important accessory. (b) Anorthosite from the Adirondack Mountains of New York consists of labradorite with very subordinate amounts of pyroxene, ilmenite, garnet, and apatite.

Anorthosites

Another variety of plutonic rocks confined almost exclusively to the Precambrian is anorthosite, a singular rock consisting primarily of plagioclase (Fig. 10-11b). Often referred to as *massif-type anorthosites*, they differ from the stratiform anorthosites in gabbroic intrusions like the Stillwater and Bushveld complexes in that they are not interlayered with more mafic rocks. Instead, they form broadly lenticular bodies of almost uniform compositions.

What makes these rocks so unusual is their total lack of aphyric volcanic equivalents. Although some very plagioclase-phyric basalts approach them in bulk composition, lavas that can be shown to have been true liquids of anorthositic compositions are unknown. Another puzzling feature is their restricted distribution in time and space. They are found almost exclusively in anorogenic terranes in a broad belt extending from the southwestern United States to Labrador and on the other side of the Atlantic, through Norway and Sweden. A smaller group has been identified in the Southern hemisphere. Without exception, the ages of all these bodies lie between 1.7 and 1.1 billion years with many close to 1.4 billion years.

Plagioclase, which accounts for at least 85 percent of their volume, is typically in the range of An-40 to An-60, in contrast to values of An-65 to An-85 in large layered intrusions. Hypersthene and augite are the most common mafic minerals, and ilmenite and apatite are important accessories. Associated rocks include gabbro, usually the hypersthene-bearing variety, norite, that typically forms gradational margins around a central core of more leucocratic rocks composed of the same minerals. In fact, most anorthosites have compositions that are consistent with simple addition of plagioclase crystals to a noritic liquid. Thus, apart from their unusual proportions of plagioclase, they are subalkaline rocks with many affinities to tholeiitic magmas derived from a shallow, depleted mantle. If they interacted with the continental crust, it was only to a minor degree, for their trace-element contents and isotopic ratios indicate no measurable amount of assimilation.

Depths of emplacement of these bodies were probably between 5 and 25 kilometers. None has been shown to have approached the surface. The magmas must have been exceptionally hot and dry, for their contact aureoles show the effects of high temperatures but little evidence of hydrothermal fluids. These properties point to what must have been a condition unique in the Earth's history, when a much steeper thermal gradient caused dry melting at shallow levels below a thin continental crust.

How this process produced rocks so rich in plagioclase is difficult to say. Most explanations appeal to unusual accumulations of plagioclase from magmas originally of more mafic compositions, but how so much plagioclase can be fractionated without corresponding amounts of mafic minerals is a mystery. The greatest proportion of plagioclase that could be precipitated from a normal mantle-derived magma is of the order of 60 percent. To produce the much greater concentrations in anorthosites requires disposal of proportionate amounts of mafic minerals in some hidden part of the system. Anorthosites deserve more study than they have received, not just because they are petrologic curiosities but because they could reveal much about the early thermal and chemical evolution of the mantle and continental crust.

■ The Granite System

The phase relations of what is loosely referred to as the "Granite System" lie within the silica-saturated part of what Bowen aptly referred to as "Petrogeny's Residua System" (**Figs.** 7-4 and **10-12**). As used by Bowen, the name was meant to imply that the phases come from the last liquid residue of crystal fractionation, not the refractory solid residue of melting. Although an oversimplification of natural systems, it is the basic starting point for understanding felsic magmas. Here we are concerned only with silica-saturated

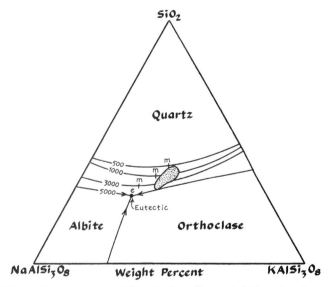

Figure 10-12 The granite system, quartz-albite-orthoclase. The quartz-feldspar boundary and temperature minimum, m, is shown for 0.5, 1, 3, and 5 kilobars water pressure. Note that at 5 kilobars the temperature minimum becomes a eutectic, and a cotectic line separates the fields of the two feldspars. The average composition of granitic rocks containing 80 percent or more of the three end-member components is indicated by the shaded oval area.

magmas that are separated by a thermal divide from nepheline-normative compositions. The latter are dealt with in Chapter 11.

Unlike mafic magmas, which evolve primarily through fractionation of anhydrous ferromagnesian minerals and calcic plagioclase, the compositions of felsic magmas are governed chiefly by crystallization of the alkali feldspars, quartz, and the hydrous minerals, mica and amphibole. In their simplest form, rocks that are loosely referred to as "granitic" can be treated in terms of three components, quartz, albite, and orthoclase, that make up 80 percent or more of their normative or modal compositions. The close correspondence of the compositions of granitic rocks to the region of minimum temperature in the phase diagram for quartz and alkali feldspar (Fig. 10-12) shows that the rocks are either the last products of differentiation or the first products of melting of a mineralogical assemblage dominated by these three minerals. These processes have important differences, depending on whether they take place under hypersolvus and subsolvus conditions.

Crystallization and Melting Under Hypersolvus Conditions

Few, if any, granitic magmas precipitate quartz before feldspar, and most crystallize plagioclase before potassium feldspar. Hence, the great majority of natural liquids will fall in the lower left part of the triangle. On crystallizing they follow a course across the feldspar field toward the quartz-feldspar boundary. A typical path for such a liquid is illustrated schematically in **Figure 10-13b**. Because both alkali feldspars form solid-solution series, the course of crystallization of liquids crossing the lower part of the diagram is not linear but curved and cannot be predicted exactly without knowing more about the composition of coexisting crystals and liquids. One can easily distinguish granites that crystallized from potassic liquids from more sodic ones by plotting the variations of Na_2O versus K_2O. For the former, the ratio of K_2O to Na_2O will decrease while the latter will have the opposite trend.

The point m where the temperature of liquid–crystal assemblages reaches a minimum is not a true eutectic for hypersolvus conditions. Only liquids with a narrow range of intermediate compositions will crystallize under equilibrium conditions at the temperature and composition of that point. We saw the reason for this when we considered binary solid-solution series in Chapter 3 and the alkali feldspar system in Chapter 4. Because it is not a eutectic, differentiating liquids reach point m only with extreme fractionation.

The same principles hold for melting. Under hypersolvus conditions, the first product of melting of crustal rocks with compositions consisting mainly of quartz and alkali feldspar may not be a liquid close to the composition m. As we noted when considering solid-solution series, the range of liquids that evolve during crystallization is not necessarily the same as those produced by

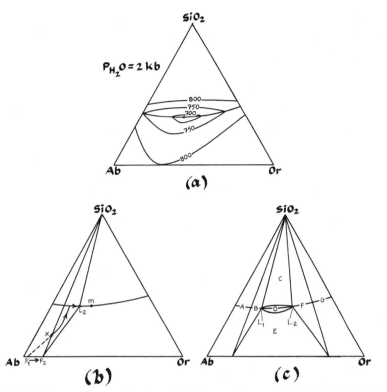

Figure 10-13 (a) The hypersolvus granite system at 2 kilobars water pressure. (b) A typical path of a liquid, X, that first crystallizes in the albite field and evolves with complete reaction. When the liquid reaches L_2, the bulk composition can be made up entirely of quartz and the equilibrium feldspar, F_2, and the liquid is consumed without reaching the temperature minimum at m. (c) A constant temperature section at 700°C. The system is divided into various fields corresponding to bulk compositions that would have the following phase assemblages at that temperature.

    A—sodic feldspar and quartz without liquid
    B—sodic feldspar, quartz, and liquid L_1
    C—quartz and liquid on the SiO_2-rich contour
    D—all liquid
    E—feldspar and liquid on the SiO_2-poor contour
    F—potassic feldspar, quartz, and liquid, L_2
    G—potassic feldspar and quartz without liquid

melting. Much depends on the degree of fractionation and equilibration of crystals and liquid as the process advances. The same principles apply to crystallization and melting in the granite system. If, on cooling, the liquid X shown in Figure 10-13b continues to equilibrate with its coexisting solid solution minerals, its course is arrested when the composition of the crystals reaches that of the original starting liquid. Without equilibration, the liquid will continue to evolve until it reaches the composition and temperature of the system at m. However, if a mineral assemblage consisting of unzoned crystals having the same bulk composition X is heated, no liquid appears at the temperature minimum at m but only at some higher temperature at which the crystals are in equilibrium with a liquid of different composition, such as L_2.

Figure 10-13c illustrates the main possibilities when crustal rocks that are being heated reach a given temperature. Depending on the bulk composition, the system can consist of liquid and quartz, liquid and Na-feldspar, liquid and K-feldspar, liquid without crystals, or crystals without liquid. The liquid composition differs according to the temperature and the mineral phases present, and the trend of liquid compositions must differ for every bulk composition, whether the process is one of crystallization or melting.

Crystallization and Melting Under Subsolvus Conditions

In a multiphase assemblage, the effect of water pressure will, as we saw in the preceding chapter (Fig. 9-17b), differ according to the nature of the individual phases, and depending on the properties of each mineral, the compositions of liquids shift to include more or less of the different crystalline phases with which they are in equilibrium. For dry conditions, the composition of melts in the granite system is not greatly altered by increased load pressure; however, with increasing water pressure, the liquidus temperature of albite is depressed more than that of quartz or orthoclase, and the composition of the minimum melting temperature (point m in Fig. 10-12) is shifted in the direction of the albite corner of the diagram.

At the same time, we have seen that increasing water pressure causes the solvus in the alkali-feldspar system to intersect the solidus and the temperature minimum to becomes a true eutectic. As a result, the minimum, m, in the three-component system (Fig. 10-12) becomes a eutectic, and a cotectic divides the lower part of the system into two fields, one for each of the alkali feldspars.

Under these conditions the behavior is more restricted, and a broader range of compositions can differentiate or melt to form liquids of the minimum-melting composition. The course of a liquid composition during crystallization or melting is simplified by the cotectic nature of the boundary between the two feldspars. A liquid initially crystallizing a sodic feldspar follows a curved path across the field of that mineral until it intersects the boundary with potassium feldspar or quartz. It then descends to the eutectic, where all three phases crystallize together. An opposite course is followed on melting.

Effects of Anorthite

The four-component system, quartz-albite-orthoclase-anorthite, can be represented by a tetrahedron (**Fig. 10-14**) having anorthite at its apex and the granite system of Figure 10-12 as its base. The front face is the familiar feldspar

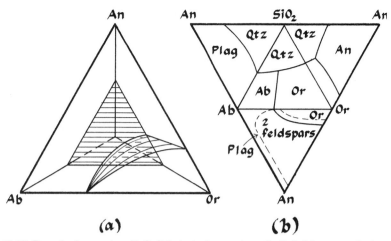

Figure 10-14 The subsolvus system Ab-Or-SiO₂-An is shown schematically in (a) as a tetrahedron with anorthite at its apex and the granite system from Figure 10-12 as its base. The surfaces bounding the volume of quartz, plagioclase, and K-feldspar are exaggerated to show their geometric relations. In (b), the tetrahedron has been unfolded to show the three component systems of the four surfaces of the three-dimensional diagram.

system, plagioclase-orthoclase. The line separating the field of quartz from that of feldspar in the granite system is now a surface extending into the interior of the tetrahedron roughly parallel to the front face. From the feldspar cotectic, a curved surface extends back into the interior until it intersects the plane bounding the quartz volume.

Because most natural magmas precipitate plagioclase before quartz or potassium feldspar, their initial compositions must lie in the volume defined by the plagioclase edge of the tetrahedron and the two saturation surfaces for quartz and potassium feldspar (**Fig. 10-15**). On crystallizing, their compositions move away from plagioclase along a curved path as the crystals become more albitic. The path of the liquid will intersect one or the other of the two saturation surfaces and begin to precipitate quartz or potassium feldspar along with plagioclase. Then, moving along one of these surfaces, it eventually reaches the intersection with the second, and all three phases—plagioclase, quartz, and potassium feldspar—crystallize together as the liquid follows the cotectic line of intersection down toward the base of the tetrahedron.

In treating felsic rocks in terms of only quartz and alkali feldspar, as we did in the preceding sections, the anorthite component is ignored, and compositions falling within the interior of the four-component system are in effect projected from the anorthite corner to the base of the tetrahedron. If the anorthite

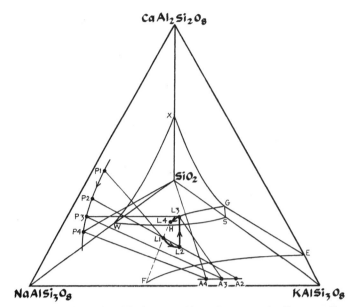

Figure 10-15 The course of crystal and liquid compositions of a magma that first crystallizes plagioclase in the system An-Ab-Or-SiO₂. The initial liquid, L_1, in the interior of the tetrahedron, first crystallizes plagioclase of composition P_1 (on the front face of the tetrahedron) and follows a curved path away from plagioclase until it intersects the two-feldspar surface at L_2. Alkali feldspar A_2 (on the front face) begins to crystallize along with plagioclase, which has now reached composition P_2, and the liquid moves away from these feldspars along the curved surface until it reaches the silica-feldspar surface at L_3. Quartz now begins to crystallize along with feldspars (now with compositions P_3 and A_3), and the liquid then descends to an end point at L_4. (After I. S. E. Carmichael, et al., 1974, *Igneous Petrology*, McGraw-Hill, 739 p.)

content is small, this may not introduce a major error, but if it accounts for more than 10 percent or so of the total, the effect can be misleading. Note that any composition on the four-phase cotectic line—that is, a liquid crystallizing quartz, orthoclase, and plagioclase simultaneously—falls within the quartz field when projected along a line from the anorthite corner through the point representing the liquid composition to the basal triangle. The apparent effect of increasing anorthite content, when projected in this way, is just the opposite of increasing water pressure on the simpler three-component system. Thus, the combined effects of water and anorthite result in a range of possible compositions for the temperature minimum, m, that is an almost linear trend across the central part of the diagram. Without knowing the anorthite content, it is difficult to say to what degree the position of m is due to one effect or the other.

Another effect of this projection is to place the composition of liquids in equilibrium with plagioclase in the quartz field of the simpler quartz-alkali feldspar system. This gives the false impression that the liquid is saturated with quartz but not with feldspar. A composition that, in the simple granite system, might appear to be the result of advanced melting of a very quartz-rich crustal rock may in fact be a normal igneous composition with enough anorthitic plagioclase to place it well above the base of the tetrahedron. This pitfall has misled more than one petrologist. It is a trap to beware of when dealing not only with granites but with any complex natural system in which one attempts to interpret compositional trends in terms of a simplified diagram that ignores other components of the rock.

Hydrous Mafic Minerals

Thus far we have assumed compositions solely of quartz and feldspar and have ignored mafic components and hydrous minerals. From what was seen in Chapter 9 when we considered the role of water in systems such as diopside-forsterite-silica (Fig. 9-19), one would anticipate that the effect of water pressure on mafic compositions will be to produce liquids that are systematically more silica-rich than under anhydrous conditions. This is no less true when amphibole or biotite are stable phases, for most of these minerals have low silica contents and in most instances are equivalent to an assemblage that includes a silica-deficient mineral, such as nepheline or leucite. Hence, any melt in equilibrium with these minerals will be more silica rich than one formed from partial melting of an equivalent anhydrous assemblage.

The stability relationships of the hydrous minerals, biotite, muscovite, and hornblende, are a function of temperature, water pressure, and oxygen fugacity, and in contrast to anhydrous minerals, which melt at lower temperatures with increasing water pressures, hydrous minerals can have the opposite relationship.

When superimposed on a single diagram, such as one for a typical granodiorite (**Fig. 10-16**), the stability limits of the hydrous and anhydrous

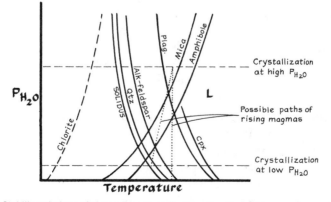

Figure 10-16 Stability relations of phases in a granodiorite as functions of temperature and water pressure. See text for a discussion of the possible courses of crystallization.

minerals form crossing curves, and the order of crystallization or disappearance of minerals with cooling or heating will differ according to the pressure, water content, and oxidation state under which the process takes place. Such a magma, if cooling at high water pressure, would first crystallize the hydrous minerals, hornblende and biotite, then plagioclase with increasing albite content, and finally the granitic end-members, quartz, orthoclase, and albite, at temperatures just above the solidus. The same liquid crystallizing at a shallow depth or with less water would first crystallize pyroxene, but, on further cooling, the pyroxene might alter to an amphibole (uralite), and the feldspars could be altered to a mica, such as sericite. This mineral assemblage is common in shallow, subvolcanic stocks.

Consider next what might happen to such a magma if it rises to erupt at the surface. The hydrous minerals formed at depth would become unstable, and exsolution of water would cause the magma to vesiculate. If the water is lost to the system, all the anhydrous minerals could crystallize, even at constant temperature (because a loss of water pressure raises their temperatures of crystallization), and the magma is likely to freeze. This probably explains why so many felsic magmas form plutons within the crust and fail to reach the surface.

Effects of a Hydrous Vapor Phase

Having said virtually nothing about the vapor phase since we last touched on it in Chapter 3, this is an appropriate place to examine some of the ways in which the volatile components of magmas behave during cooling and crystallization. To do this, it will be helpful to look at a few basic principles of liquid-vapor systems and how they apply to magmas.

At the outset, it must be emphasized that when we speak of a volatile component, such as water, no distinction can be made between the vapor and a hydrous liquid phase at the temperatures and pressures of subsurface magmatic systems. Because the effect of increasing temperature on the liquid phase is to decrease its density, and the effect of pressure on the vapor phase is the reverse (**Fig. 10-17a**), the densities of the liquid and vapor converge, and beyond a certain condition, known as the critical point, the liquid and vapor are no longer separated by a sharp difference in properties. The critical point of pure H_2O is only 221 bars and 374°C, and those of other magmatic gases are even lower; thus, it is clear that a high-temperature gas phase at any depth greater than a kilometer or so must be *supercritical*. That is to say, it has no well-defined boiling temperature but rather a continuous range of density. We may still refer to this fluid as a

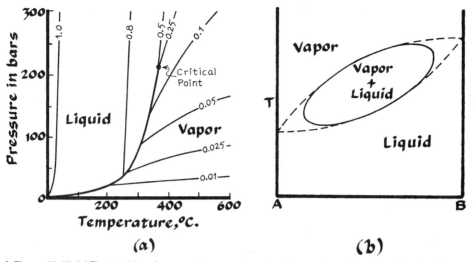

Figure 10-17 (a) The densities of water and steam, shown by light, numbered lines, vary directly with pressure and inversely with temperature in such a way that the densities of the two phases converge on common values, and above the critical point (C.P.) at 374° and 221 bars the distinction between the liquid and vapor phases disappears and the properties are gradational. (b) Vapor–liquid relationships for two components with differing vapor pressures resemble those for liquid–solid relations. The component with the low boiling temperature and high vapor pressure is more concentrated in the vapor than in the liquid phase. If both components are above their critical temperatures, the distinction between the vapor and liquid is gradational for compositions near the end-members.

vapor phase, however, in order to distinguish it from silicate liquids, which, under most conditions, have physical properties different from those of H_2O-rich fluids.

In granitic systems, these differences are not great. Supercritical hydrous fluids are excellent solvents, capable of dissolving silica, alkalies, and other components, so that the vapor phase is composed of more than the components one normally thinks of as volatiles. Fluids containing large amounts of these components have properties approaching those of silicate melts, and the difference between a hydrous fluid with dissolved silicates and a silicate melt with dissolved volatiles becomes rather narrow.

A magmatic vapor phase can be thought of as consisting of two sets of components: those like water that have low boiling temperatures and high vapor pressures and those like silica that have very high boiling temperatures and low vapor pressures. Their phase relationships are analogous to those of two miscible liquids of different volatilities, say water and alcohol (**Fig. 10-17b**). Their phase diagram at constant pressure resembles the liquidus–solidus loop of solid solution series. The component with the low boiling temperature (high vapor pressure) is preferentially enriched in the vapor. With falling temperature, a given intermediate composition does not change directly to a liquid but passes through an intermediate range of vapor + liquid, just as a liquid passes through a region of liquid + solid in crystallizing a solid solution in a "condensed" system.

The liquid-vapor loop in Figure 10-17b differs from that of a solid solution in that the curves do not converge at the end-member compositions. Instead, the vapor passes gradually into the liquid at the two ends because in those regions the critical conditions are exceeded for corresponding end-member components, and the boundary between the liquid and vapor is entirely gradational. Because each of the components has its own critical temperature and pressure, the system does not have a unique critical point but rather a curved boundary at each end. The student may find it instructive to show how this can be deduced from the Phase Rule.

With this general background, we can turn now to a system closer to natural melts and see how a vapor phase may behave during crystallization of a magma. The system water-albite (**Fig. 10-18**) will serve to show most of the important relationships. Taking water as the component with the low boiling temperature and high vapor pressure and albite as the one with the opposite properties, the vapor–liquid relationships would be like those in Figure 10-17b. (The boiling temperature of albite is so high that it is meaningless in terms of natural conditions and is not shown.) The liquid–solid relationships are depicted schematically as a sloping curve. Theoretically, there could be a eutectic close to the H_2O end of the diagram, where a small amount of albite would lower the freezing temperature of ice, but the solu-

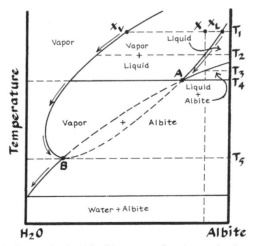

I Figure 10-18 Schematic diagram for the H_2O-albite system. See the text for discussion.

bility of albite in water is so low that this can be ignored. The middle part of the curve overlaps with the vapor–liquid region so that superposition of the diagrams results in a combined set of phase relations, which shows the basic elements of the entire system.

Take, for example, a magma of composition X at an initial temperature, T_1. Two phases are present, a liquid of composition X_L and a more H_2O-rich vapor of composition X_V. As the magma cools, the solubility of water in the liquid increases, and the proportion of vapor will diminish at the same time that the amount of liquid increases. At temperature T_2, all the vapor has disappeared, and only a single phase, a liquid of composition X, remains. On further cooling, crystals of albite appear at T_3, and at temperature T_4 an H_2O-rich vapor phase reappears. Note the three-phase invariant point at A. The system cannot decline in temperature below this point until all of the liquid is gone, but after that happens, albite continues to crystallize from the vapor phase. (The amount of albite crystallized in this interval is less than indicated by the diagram, which exaggerates the compositional differences in order to show the relations more clearly.) Note that in the lower part of this interval albite is dissolved, and the concentration in the vapor phase increases again until a second invariant point, B, is reached at T_5. Thereafter, albite crystallizes again.

Many pegmatites are thought to crystallize under conditions similar to those portrayed by the albite-water system, particularly in the lower temperature range where the vapor phase is rich in H_2O. The low viscosity and great solvent power of the fluid enhance its mobility and ability to transport components that have been concentrated in the late stages of crystallization of large bodies of magma. The high water content also facilitates growth of very large crystals.

■ Generation and Rise Through the Crust

The effect of increasing load pressure on melting or crystallization is, as we saw in the case of basalts, to raise the temperature of crystal–liquid equilibrium by amounts that are mainly a function of the difference between the specific volume of the solid and liquid. Because the specific volume of liquid is greater than that of the solid, the change of volume on melting, ΔV, is positive, and the Clapeyron equation

$$\frac{dT}{dP} = \frac{T\Delta V}{\Delta H}$$

shows that the slope of the liquidus curve is also positive, and its temperature increases with pressure (**Fig. 10-19b**). Increasing water pressure has the opposite effect. It lowers the melting temperature because the combined volume of an anhydrous solid and water vapor is greater than that of the liquid with water in solution, and ΔV is negative. The temperature of the liquidus curve must therefore decrease with increasing water pressure, and two limiting cases can be defined, one for dry melting or crystallization in which the T-P curve is positive, and another for wet conditions that has a negative slope. To see how melting might proceed under such conditions, consider a hypothetical crustal source with a composition within the middle range of the granite system so that the first melt has a composition at the eutectic in Figure 10-12.

The relative amounts of melt and crystals during melting or crystallization are strongly dependent on the solubility of water in the liquid. The amount of water the liquid can dissolve increases with increasing water pressure (again because the volume of melt plus water vapor is much greater than that of the melt with water in solution). A typical solubility curve is shown in **Figure 10-19a**. The amount of water dissolved in the liquid on the wet-melting curve in Figure 10-19b corresponds to saturation for the prevailing pressure, while on the dry curve, of course, it is nil, and between the two curves is a region in which the liquid has an intermediate temperature and water content.

A rock having a water content equivalent to its solubility at the prevailing pressure would melt completely at the temperature corresponding to the

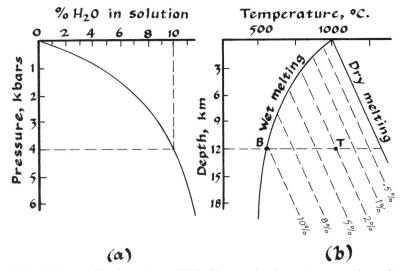

Figure 10-19 (a) The solubility of water in granitic liquids, assuming that water pressure is equal to load pressure at the depths indicated on the vertical axis. (b) The melting curves of granite as functions of pressure, temperature, and water content. See text for discussion.

water pressure on the wet-melting curve. But if it has less water—the more likely situation in nature—it will begin to melt on the wet curve (where the first liquid would be saturated), but it will melt only as far as is required to saturate the liquid at that temperature and eliminate the vapor phase. At this point, ΔV changes from negative to positive, and to attain further melting, the temperature must rise and the liquid must become poorer in dissolved water. For example, a rock with 2-percent water will begin to melt at a temperature of about 600°C at a water pressure of roughly 4 kilobars. Suppose that the solubility at this pressure is 10 percent. Melting will therefore proceed at the minimum temperature at this depth until 20 percent of the rock is melted and the liquid is saturated. The rock would only be completely melted if it were heated to a temperature of about 1000°C.

A wide range of melting could result if the pressure of water increases with depth and the geothermal gradient rises above the melting curves in the manner shown schematically in **Figure 10-20a**. The range of melting starts at the uppermost intersection of the geothermal gradient with the wet melting curve and extends down the gradient with increasing amounts of liquid until the entire rock would be melted at the depth where the temperature reaches the melting curve corresponding to the total water content of the rock. (These relationships apply only if the water is a separate phase and is not locked up in hydrous minerals. The water in micas or amphibole has little effect on melting temperatures unless these minerals become unstable and a separate vapor phase is created. We shall see shortly that this difference has a strong effect on how magmas are generated under natural conditions.)

If a parcel of granitic magma were saturated with water at a temperature corresponding to the wet melting curve at some depth and then were to rise to a level of lower pressure, the solubility of water would decline, the magma would exsolve water vapor, and the melt would begin to crystallize. It can be seen from the negative slope of the wet-melting curve that if such a magma did not rise in temperature during ascent, it could crystallize completely. The heat of crystallization might help sustain a somewhat higher temperature and enable the liquid to follow the melting curve to lower pressures, but it is doubtful whether this would be enough to offset the strong quenching effect of exsolving water. A second magma starting at the same depth but on the dry melting curve would have little or no water and an initially higher temperature. On rising to a level of lower pressure, it enters a field above its liquidus and would be superheated. Adiabatic cooling (about 0.2 or 0.3 degrees per kilometer) would be insufficient to prevent the magma from becoming superheated, but loss of heat to the surroundings becomes increasingly important as the magma enters cooler rocks of the shallow crust.

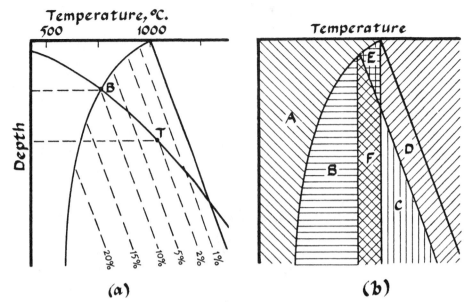

Figure 10-20 (a) Schematic illustration of the melting relations of granite with 2-percent H_2O in a geothermal gradient. Melting begins at point B, where the geothermal gradient first intersects the curve for wet melting, and increases with depth until the rock is totally melted at point T, where the thermal gradient reaches the melting curve for the water content of the rock. (Adapted from O. F. Tuttle and N. L. Bowen, 1958, *Geol. Soc. Amer. Mem.* 74.) (b) Regimes in which differing effects would be observed when magmas generated from a rock with 2-percent H_2O rise toward the surface.

A. No melting.
B. Partial melting; on rising, a crystal–liquid mixture first absorbs crystals and then grows larger ones.
C. Partial melting; on rising, the crystal–liquid mixture would be entirely liquid and superheated at the surface if no heat is lost en route.
D. Total melting; the liquid would reach the surface superheated if no heat were lost en route.
E. Total melting; the liquid begins to crystallize before reaching the surface.
F. The relationships in field F are left for the student to deduce.

In nature, the presence of other components and the cooling effects of decompression and heat losses to the surroundings will modify this basic outline.

Between these limiting cases we can distinguish several types of intermediate behavior. The diagram in Figure 10-20b is divided into six general zones, each characterized by a particular sequence of melting or crystallization that could result in different petrographic features if the magmas rose to a shallow level or erupted volcanically.

These examples assume the pressure of water to be equal to the total load pressure at any given depth and have been highly simplified in order to illustrate certain basic principles. In reality, conditions must be more complex. In nature, the pressure of water is probably less than the total load pressure, and as we saw in Chapter 5 (Fig. 5-14), under such conditions, the solubility of water is reduced. The amount of water a magma contains under such conditions is a function not only of the pressure of water, which tends to cause water to enter the magma, but also of load pressure, which tends to squeeze it out. Thus, the water content can increase with falling load pressure, and magmas saturated at depth may become undersaturated and absorb water as they rise.

Melting of Sedimentary and Meta-Sedimentary Rocks

The problem of generating large volumes of siliceous magmas is greatly reduced if the source rock has a composition that is already close to that of granite. This, together with the common association of orogenic plutons with thick, deeply buried, and intensely metamorphosed sedimentary rocks, supports the notion than many granites are the products of *anatexis*, as the process of melting continental crust is called. Although the principles of melting such rocks are often thought of as the reverse of crystallization, the process is altered somewhat when the compositions and mineral assemblages differ from those of the simple system. Most thick accumulations of continental sediments are *pelitic*—that is, they contain large proportions of clay and quartz. Most have such large amounts of these minerals that they are at least as rich in silicon, aluminum, and potassium as are granites. Unlike most granites, however, they tend to have less sodium than potassium.

Most sediments and their metamorphic equivalents contain large proportions of hydrous minerals, such as muscovite, biotite, and amphibole, that have a strong effect on their melting behavior. Depending on their compositions and depths of burial, melting proceeds in multiple stages each of which is governed by the effects of increasing temperature on a particular assemblage of minerals. The way in which melting advances is illustrated schematically in **Figure 10-21a**, which shows the sequence of changes that accompany an increase of temperature at a given pressure. Curve (1) depicts the upper stability limit of an assemblage that includes a hydrous mineral, H, in the presence of water vapor. On crossing this curve the assemblage goes to a melt, M_S, that is saturated with water. The hydrous mineral disappears, and the water vapor is completely dissolved in the melt, but if water is present in excess of the amount that can be dissolved in the melt, a vapor phase will still remain.

The same assemblage with a hydrous mineral but without a separate vapor phase will be stable to a higher temperature corresponding to curve (2) and will break down to one or more anhydrous minerals, A, and a melt, M_U, that is normally undersaturated with water. Melting of this kind is sometimes referred to as *dehydration melting* and is probably most common in nature.

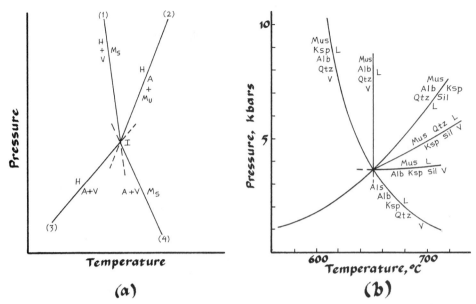

Figure 10-21 Stability limits and melting curves for hydrous and anhydrous minerals in the crust. The four curves in the schematic diagram (a) illustrate the principles explained in the text. H stands for a hydrous mineral, A for an anhydrous one, M_s for a melt saturated with water, and M_u for an undersaturated one. V is water vapor. The more complex natural system (b) has basically the same form but is complicated by the greater number of possible minerals. Mus is miscovite. Ksp is potassium-rich alkali feldspar. Alb is albite. Qtz is quartz, Sil is sillimanite, and L is liquid. [Part (b) is adapted from Thompson and Algor, 1977, *Contrib. Miner. Petrol.* 63:247–269.]

Conditions differ at pressures below that of the invariant point, I, because in this region the temperatures at which hydrous minerals break down are below the melting curve. In fact, the pressure of point I is the minimum pressure at which the hydrous mineral can coexist with any melt, whether it is saturated with water or not. Note that on curve (3), the low-pressure equivalent of (2), the hydrous mineral goes to an anhydrous equivalent plus water vapor, but no melt appears. Only if these products continue to rise in temperature and reach curve (4) can they then produce a melt that is saturated with water. The form of these curves for common minerals in pelitic rocks is shown in part (b) of the same figure.

Melting of Mafic Igneous and Meta-Igneous Rocks

The long-held belief that the only way to produce significant amounts of siliceous magma is to melt siliceous rocks is not strictly true. Under the right conditions, a silica-rich melt can be produced from almost any common multicomponent crustal rock, even a basalt, if the rocks are rich in hornblende or biotite. The amounts of such liquids, of course, are severely limited by the bulk composition of the source.

Typical liquids produced experimentally by partial melting of various types of basalts under elevated water pressures are shown on the diagram for the granite system at 5-kilobars water pressure (**Fig. 10-22**). Note that most first melts fall in the quartz field, and their positions are more or less where one would expect them to be if the mineral assemblage includes abundant plagioclase. Moreover, basalts of different compositions ranging from quartz tholeiites to nepheline basanites would have initial melts with compositions falling close to the line defined by the temperature minimum with varying water pressure. It is curious, and somewhat frustrating, that so many of the effects we have examined, including fractionation of plagioclase, varying water pressure, differing anorthite contents, and now the composition of mafic assemblages, all produce trends of liquid compositions that are almost indistinguishable, at least in terms of the end-members of the granite system.

With progressive melting of an amphibole-rich source rock, liquid compositions first become more silica rich and then poorer in normative orthoclase. This is because so much of these component enters the first melt that little is left in the solid, and further melting simply dilutes the initial concentration. Thereafter, silica contents decline as more amphibole enters the melt.

The quantity of liquid produced by melting increases sharply at temperatures high enough for the hydrous minerals to break down, but this

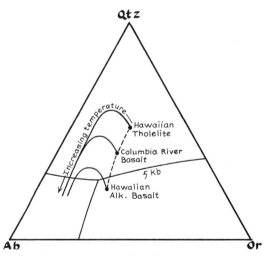

Figure 10-22 Liquids produced by partial melting of basalt at 5 kilobars water pressure. Compositions have been plotted in terms of their normative quartz, albite, and orthoclase. At the conditions of melting, the residue is mainly amphibole and plagioclase. The solid curves show the effects of progressive melting of basalts of differing degrees of silica saturation. (After R. T. Helz, 1976, *Jour. Petrol.* 17:139–193.)

requires higher temperatures for amphibole than metamorphosed pelitic sediments. For example, melting of water-rich metamorphosed pelitic rocks yields about 20-percent melt at 800°, in contrast to amphibolites which require temperatures close to 950° to produce comparable amounts of melt. Such temperatures are beyond the range of normal conditions of regional metamorphism and can be produced only from a high-temperature heat source, such as a nearby mafic intrusion. Because of their higher temperatures and greater water contents, liquids produced in this way have much lower viscosities and are capable of rising more rapidly through the overlying crust.

■ Crystallization and Differentiation in the Crust

By the time plutons are uplifted and exposed by erosion, all traces of magma have long since solidified, and all that one can examine are the cold, dissected remains of what was once a large magmatic intrusion. A petrologist attempting to reconstruct the life of such a body is in much the same position as a surgeon would be if he had studied only cadavers without ever having seen a healthy living organism.

The task of the volcanologist is no less difficult. He must interpret the origins of large masses of siliceous magma erupted from a chamber that he has no hope of observing directly. In a sense, ignimbrites are simply "unroofed batholiths" in that they are derived from the gas-rich upper levels of plutons that have breached the surface. Because they cool too quickly to re-equilibrated at lower temperatures, they preserve a record of the crystal–liquid relations at the instant the magma reached the surface. By integrating information from these volcanic magmas with observations of granitic plutons, one can reconstruct a composite picture of how siliceous magmas evolve.

The Long Valley Caldera of southeastern California (**Fig. 10-23a**) is an exceptionally well-studied and instructive example in which this has been done. One of a series of volcanic centers close to the eastern front of the Sierra Nevada Range, Long Valley, has long been the scene of recurrent activity, including an eruption 700,000 years ago of 170 km^3 of rhyolite known as the Bishop Tuff.

Figure 10-23 (a) Simplified geologic map of Long Valley Caldera, California. (b) Schematic section through the magma body under Long Valley as it has been inferred from the ages, spatial relations, compositions of eruptive units, and geophysical studies. The vertical dimension is exaggerated in order to show the various units more clearly. (After R. A. Bailey, G. E. Dalrymple, and M. A. Lanphere, 1976, *Jour. Geoph. Res.* 81:725–744, and R. A. Bailey, personal communication.)

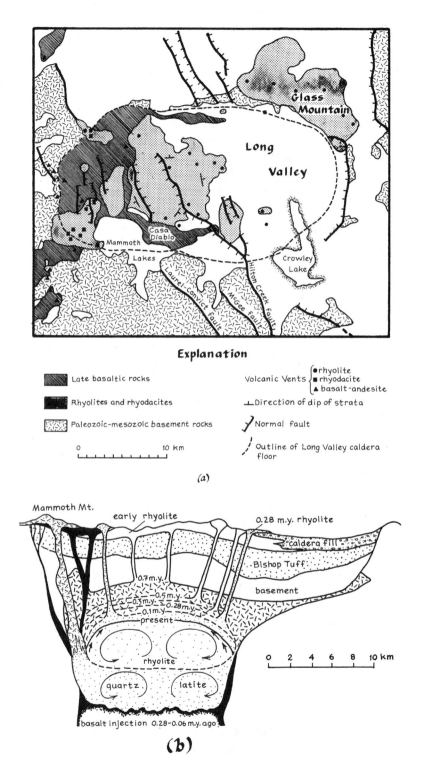

Explanation

Late basaltic rocks

Rhyolites and rhyodacites

Paleozoic-mesozoic basement rocks

0   10 km

Volcanic Vents ⎰• rhyolite
⎱■ rhyodacite
▲ basalt-andesite

⊥ Direction of dip of strata

⅄ Normal fault

Outline of Long Valley caldera floor

(a)

(b)

Some idea of the size and form of the magma chamber can be inferred from geophysical measurements and the surface expression of the caldera. The zoning and temporal variations of the magma have been deduced from the ages, compositions, and spatial distribution of its eruptive units. The picture that emerges is one of a large, compositionally stratified intrusion with the form shown in **Figure 10-23b**.

Using the temperature-dependent compositions of iron-titanium oxide minerals (Fig. 4-16), the temperature of the magma has been shown to have increased from about 720°C for the first magma erupted to about 790°C for the last. Although it varied by only about 2 weight percent SiO_2, the magma was compositionally zoned from the first to last erupted liquids. The first magma, which presumably occupied the uppermost part of the reservoir, was enriched in most incompatible elements, such as Rb, U, Th, Nb, Cs, Ta, Y, and the heavy rare-earths, relative to the last magma discharged from lower levels. However, it is poorer in some incompatible elements, such as Ba, Zr, P, Ce, and the light rare-earths, which are normally enriched by crystal fractionation. Although the liquid fraction of the magma was isotopically inhomogeneous, the Sr 87–86 ratio is in the range of only 0.706 to 0.709. It seems unlikely, therefore, that the trace-element variations could be due to large amounts of assimilation. Ba was probably depleted by crystallization of sanidine, which takes in large amounts of that element. The accessory minerals, allanite, apatite, and zircon, although present in only small amounts, take in large amounts of other incompatible elements and deplete the liquid in the light rare-earths more than the heavy.

Part of the compositional zoning can probably be attributed to inward crystallization and fractionation of light liquids during sidewall crystallization (Fig. 9-14), but density stratification of successive intrusions may have been at least equally important. As the magma evolved, new injections of denser, less-differentiated liquids would tend to pond at the base. Although it is difficult to know how much each of these mechanisms contributes to the zoning, there is little doubt that periodic intrusions of fresh magma prolong the life of the magma by balancing losses to cooling and volcanic eruptions.

■ Origins of Granitic and Rhyolitic Magmas

The "Granite Problem"

Granitic and siliceous volcanic suites that are notably deficient in intermediate compositions but have disproportionately large volumes of extreme differentiates present an obvious problem if one is to attribute their huge volumes to conventional crystal fractionation of basaltic magma. They would require crystallization of parent bodies of impossibly large dimensions.

It is mainly to surmount this volume problem that most explanations for large masses of felsic magma dismiss the possibility of differentiation from mafic magmas and, instead, call on melting of crustal rocks that are already enriched in silica, alumina, and alkalies by surficial processes. This inference is logical, in view of the almost total restriction of large bodies of siliceous rocks to continental settings and their scarcity in regions of oceanic lithosphere. As mechanisms for concentrating lithophile elements, the processes of weathering, erosion, and sedimentation, although slow, are far more efficient than those of magmatic differentiation. By postulating that these crustal rocks are the sources of siliceous magmas, the problem of generating large volumes of granite and rhyolite is greatly reduced.

Geological support for this reasoning is not hard to find. Orogenic granites are typically associated with thick piles of sediments that accumulated in rapidly subsiding geosynclines where they seem to result from deep burial and heating that culminates in upheaval and large-scale plutonism. Crustal melting and generation of magmas has been well documented in many high-grade metamorphic terranes. In some instances, this process results in mobilization of what was clearly a magma; in others, crustal rocks seem to have been converted to granites, more or less in place, by a process of "ultrametamorphism" that transforms meta-sediments of varied compositions into homogeneous, coarse-grained, felsic rocks with textures and compositions very similar to those of magmatic granites. This process of *granitization* involves wholesale compositional changes and exchange of mafic and felsic components over distances of tens or even hundreds of meters. In the past, some of the more extreme advocates of this interpretation maintained that the process takes place entirely in the solid state, but more recent work indicates that a melt of some sort is indispensable.

Two arguments have been offered in support of "granitization" by ultrametamorphism. The first is based on examples of metamorphic rocks that can be seen to grade imperceptibly into rocks that to all intents and purposes are granites but retain a "ghost stratigraphy" inherited from distinctive horizons in the layered rocks from which they are derived. In some cases, distinctive layers of amphibolite, quartzite, and marble project along the strike from a metamorphic sequence into the interior of an otherwise homogeneous granite. Because of their extreme compositions, these metamorphic remnants are more resistant to granitization and persist long after more pelitic rocks have been totally transformed. In other instances, the granitic rocks have only subtle textural differences but retain zones with anomalous concentrations of trace elements inherited from an earlier, metasedimentary protolith.

A second argument was a geometrical one based on the so-called room problem. Because granitic bodies occupy such enormous volumes, they could not be intruded without a corresponding displacement of surrounding rocks. Assimilation and stoping can account for only a minor part of this volume and does not really create new space. It seemed logical, therefore, to explain the emplacement of large plutons, not as introductions of great volumes of new magma, but as a simple redistribution of the components of the crust. This argument lost much of its strength when it was shown that plutons are not steep-sided diapirs of greater vertical extent but sill-like bodies that were intruded from dikes and simply lifted their roof.

The fact that granites have compositions consistent with experimentally determined phase relations implies that they crystallize under conditions approaching equilibrium, but it says nothing about how they reach that state. As we have seen time and again in earlier chapters, a magma can arrive at a particular mineralogical composition by any of a number of routes, including crystal fractionation, partial melting, assimilation, and mixing. To resolve this question, one must rely on a variety of clues, both geological and geochemical.

■ Geological and Geochemical Evidence

Of the many detailed field studies that have contributed to our understanding of granite, one of the most noteworthy was focused on an area in County Donegal in northwestern Ireland, where an extraordinary cluster of well-exposed granitic bodies spans the entire spectrum from catazonal plutons associated with high-grade metamorphism to epizonal intrusions and subvolcanic stocks (**Fig. 10-24**). H. H. Read and Wallace Pitcher, together with their students and co-workers, recognized an evolutionary sequence or "granite series" that encompasses a continuous spectrum of structural and mineralogical relationships. With increasing degrees of melting, magmas were generated, mobilized, and separated from their refractory residue and then rose diapirically to shallower levels. Examples of all stages of this process were identified within a remarkably small area.

The oldest of the Donegal granites is the Thorr Tonalite, a large irregularly shaped body that intruded metasedimentary rocks of diverse lithologies. It contains rafts and xenoliths of all sizes torn from its roof and walls and has converted many of these older rocks to migmatites with differing degrees of textural and compositional homogenization. The Main Donegal Granite (Fig. 10-24b), more properly a granodiorite, is structurally gradational into the surrounding metamorphic rocks and retains a remarkable

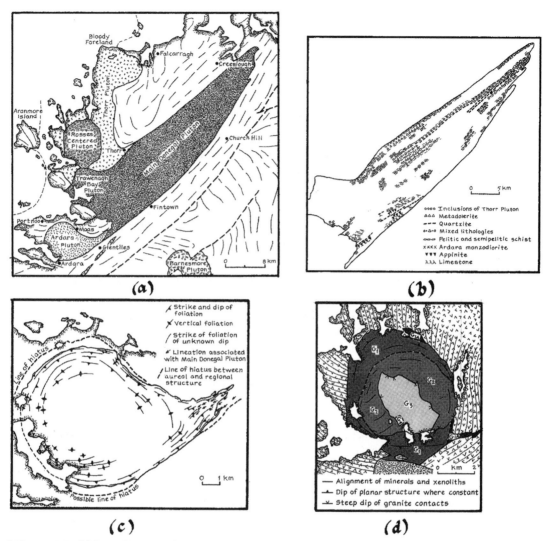

Figure 10-24 (a) Generalized map of Donegal, northwestern Ireland, showing the spatial relations of the main granites. (b) A more detailed map of the Main Donegal Granite showing ghost stratigraphy inherited from granitized metasedimentary rocks. (c) The Ardara pluton illustrates the early stages of mobilization of granitic magma in a zone of plastic deformation. (d) The Rosses Centered Complex is interpreted as a nested set of shallow intrusions. (After a compilation by W. S. Pitcher and A. R. Berger, *Geology of Donegal*, Wiley Interscience, 435 p.)

ghost stratigraphy in the form of layers of quartzite, metadiorite, and schist extending along the strike from the granodiorite into adjacent metamorphic units. Because of their more refractory compositions, these layers were more resistant to melting by wedges of magma that appear to have intruded them from greater depths. The Ardara Pluton (Fig. 10-24c) contains fewer inhomogeneities and, from its curious pod-like form and flow-induced foliation, can be seen to have been intruded plastically, pushing aside the adjacent rocks to take on a tadpole-like shape. The Rosses Ring Complex (Fig. 10-24d) illustrates a more advanced stage of rise into shallower structural levels. It truncates adjacent units and has sharply defined concentric zones produced by successive intrusive pulses.

Those parts of this series that grade into metamorphic rocks were melted and partly mobilized, but the melting did not result so much from deep burial as from heat supplied by magmas rising from deeper levels. The thermal gradients inferred from the widths of the metamorphic zones around deep-seated granites are much steeper than they would be if they resulted from deep burial. Moreover, some of the intrusive bodies are geochemically distinct from the rocks they intrude and must come from melting of a deeper source in the lower crust or upper mantle. Thus, only a small part of the granite has a local origin in the metamorphic rocks; by far the greater part comes from greater depths.

Sorting out the relative contributions of crustal and mantle components to such rocks is a formidable task requiring all the resources of igneous petrology. Studies of the Scottish Tertiary igneous centers discussed in Chapter 7 (Fig. 7-10) are a particularly instructive example because they illustrate the various geochemical tools that can be brought to bear on the problem.

Rare-Earth Elements

Among the trace components of granites, the REEs are especially useful because they vary in abundance by very large factors and are fractionated in different ways by individual minerals. The range of rare-earth abundances of eleven Scottish granites are shown in **Figure 10-25**, together with corresponding data for basement rocks and basalts from the same region. If we compare the granites with the Lewisian gneiss and Torridonian sandstones making up the thick basement series, the two groups have no overlap; the abundances in the granites are uniformly higher. This is what we would expect if the granites are partial melts because in such a process the REE would be enriched in the liquid fraction by an amount that varies inversely with the proportion of melt. Note, however, that the amount of enrichment is not uniform; the light REE are enriched much less than the heavier elements. The relative degrees of enrichment of the light and heavy elements depend on the identities of the residual crystalline phases, which have differing distribution coefficients relative to a liquid. All the common residual minerals that are likely to be left by partial melting of felsic crustal rocks would tend to increase the light elements in the liquid more than heavy ones

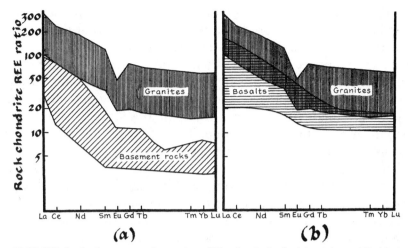

Figure 10-25 REE distribution patterns for granites of Skye (vertical ruling), Lewisian and Torridonian basement rocks (diagonal ruling), and basaltic lavas and dikes (horizontal ruling). (After *Igneous Case Study*, The Open University, 1980.)

(Fig. 5-24). The opposite seems to be true of the rocks of Skye. Hence, if crustal rocks contributed anything to the granites, it was either a very small amount, or it was a material unlike the exposed basement and had an anomalous REE pattern.

Comparing the granites with their associated basalts (Fig. 10-25b), we note that although the granites are richer in the REE, there is a substantial range of overlap between the two sets of data. This lends support to the interpretation of the granites as differentiates of more basic magmas. Just as in partial melting, liquids produced by crystal fractionation tend to be enriched in the REE, particularly at the light end of the series. The two sets of data could be interpreted as a continuum in which the REE in the granites have been enriched by differentiation of more mafic magmas. Moreover, the proportion of light REE increases with progressive differentiation, just as one would expect if pyroxene or hornblende had been fractionated. The negative Eu anomaly indicates that plagioclase was among the fractionated phases.

Strontium Isotopes

We have already seen from our consideration of calc-alkaline volcanic rocks in Chapter 9 how the Rb-Sr system has been used to evaluate the contribution of crustal material to andesitic magmas. The same system can be used to even greater effect in studying granites and rhyolites. The Rb-rich Precambrian basement rocks underlying the Tertiary igneous centers of Skye have $^{87}Sr/^{86}Sr$ ratios of the order of 0.720 or more. Thus, their strontium is much more radiogenic than that of the basalts and should produce a marked contrast between magmas produced by melting of crustal rocks and those coming from the mantle. When the various igneous rocks of Skye were analyzed and the ratios recalculated to their initial values at the time of crystallization about 53 million years ago, they were found to fall into two distinct groups. Granites, granophyres, and rhyolites have larger ratios than the basalts and ultramafic rocks, and a histogram (**Fig. 10-26**) shows two separate populations with virtually no overlap. The difference is much greater than the possible margin of error in the analytical data.

What is surprising is not that the isotopic ratios differ but that they differ so little. The Sr ratios of the granites, although larger than those of the basalts, are much less than would be expected from source rocks with ratios of 0.720 or more. If they are partial melts their ratios should be at least as large, and under some conditions, they could be much greater. The reason for this can be seen by examining where the radiogenic Sr resides in the rocks and how it is partitioned into early melts.

When the individual minerals in ancient metamorphic and plutonic rocks are analyzed separately, they are found to have Sr ratios that differ greatly according to their individual Rb/Sr ratios (Fig. 9-22). Micas and potassium feldspars, which have proportionately more Rb and less Sr, evolve proportionately more radiogenic Sr. We saw in an earlier section (Fig. 10-16) that these same minerals are among the first phases to break down or melt with rising temperature at relatively low pressures. Hence, the first melts

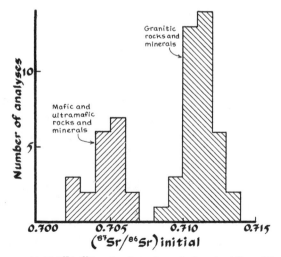

Figure 10-26 Histogram of initial $^{87}Sr/^{86}Sr$ ratios for rocks and minerals of Skye. (Modified from a figure in *Igneous Case Study*, The Open University, 1976.)

produced by rapid melting of a source containing rubidium-rich minerals could have strontium ratios that are more radiogenic than that of the original source rock.

This effect can be seen in coarse-grained xenoliths that have been partly remelted in basalts. An example from a Late Cenozoic basaltic lava in Arizona contains several percent of black, magnetite-rich glass formed by partial melting breakdown of hydrous minerals and when the felsic xenolith was immersed in the basalt. While the Sr ratio of the entire xenolith was found to be only 0.7060, that of the glass was 0.7231. This impressive enrichment, together with the large amount of magnetite in the glass, suggests that the glass was an early liquid produced by breakdown of biotite.

It is doubtful whether such large differences would be preserved during a prolonged period of heating leading up to melting of the basement. More likely, the minerals would re-equilibrate before large amounts of melt could be produced. If micas break down early in the heating process, however, the water-rich solutions released from the rock could carry large concentrations of soluble components, including Sr, into any nearby magma. Because granites and rhyolites have very low absolute abundances of strontium, a small amount of contamination with ^{87}Sr from such a source would cause a marked increase in the isotopic ratios of the magmas. Other isotopic systems seem to confirm that this is indeed what happened in the case of the Skye granites.

Neodymium Isotopes

As we have noted earlier, basalts and mantle rocks tend to have relatively large ratios of Sm to Nd (i.e., they are relatively depleted in light REE). Crustal rocks, such as the Lewisian gneiss of Scotland, are quite different; they have relatively small Sm/Nd ratios and are enriched in the light REE. As a consequence, the $^{143}Nd/^{144}Nd$ ratio of granite derived from felsic continental basement rocks should be relatively small compared with contemporaneous mantle-derived rocks. Conversely, a large ratio would indicate derivation from a mantle source or basalt. When the ratios of Nd of the Skye rocks are plotted against those of Sr (**Fig. 10-27**), most suites of related rocks define lines with negative slopes. This is why mantle-derived volcanic rocks of the oceans fall on the "mantle array" illustrated in Chapter 8 (Fig. 8-2). Oceanic rocks have a relatively small amount of variation along this line because they differ little in their Rb/Sr and Sm/Nd ratios, but a much wider range is found in crustal rocks, such as those forming the metamorphic basement under western Scotland.

The $^{143}Nd/^{144}Nd$ ratios of the granites of Skye are about 0.5115 and fall between those of the Lewisian gneiss (about 0.5110) and young basalts

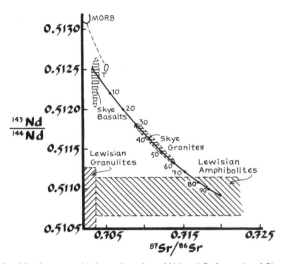

Figure 10-27 Relationships between the isotopic ratios of Nd and Sr for rocks of Skye. Also shown is a typical trend for basalts of mid-ocean ridges (MORB) and other oceanic rocks (T). (After a compilation by R. S. Thorpe and P. W. Francis, 1979, *Origins of Granitic Batholiths: Geochemical Evidence*, M. P. Atherton and J. Tarney, eds., Shiva Publ. Co., pp. 65–75.)

(about 0.5125), and when plotted against the ratios of Sr isotopes in the same rocks, they define distinct fields. Note that the spread of values for the granites is consistent with mixing of the basalt with Lewisian basement rocks having relatively large $^{87}Sr/^{86}Sr$ ratios. The proportions contributed by the basement, calculated by a simple mass balance, are indicated by the numbers along the mixing curve. Note that between 30 and 60 percent of the Nd and Sr in the granite appears to be from the basement rock.

This conclusion based on the radiogenic isotopes seems to conflict with the deduction, drawn from the REEs alone, that the granites have little if any crustal component. The apparent anomaly might be explained by postulating fractionation of a mineral relatively rich in the light REE. Although none of the common minerals of granites have the appropriate patterns to do this, certain REE-rich accessory minerals, such as sphene, apatite, and allanite, could effect these changes, provided they were fractionated in large amounts. Another, more plausible possibility is the one suggested by the strontium, namely that the isotopic ratios of the granites result from contamination by fluids released by breakdown of hydrous minerals in the basement wall rocks.

Lead Isotopes

The isotopes of lead afford yet another way of detecting crustal contamination, but they can yield much more information than either strontium or neodymium. The isotopes of most use are those produced by radioactive decay of uranium (Appendix D). Like rubidium, uranium and lead are strongly concentrated in the continental crust. The two important isotopes, ^{235}U and ^{238}U, decay with long half lives to ^{207}Pb and ^{206}Pb, respectively. Just as the isotope ^{86}Sr is used as a reference for the abundance of radiogenic ^{87}Sr, ^{204}Pb, a nonradiogenic isotope of lead, is used as a reference for radiogenic ^{207}Pb and ^{206}Pb.

Because ^{207}Pb and ^{206}Pb are produced at different rates and the rates change as the amounts of the parent isotopes decline, a curved line will result when their ratios to ^{204}Pb are plotted against each other, as they are in the growth curve shown in **Figure 10-28a**. Note that the ratio $^{207}Pb/^{204}Pb$ increases at a diminishing rate compared with $^{206}Pb/^{204}Pb$. This change reflects the

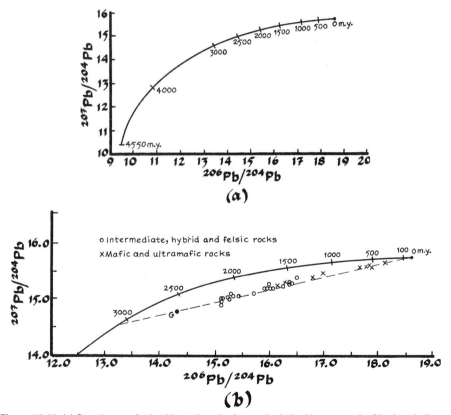

Figure 10-28 (a) Growth curve for lead isotopic ratios in mantle-derived igneous rocks. Numbers indicate ages in millions of years. The leads from basalts differ slightly from one province to another. (b) An enlarged section of the growth curve in (a) showing the principal rocks of Skye. (After *Igneous Case Study*, The Open University, 1976.)

different half-lives and initial abundances of the uranium isotopes from which the leads evolve. When the lead ratios of rocks of mantle origin are corrected for age and plotted against one another, they fall close to a curve of this form. Igneous rocks of different provinces have slightly different growth curves, indicating that their mantle sources are not entirely homogeneous.

These relationships can be used in a number of ways. First, one can measure the ages of certain types of igneous rocks by simply determining where their lead ratios fall on the growth curve. Second, they can be used to determine whether the lead in a given rock or mineral has come directly from depleted mantle or has evolved in the upper crust or lithosphere. Once separated from the mantle, the proportions of uranium and lead in a rock produced by partial melting and other processes, including weathering, erosion, and mixture with other rocks, may differ from those responsible for the normal growth curve. Hence, the lead in such a rock will evolve in different ways and will not fall on the same curve as rocks derived directly from the mantle.

If an ancient crustal rock with a composition displaced from the growth curve is incorporated into a younger magma, the lead of the contaminated magma no longer corresponds to that of the mantle. Its composition is the result of mixing two leads, one from the mantle and another from the contaminant. **Figure 10-28b** illustrates the measured ratios for a variety of Tertiary igneous rocks from Skye. Note that they fall on a remarkably linear cord, one end of which intersects the growth curve at a point corresponding to the time of igneous activity 60 million years ago. The average composition of the metamorphic Lewisian complex also falls on this line at point G. (The position of this point on the concave side of the growth curve is thought to result from deep-seated crustal processes that removed large amounts of the lithophile elements, including U, about 3000 million years ago.)

Skye granites have lead with compositions closer to that of the average Lewisian gneiss than do intermediate and mafic rocks. Thus, the spread of compositions is consistent with differing contributions of lead from the metamorphic complex to lead derived from the mantle at the time the magmas rose through the crust.

Note that mafic and even ultramafic rocks have also been contaminated with lead from a U-depleted source, even though they show no other evidence of assimilation. Because their lead contents are very small, the amounts required to change the ratios in these rocks are also small. The granites, however, have much greater absolute concentrations of lead, and to attain their present ratios they would have to gain proportionately greater amounts from the metamorphic crustal rocks.

Oxygen Isotopes

In weighing data such as these, it is important to bear in mind that the rocks have had complex histories, and the magmatic processes ending with crystallization were not necessarily the last chapter of their life story. Interpretations of isotopic relations of the granites of Skye illustrate this lesson very well.

Shortly after the results obtained from lead and strontium had been presented and petrologists had accepted the conclusion that crustal rocks had played an important role in the genesis of the granites of Skye, another study was initiated to determine the effects of hydrothermal alteration on the isotopic character of oxygen in the same rocks. We have already noted in Chapter 6 that circulation of meteoric water through shallow intrusions has the effect of exchanging oxygen and that this results in a lowering of the ratio of ^{18}O to ^{16}O. This same effect was found in the rocks of Skye (**Fig. 10-29**). Large quantities of hot, meteoric water must have moved through the rocks. Hydrothermal waters are known to carry as much as 100 ppm Pb and 750 ppm Sr scavenged from the rocks through which the water flows. Hence, if oxygen, the most abundant element in the rocks, is exchanged with hydrothermal solutions, why not lead and strontium as well? Only minute amounts would have to be selectively dissolved from the metamorphic and sedimentary basement and deposited in the igneous rocks to produce the observed isotopic compositions.

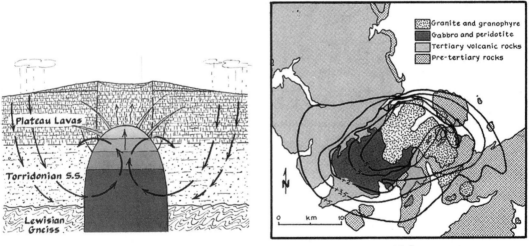

(a)   **(b)**

Figure 10-29 (a) A schematic section illustrating subsurface structures of the Skye granites as inferred from surface exposures and geophysical measurements. Also shown is the pattern of convective circulation of meteoric water. (b) Zoning of oxygen isotopic ratios resulting from circulation of meteoric water in a hydrothermal system that continued long after the main bodies of magma had solidified. (After R. W. Forester and H. P. Taylor, Jr., 1977, *Amer. Jour. Sci.* 277:136–177.)

If hydrothermal solutions had contributed lead to the rocks we would expect the effect on the Pb ratios to be greatest in rocks with the smallest initial lead contents, namely rocks of mafic composition and with the smallest oxygen ratios. The fact that no such correlation is seen suggests that the amount of contamination with hydrothermal lead has not been great. Nevertheless, the problem of subsolidus alteration is one that should always be kept in mind, especially in dealing with large intrusions that have cooled in or near a zone of hydrothermal circulation.

■ Granites and Crustal Evolution

The conclusion that we are left with after this examination of the rocks of Donegal and Skye is that the geochemical character of the magmas is the result of differing contributions from the mantle-derived magmas and the continental crust through which they rose. Moreover, such liquids could not have risen from their source without the additional heat gained from the mafic, mantle-derived magma.

It was long assumed that voluminous felsic magmas could only be produced by melting of crustal rocks of similar composition. While many are indeed directly traceable to crustal sources, others clearly are not. Although all are rich in silica and other lithophile components, many have isotopic ratios that are not notably radiogenic, and minor amounts of contamination could account for any small excess over the ratios of mantle-derived basalts.

One might well ask why, if siliceous magmas are not necessarily products of anatectic melting of continental sediments, are they confined to continental settings? The answer, of course, is that they are not. The rhyolites of Iceland, the Galapagos, and other oceanic islands, though rare, are highly instructive, for they demonstrate that such magmas can be produced in the absence of continental crust provided certain special conditions prevail.

The amount of rhyolite that can be derived directly from a mantle peridotite is, of course, trivial. We have seen, however, that rhyolitic liquids can be produced in substantial volumes by melting of basalts at elevated water pressures. Thus, an appropriate lithosphere, namely one of amphibolitic character, could be a suitable source rock, whether the overlying crust is oceanic or continental. We noted in the opening chapter that the long-term compositional stability of the continents could not be maintained unless the contribution of mantle-derived mafic magmas were balanced by comparable additions of more felsic material. Although the process is indirect and much slower than it was during the early history of the earth, the mantle continues to contribute felsic material to the continents. In the chapter that follows, we examine one final group of continental igneous rocks that are another important element in the long-term evolution of the Earth.

Selected References

Clemens, J. D., N. Petford, and C. K. Mawer, 1997, Ascent mechanisms of granitic magmas: causes and consequences. In: Holness, M. B. (ed.). *Deformation-Enhanced Fluid Transport in the Earth's Crust and Mantle.* Chapman Hall, London, pp. 145–172. An excellent treatment of emplacement mechanisms.

Petford, N., A. R. Cruden, K. J. W. McCaffrey, and J. L. Vigneresse, 2000, Granite magma formation, transport and emplacement in the Earth's crust. *Nature* 408:669–673. A concise summary of new interpretations of the generation, rise, and emplacement of plutonic rocks with references to recent work on these subjects.

Pitcher, W. S., and A. R. Berger, 1972, *The Geology of Donegal: A Study of Granite Emplacement and Unroofing*, Wiley Interscience, 435 p. A summary of detailed studies of a series of plutons illustrating a wide range of structural and mineralogical relations ranging from ultra-metamorphism to a high-level ring complex. The work in Donegal had a great influence on many basic concepts of granitic magmatism.

Smith, R. B., and L. W. Braile, 1994, The Yellowstone hotspot, *Jour. Volc. Geoth. Res.* 61:121–187. A very comprehensive paper on the structural and petrologic aspects of a continental hotspot and the origins of large siliceous ignimbrites.

Tuttle, O. F., and N. L. Bowen, 1958, Origin of granite in the light of experimental studies in the system $NaAlSi_3O_8$-$KAlSi_3O_8$-SiO_2-H_2O, *Geol. Soc. Amer. Memoir* 74:153. An early account of experimental work on the simple "Granite System" and how it pertains to plutonic rocks. This classic work had a major influence on interpretations of granitic rocks.

11 Magmatism of the Continental Interiors

The previous chapter dealt with a wide range of felsic plutonic and volcanic rocks, mainly in orogenic settings. We turn now to a more diverse group of continental rocks, mostly of alkaline compositions, that occur in the interior regions of continents, far from the orogenic belts of convergent plate boundaries. Although some are associated with continental rifting, many of these rocks have no distinctive tectonic environment. Indeed, if they share anything in common, it is their tendency to appear in widely differing settings that seem to have little relationship to their unusual compositions.

For this reason, it is easier to group these rocks according to their compositions than to their tectonic setting. The main types are best distinguished by their relative concentrations of three components, Na, K, and Ca, and each of these groups may be relatively poor in SiO_2, Al_2O_3, or both.

Unlike oceanic suites in which sodium is the dominant alkali and Na_2O/K_2O is rarely less than 2.0, continental suites have a much greater range of alkali ratios, and either sodium or potassium may be more abundant. Moreover, some of these rocks, especially sodic types, have abnormally large amounts of calcium, which may be in the form of calcite or one of a group of calcium-alumina silicates, such as melilite or the calcium-bearing olivine, monticellite.

When these Na-, K-, or Ca-rich types are deficient in SiO_2, they are *alkaline* in the sense that their composition is "critically undersaturated" with silica and yields Ne in the norm. If they are deficient in Al_2O_3, they are *peralkaline* in the sense that alkalies are exceptionally enriched with respect to Al_2O_3 but not necessarily SiO_2. That is, they have a molecular excess of $Na_2O + K_2O$ over Al_2O_3 so that the alumina content is insufficient to accommodate all the alkalies in normative feldspars.

Taken as a whole, the rocks span an especially wide range between felsic and ultramafic end-members, but individual groups may have a very limited range of differentiation. One group, the *lamprophyres*, includes a great variety of very mafic alkaline compositions that fit no scheme of differentiation or genetic lineage. Some are associated with calc-alkaline plutonic rocks, others with alkaline and peralkaline series, and still others have no clear petrologic or tectonic association of any kind.

It must be emphasized that the various divisions of these rocks are gradational one into the other, and any classification, either compositional or genetic, is simply an artificial device to aid in recognizing certain common associations. The limitations of these classifications will become evident as we consider each of the various groups in turn.

■ Alkaline and Peralkaline Series

Alkali-Silica relationships

Many continental alkaline series share certain of the distinguishing marks of their oceanic counterparts. Some resemble the familiar Hawaiian alkaline series of alkali olivine basalt-hawaiite-mugearite-benmoreite-trachyte in that they straddle the division between hypersthene-normative and nepheline-normative compositions. In the more alkaline series, basanites have differentiated through feldspathoidal intermediate rocks to phonolites not unlike those of Tahiti described in Chapter 7. Continental suites tend to include rocks of more extreme compositions, such as *nephelinites* and *leucitites*, with very abundant Na or K feldspathoids but little or no feldspar.

Thus, one can think of a continuum including sodic rocks of various degrees of alkalinity or peralkalinity at one end and a corresponding potassic series at the other. Between these two extremes, a large intermediate group has roughly equal proportions of the alkalies and two separate feldspars, one plagioclase and the other anorthoclase, sanidine, or orthoclase. These two-feldspar volcanic rocks, loosely referred to as *trachytic*, include rocks that have been called *trachybasalts*, *latites* (or *trachyandesites*), *tristanites*, and *trachytes*. (Note that the use of the term trachyte in this sense implies nothing about the degree of silica saturation, as it does when used to refer to differentiated rocks with less than about 10-percent normative Q and no normative Ne.) Some varieties contain small amounts of modal leucite.

Alkali ratios tend to vary widely with differentiation, even in a single series (**Fig. 11-1**). A common tendency is for series initially rich in either sodium or potassium to converge with differentiation on felsic end-members with more nearly equal amounts of the two alkalies, as would be expected if the liquids were descending toward a eutectic-like end-member. Looking again at "Petrogeny's residua system" (**Figs. 11-2 and 11-3**) and noting especially the range extending from the granitic rocks of the preceding chapter into that of more silica-poor compositions considered here, we find that the felsic differentiates of alkaline magmas fall in a thermal valley crossing the central

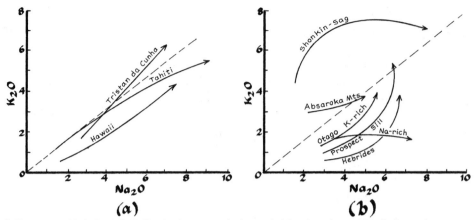

Figure 11-1 Variations of alkali ratios in some typical oceanic (a) and continental (b) alkaline series (Hawaiian curve after H. S. Yoder, Jr., and F. Chayes, 1982, *Carnegie Inst. Washington Yrbk.* 81:309–314).

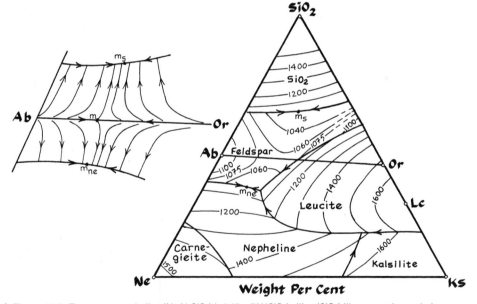

Figure 11-2 The system nepheline (NaAl-SiO₄)-kalsilite (KAlSiO₄)-silica (SiO₂) illustrates the evolution of felsic differentiates toward compositions enriched or depleted in silica. The insert on the left shows paths of cooling liquids in the low-temperature region of the albite-orthoclase join. Orthoclase melts incongruently to leucite and liquid but only at low pressures. Carnegieite is the high-temperature form of nepheline; it is not found in nature. (After J. F. Schairer, 1950, *Jour. Geol.* 58:512–517. The small diagram at left is adapted from a similar figure by S. A. Morse, 1980, *Basalts and Phase Diagrams,* Fig. 15-11.)

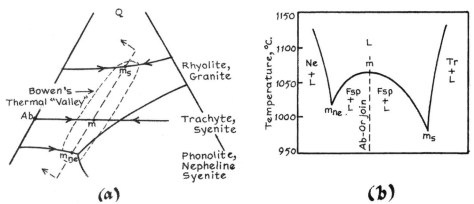

Figure 11-3 (a) The thermal valley defined by the minima in the system Ne-Ks-SiO_2 in Figure 11-2. The section shown in (b) passes through the compositions of felsic differentiates of alkaline, transitional, and subalkaline series.

part of the diagram. Ignoring for the moment the field of leucite, consider only the region of the thermal trough. Note that syenites and trachytes occupy a region near the feldspar join where compositions have little or no normative quartz or feldspathoid. Silica-deficient nepheline syenites and phonolites have compositions below this join close to a minimum on the cotectic boundary between the feldspars and feldspathoids.

The course of a differentiating liquid that crystallizes alkali feldspars along the Ab-Or join should diverge from that line and descend along curved paths away from feldspar and toward one side or the other of a thermal trough. The nature of this divide is shown more clearly by a section down the axis of the thermal trough (Fig. 11-3b). Any liquid with an initial composition on the silica-saturated side of the divide should become richer in SiO_2 and evolve toward a granitic composition; liquids even slightly undersaturated should descend toward the low-SiO_2 minimum. In nature, however, many series straddle this divide and stubbornly refuse to descend to either of the two minima.

Alkali-Alumina Relationships

The peralkalinity that is so characteristic of many continental series results from a relationship between alkalies and alumina that is in some ways similar to that between alkalies and silica. Depending on how the amount of alumina is related to that required for the feldspars or feldspathoids, a rock is said to be *peraluminous, metaluminous,* or *peralkaline.*

In *peraluminous* rocks, molecular $Al_2O_3 > Na_2O + K_2O + CaO$. This excess is expressed in the norm as corundum, Al_2O_3, and in the rocks by minerals such as mica or garnet. In *metaluminous* rocks, molecular $Na_2O + K_2O + CaO > Al_2O_3 > Na_2O + K_2O$, and the norm has both anorthite and diopside. This group includes most of the common subalkaline magmas as well as the great majority of oceanic alkaline rocks. In *peralkaline* rocks, molecular $Na_2O + K_2O > Al_2O_3$, and as a result, part of the Na_2O is assigned to normative acmite ($Ac - NaFeSi_2O_6$) and, in rocks very rich in Na or K, to sodium or potassium metasilicate ($Ns - Na_2SiO_3$ and $Ks - K_2SiO_3$, respectively). In natural assemblages, the excess alkalies enter pyroxene, amphibole, and other minerals in which iron and titanium take the place of aluminum.

Peraluminous, metaluminous, or peralkaline rocks may be saturated or undersaturated with silica and may have normative Ne, Hy, or Q. Q-normative felsic peralkaline rocks are sometimes referred to as *comendites* or *pantellerites*, but the use of these names to characterize such a rock series can be somewhat misleading. Defined originally on the basis of petrographic characteristics, the names have been applied so loosely to rocks of various chemical compositions that modern usage has tended to avoid them in favor of the more general term, peralkaline rhyolite.

Differences in the amount of alumina are closely reflected by the proportions of feldspars and mafic minerals in modal assemblages. Note that anorthite, $CaAl_2Si_2O_8$, contains twice as much aluminum as the alkali feldspars, $(Na,K)AlSi_3O_8$. Because all liquids crystallize a feldspar that is relatively richer in anorthite, and hence Al, than their coexisting liquid, it follows that at the same time the liquid is enriched in Na and Si it is also impoverished in both Al and Ca. Hence, crystallization of plagioclase should cause derivative liquids to become relatively depleted in Al and hence potentially peralkaline. This is true, of course, only if feldspar is the dominant crystallizing mineral. If ferromagnesian minerals lacking Al are also crystallizing in abundance, their removal may offset the effect of plagioclase and impede depletion of Al, as was pointed out in Figure 5-13. Thus, whether a given magma follows a peralkaline trend depends chiefly on the proportions of feldspar and ferromagnesian minerals in the fractionating assemblage.

■ Alkaline and Peralkaline Series of Continental Rifts

A few examples of alkaline and peralkaline rocks of widely differing ages will serve to illustrate the salient features of these magmas in continental extensional environments. Starting with the presently active East African Rift and then turning to the progressively older Oslo Graben and Gardar Province of Greenland, the characteristics these rocks share can be traced through almost the entire span of Phanerozoic time.

East Africa

The largest and perhaps the most varied province of Cenozoic continental volcanism is the one extending from the Red Sea and Afar Depression through Ethiopia and the rift valleys of East Africa as far as southwestern Tanzania (**Fig. 11-4**).

The most active part of this region today is that of the Afar Depression in Ethiopia where floods of alkaline basalt issue from fissures along an active rift extending southward from a junction with the Red Sea. Starting in

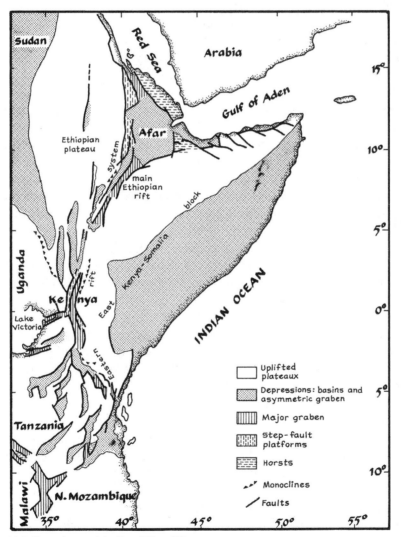

I Figure 11-4 General map of the East African Rift zone.

Eocene time and continuing with few interruptions down to the present, floods of magma covered what are now the broad slopes of the Ethiopian Plateau. Most of the activity before the main period of uplift took the form of fissure eruptions of both basalt and siliceous ignimbrites but changed during the late Miocene and Pliocene to shield-building eruptions of basalt and closely associated rhyolite and phonolite. In the Plio-Pleistocene, explosive eruptions of siliceous magma were confined mainly to the axis of the rift, while basalts poured out on the adjacent plateau. The current activity in the rift produces both basalt and rhyolite, with the latter accounting for an unusually large proportion, perhaps as much as 25 percent of the volume of eruptive rocks. Representative compositions are given in **Table 11-1**.

This activity in the northern part of the system differs from that farther south in Kenya and Tanzania, where the Tertiary activity produced much less basalt but relatively greater proportions of differentiated magmas, including very fluid sheets of flood trachytes and phonolites that spread from vents on the floor and margins of the rift. In the eastern or Gregory Rift, the rocks tend to be sodic, whereas volcanoes along the western branch have produced some of the most potassic lavas known. Of the total of 220,000 cubic kilometers of Cenozoic volcanic rocks in the region of the Kenya Rift, about half consist of phonolite, trachyte, and rhyolite, and many of the central volcanoes consist almost exclusively of these highly differentiated rocks. Basalts are rare throughout the late Cenozoic series, especially in the central part of the rift.

The various types of lavas, ignimbrites, and tephra have a range of compositions rarely found in such close association. Individual Quaternary volcanoes along the axis of the Gregory Rift of Kenya and Tanzania are composed almost entirely of phonolite (e.g. Suswa), trachyte (Menengai), or peralkaline rhyolite (near Lake Naivasha). Still others, such as Kilimanjaro, Mt. Kenya, and the large Pliocene center of Mt. Olokasalie in south-central Kenya, have produced a variety of series mixed in seemingly random fashion throughout the eruptive sequence. The rocks of Mt. Olokasalie form at least three distinct trends with differing degrees of silica saturation: one hypersthene normative, one mildly alkaline (with a few percent normative Ne), and a third that is much more strongly undersaturated in silica and includes rocks with more than 7-percent normative nepheline.

The petrologic identity of each these series is well defined by the different trends of their large-ion lithophile elements (LILE) with differentiation (**Fig. 11-5**). Silica, which is such a useful measure of differentiation in most other types of rocks, can be misleading in the highly alkaline series, for it may increase nonlinearly or even reverse its trend and decline in the latest stages of differentiation. The incompatible elements, which are so plentiful in these rocks, provide much better indices of differentiation.

Table 11-1 Chemical compositions of some characteristic rocks of a peralkaline series from the Afar Rift, Ethiopia.

	1.	2.	3.	4.	5.	6.	7.
SiO_2	48.36	47.00	56.81	59.44	65.02	69.91	72.11
TiO_2	1.92	3.10	1.76	1.82	0.36	0.43	0.38
Al_2O_3	15.34	13.26	13.88	15.00	14.88	13.14	9.35
Fe_2O_3	2.42	7.82	0.70	2.02	1.75	2.83	2.30
FeO	7.75	8.22	9.37	5.45	3.48	1.53	3.80
MnO	0.17	0.19	0.29	0.21	0.13	0.14	0.21
MgO	7.83	4.84	2.13	1.76	0.04	0.01	<0.01
CaO	12.04	8.62	5.04	4.00	1.34	0.71	0.34
Na_2O	2.42	3.70	5.00	4.97	5.90	5.95	5.74
K_2O	0.63	1.55	2.15	3.27	4.30	4.52	4.40
P_2O_5	0.32	0.59	0.72	0.48	0.04	0.01	0.02
Trace Elements (ppm)							
Ni	33	10	6	9	2	2	8
Cr	291	18	6	49	56	–	77
Ba	–	541	408	715	736	652	<10
Rb	13	33	49	79	106	111	147
Sr	343	462	360	323	60	24	3
La	19.9	48.7	60.0	80.8	111.5	–	159.5
Sm	–	10.6	16.6	13.8	17.4	–	27.0
Lu	0.3	0.4	1.0	0.9	1.4	–	2.0
Molecular Norms							
Ap	0.68	1.28	1.55	1.02	0.09	0.02	0.04
Il	2.71	4.48	2.53	2.59	0.51	0.60	0.54
Mt	2.56	8.49	0.75	2.15	1.87	0.44	–
Or	3.77	9.51	13.10	19.70	25.97	26.86	26.57
Ab	21.99	34.50	46.29	45.51	54.15	45.28	25.59
An	29.49	15.57	9.36	9.14	1.45	–	–
Di	23.08	19.86	9.25	6.32	4.06	2.78	1.27
Hy	7.81	3.22	13.84	6.72	2.04	0.35	5.18
Ol	7.90	3.09	–	–	–	–	–
Q	–	–	3.33	6.84	9.86	16.90	28.58
Ac	–	–	–	–	–	6.77	6.55
Ns	–	–	–	–	–	–	5.67

1. Olivine basalt. 2. Ferro-basalt. 3. Dark trachyte. 4. Oversaturated trachyte. 5. Alkali rhyolite. 6. Pantellerite or peralkaline rhyolite. 7. Comendite or peralkaline rhyolite.

Source: F. Barberi, et al., 1975, Jour. Petr. 16:22–56.

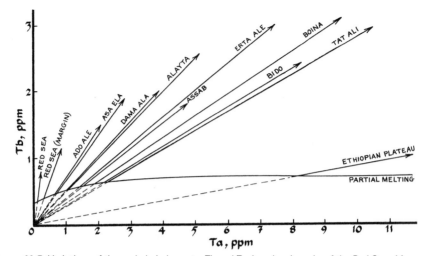

Figure 11-5 Variations of the excluded elements, Tb and Ta, in volcanic rocks of the Red Sea, Afar Depression, and Ethiopian Plateau. Rocks erupted from each volcanic center show linear correlations between the two elements, as would be expected for fractional crystallization of a magma having the Tb/Ta ratios along a curve defined by differing degrees of partial melting of a source that is common for the entire region. Compare with Figure 5-23. (After G. Ferrara and M. Treuil, 1974, *Bull. Volc.* 38:548–574.)

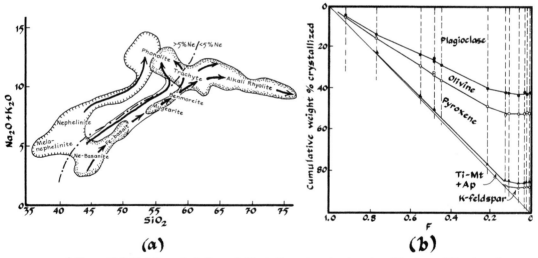

Figure 11-6 Variations of alkalies and silica in the major volcanic series of the Kenya Rift system of Figure 11-4. The heavy dashed line with dots separates rocks with greater than and less than 5-percent normative nepheline. (After B. H. Baker, 1987, Alkaline igneous rocks, *Geol. Soc. Spec. Publ.* No. 30, 293–312.) (b) Cumulative weight percentages of minerals fractionated from a parental basaltic magma to produce the various differentiated rocks of the basalt-benmoreite-trachyte series in (a). F is the fraction of liquid represented by individual rocks. Calculations are based on the method illustrated in Table 7-4. (After B. H. Baker, *et al.*, 1977, *Contr. Mineral. Petr.* 64:303–332.)

The parental magmas of each individual series can be explained as products of differing degrees of partial melting, whereas their differentiated series evolve along separate trends governed by crystal fractionation (**Fig. 11-6**) The huge volumes of phonolitic and trachytic magma and paucity of basalts seem to demand a special mechanism of differentiation. We return to consider this problem in a later section.

The Oslo Graben

A close plutonic counterpart to the East African volcanic series can be seen in the Oslo Graben of southern Norway (**Fig. 11-7**). Although volcanic rocks are still preserved in a number of thick sections of felsic ignimbrites, the roots from which they were erupted are also visible in a cluster of calderas and subvolcanic intrusions.

The volcanic series consists of an early group of basaltic lavas, apparently the products of fissure eruptions, followed by large volumes of differentiated lavas and ignimbrites. The lavas were given the local name *rhomb porphyries* because of their prominent feldspar phenocrysts but are more generally classified as trachyandesites and latites. Their moderately high potassium content, which in some of the more felsic rocks exceeds that of sodium, is responsible for the prominence of the alkali feldspar that is so characteristic of these rocks. A few rhyolitic lavas are found in the upper parts of the section, but like basalts, they are very subordinate in volume to intermediate rocks.

The intrusive rocks have been divided into two groups: one making up a series of subvolcanic gabbroic and dioritic sills and pipes that must have fed many of the lavas erupted early in the history of the region and the other a cluster of plutonic stocks and ring dikes of syenitic and granitic compositions. Various forms of potassium-rich feldspar appear in all but the most basic compositions. Even some of the gabbros contain enough potassium feldspar to deserve the name *essexite*, and in the felsic plutonic rocks, alkali-feldspar is dominant over plagioclase and may even be the sole feldspar. By far the most voluminous member of the series is a varied group of augite syenites (*larvikites*), which are plutonic equivalents of the rhomb porphyries. Some of these contain quartz, others nepheline.

Depending on whether the feldspars are a subsolvus assemblage of two feldspars, oligoclase and alkali feldspar, or a single hypersolvus micropherthite, these rocks have very different appearances, both in hand specimen and under the microscope (**Fig. 11-8**). The perthitic variety is a beautiful, dark-bluish rock in which the feldspar has a *schiller* structure, giving it a spectacular satiny sheen. A popular ornamental stone, it is commonly seen in the facing of hotels and bars, particularly those aspiring to an affluent clientele.

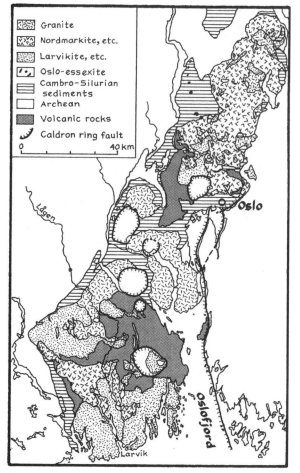

Figure 11-7 Generalized map of the Oslo region of southern Norway. (After C. Oftedahl, 1978, *Jour. Volc. Geoth. Res.* 3:343–371, with modifications from other work by the same author.)

With progressive differentiation, larvikites give way to two divergent series. One becomes progressively oversaturated with SiO_2 as it follows a trend through single-feldspar syenites (*nordmarkites*) close to the boundary of silica saturation before passing into biotite-bearing hypersolvus granites. A second branch grades into nepheline-rich rocks that, with increasing amounts of alkali feldspar, reach the composition of nepheline syenite. The silica-rich branch is volumetrically most important; nepheline syenite amounts to less than a percent of the total mass of exposed rock. The divergent trends are closely related to the thermal divide in the residual system illustrated in Figures 11-2 and 11-3. Magmas that were initially saturated

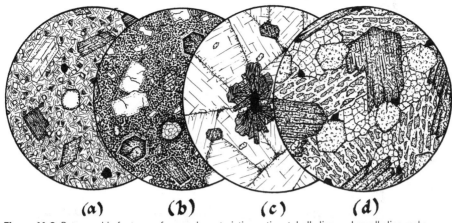

Figure 11-8 Petrographic features of some characteristic continental alkaline and peralkaline rocks. (a) A phonolite of Puy de Dôme, central France, consists of phenocrysts of albite, sodalite, Na-pyroxene, amphibole, biotite, and titaniferous magnetite in a groundmass of sanidine, albite, clinopyroxene, apatite, magnetite, and analcite. (b) A nephelinite lava from Tanzania contains phenocrysts of nepheline, Na-pyroxene, melanite garnet, and apatite in a nearly opaque groundmass. Vesicles are filled with zeolites and calcite. (c) Larvikite from Larvik, Norway, consists of about 80-percent combined alkali feldspar and oligoclase, the remainder being made up of clusters of Ti-augite, biotite, amphibole, and unusually abundant apatite. (d) Kakortokite from the Ilimaussaq Intrusion of southern Greenland contains the zirconium silicate, eudialyte (nearly equant grains of high relief), together with alkali feldspar, Na-amphibole, augite, and nepheline (diameter of fields 3 mm).

with silica descended toward the granitic minimum, while those that were more silica deficient followed the opposite course.

The Gardar Province of Southwestern Greenland

A similar structural setting seems to have been the focus of another complex of highly alkaline rocks near the southwestern edge of the Greenland icecap. A northeast-trending graben cutting very ancient felsic plutonic and metamorphic rocks still preserves remnants of a volcanic series, mainly basaltic, that appears to have been fed by a group of dikes trending parallel to the boundary faults of the down-dropped block. Cutting across these structures are several large dikes and irregularly shaped stocks composed of an extraordinary group of plutonic rocks ranging in composition from granite to nepheline syenite.

Although most of the volcanic series has been stripped away by glacial erosion, the subvolcanic feeders of the system are still preserved. They extend from basanites through hawaiites and mugearites into a divergent series of phonolite, trachyte, and sodic rhyolite (**Fig. 11-9**). A few small intrusions of lamprophyre, carbonatite, and related rocks came in, apparently at a late stage.

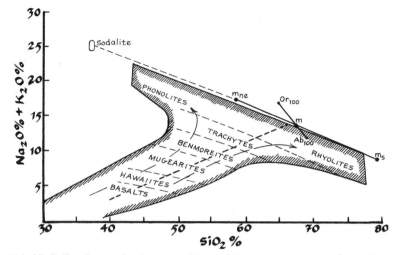

Figure 11-9 Alkali-silica diagram showing compositional relationships of rocks of the Gardar Province of southern Greenland. Note that the felsic differentiates diverge into a broad series ranging from nepheline phonolites and syenites to rhyolites and granites. As in the African rocks (Fig. 11-6), the silica contents of the phonolitic series declines in the later stages of differentiation while that of the rhyolitic series increases. The upper compositional limit follows the alkali-silica ratio of the thermal valley in the residua system, indicated by the line $m_{ne} - m_s$ from Figure 11-3. The dashed diagonal line down the axis of main trend of compositions separates compositions that are critically understaturated (above) from those with normative hypersthene (below). (After B. G. J. Upton, 1974, *The Alkaline Rocks,* H. Sorensen, ed., pp. 221–238.)

Many of the phonolites and rhyolites are peralkaline. All are hypersolvus in that they contain only a single feldspar. Mafic rocks are distinctly subordinate and are common only in early dikes. Some contain megacrysts of plagioclase and xenoliths of anorthosite, indicating that plagioclase crystallized in abundance and must have had a strong influence on the trends of differentiation, but the way in which critically undersaturated mafic liquids differentiated to silica-saturated rhyolites and granites is unclear. Small bodies, such as dikes and volcanic necks, have relatively greater alkali contents than larger, more slowly cooled intrusions, indicating that volatile components were retained under conditions of rapid cooling but, during slow crystallization, escaped into the surrounding rocks to produce broad metasomatic aureoles.

Perhaps the most remarkable feature of the peralkaline intrusive rocks, apart from their rich concentrations of lithophile elements, is the structural relations of the differentiated plutonic series. The Ilimuassaq intrusion, for example, consists of a group of syenites that appear to have differentiated from a single body of magma having a form somewhat like that of the Skaergaard intrusion. The original magma had the composition of a peralkaline phonolite and crystallized as augite syenite in a border group along

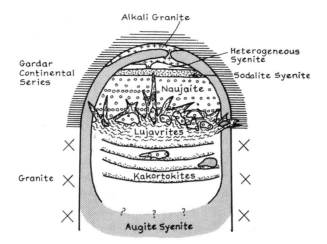

Figure 11-10 Schematic diagram showing the inferred relationships between the principal units of the Ilimaussaq Intrusion of southern Greenland. The augite syenite encasing the main body was the first unit to crystallize. The heterogenous syenites, along with a sodalite-rich variety (known as *foyaite*), seem to have accumulated buoyant crystals by flotation at the same time that the lower kakortokites formed on the floor. The naujaites of the upper series then crystallized, possibly from a liquid overlying the denser magma from which the kakortokites were formed on the floor, but fragments of solid naujaite were dense enough to settle into the lower units along with blocks of augite syenite from the roof. The last liquid to crystallize in the core formed lujavrites, which have extensively invaded the overlying rocks. At some time during this long sequence, alkali granites of unknown origin intruded the upper margins of the intrusion. The vertical scale has been exaggerated by about three times to show the relationships more clearly. (After J. Ferguson, 1964, *Medd. om Gron.* 172:5–82.)

the roof and margins of the intrusion. The interior of the body differentiated into three main divisions analogous to the Upper Border Series, Layered Series, and Sandwich Horizon of the Skaergaard (**Fig. 11-10**). Composed mainly of differing proportions of sodalite, aegirine, arfvedsonite, alkali feldspar, and the bright red zirconium silicate, eudialite, the rocks have suitably improbable names: *naujaite, kakortokite,* and *lujavrite*. The first of these forms the Upper Border Series and consists of up to 80-percent sodalite that accumulated by floatation. The kakortokite, a spectacular red and black rock consisting of alkali feldspar, aegirine, amphibole, and eudialite, has very well-developed layering and is thought to have accumulated on the floor. At middle levels are the lujavrites, representing the last residual magma left by crystallization of the other two groups, much as the Sandwich Horizon of the Skaergaard was but forming a much thicker unit. The mechanism of crystal fractionation is uncertain. Some of the structural features, such as spectacular layering and foundered blocks from the roof, point to gravitational segregation, but conflicting density relationships and anomalous compositional variations indicate that other processes may have been important. Feldspar and feldspathoids were certainly the most important fractionating minerals, and as we note shortly, could account for at least some of the compositional variations of the major units.

Mantle Origins of Rift-Related Rocks

Continental rifts resemble midocean ridges in the sense that both are extensional zones where magma can be produced by decompression of an upwelling asthenosphere (Fig. 8-9). Although the asthenosphere is overlain by thick continental lithosphere, the compositions of the primary rift-related basalts seem to be independent of the continental environment.

This is shown most clearly in a chain of small islands extending from the Gulf of Guinea directly across the continental shelf into the active Cameroon line on the West African mainland (Fig. 11-11). The chain is thought to be related to the Benue trough, a Cretaceous rift that has drifted about 300 km northwest of the presently active zone. The basalts erupted along the oceanic sector of this line are essentially identical to those erupted on the continent. Although they vary widely in composition, the variations are consistent from one end to the other, and any basalt found in one sector can be matched by an identical one in the other.

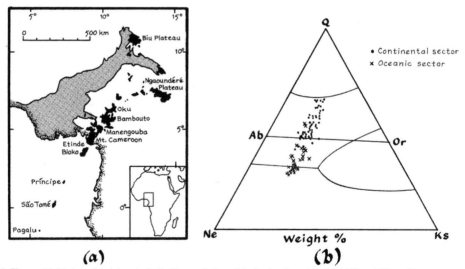

(a)   **(b)**

Figure 11-11 Location (a) and alkali-silica variations (b) of volcanic rocks of the West African rift system passing from the Cameroon line across the continental shelf and into the deep ocean basin. Although the basalts of the oceanic and continental sections are essentially the same, the oceanic magmas follow a trend toward phonolite, while those of the continental setting diverge toward increasing silica enrichment. (After J. G. Fitton, 1987, Alkaline igneous rocks, *Geol. Soc. Spec. Publ.* No. 30, 273–291.)

If the basalts are products of the same asthenosphere as midocean ridge basalts, they must be the result of much smaller degrees of melting. To obtain basalts with the observed concentrations of alkalies and incompatible elements from such a source, the fraction of melt could not be more than 1 percent. The problem is alleviated somewhat if we accept a certain amount of inhomogeneity in the asthenosphere. As we noted in Chapter 8, the basalts of oceanic ridges owe their uniformity to collection and homogenization of large amounts of magma; where smaller volumes are discharged from off-axis volcanoes, the compositions are less uniform. Under these conditions, the enriched parts of the asthenosphere are the first to melt, and the first liquids are richer in incompatible elements than a larger volume of melt at the ridge axis where these components are diluted and local variations are averaged out. The small volumes of highly varied basalt erupted on the islands of the Cameroon line may have a similar explanation.

Throughout the chain, differentiated lavas make up a large proportion of the volume. They define a forked trend on an alkali-silica diagram very much like that of East Africa and other purely continental suites. With few exceptions, the oceanic magmas evolve to phonolite and the continental magmas to rhyolite (Fig. 11-11b). The normal trend seems to have been toward increasing alkalinity, but contamination of the early and intermediate magmas with siliceous crust appears to have deflected the liquid toward the silica-enrichment side of the thermal divide.

Differentiation of Peralkaline Magmas

In each of these examples, the late-stage divergence toward alkaline, subalkaline, or peralkaline liquids conforms to what might be expected from a system with a thermal divide, such as that of Q-Ne-Ks (Figs. 11-2 and 11-3). In any system of this kind, small differences in the compositions of parental magmas and the conditions of crystal fractionation are greatly magnified in the final products of strong differentiation. Feldspar, because it is the most important phase in felsic liquids, has the greatest effect on concentrations of alumina, alkalies, and silica and must play a major role in determining whether individual series are enriched or depleted in these elements.

The importance of feldspars in the late stages of evolution of peralkaline magmas suggests that water pressure was initially large but at some stage fell sharply, causing disproportionate amounts of that phase to crystallize. Although we do not known exactly how this came about, we can draw certain inferences about the physical mechanisms of differentiation.

The mineral assemblages responsible for the observed trends are those that would be stable at relatively shallow depths in the crust. Little deep-seated differentiation is called for, but the large proportions of felsic magmas

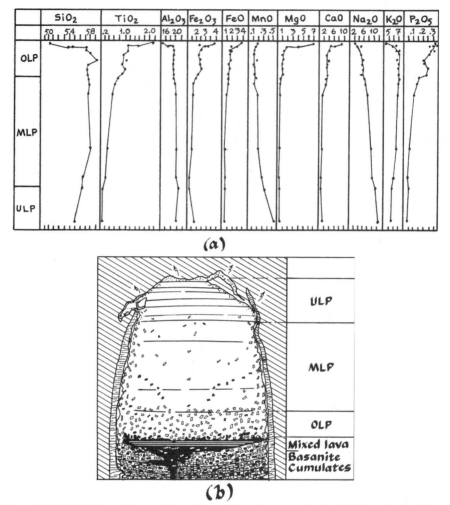

Figure 11-12 (a) Variations of compositions of phonolitic tephra erupted explosively from the zoned magma chamber of Laacher See in Western Germany. (b) A schematic illustration of what the zone magma chamber may have been like immediately prior to eruption. (Adapted from G. Wörner and H. U. Schmincke, personal communication.)

and their contemporaneous association with more basic magmas seem to require a mechanism similar to that proposed for similar relations in calc-alkaline differentiates and siliceous ignimbrites. The effects of these processes are seen in volcanic complexes in which the eruptive sequences and spatial relationships of the magmas have been determined from the field relations of eruptive units. Two examples from western Europe are especially instructive.

The Quaternary volcanic center of Laacher See produced some of the most extreme alkaline compositions of all the many vents along the axis of the Rhine Graben. A large, explosive eruption about 11,000 years ago laid down 5 cubic kilometers of pumice over an area extending from the Baltic Sea to northern Italy. The first phase of the eruption produced a crystal-poor phonolite rich in volatile and incompatible elements and accounting for about 85 percent of the total erupted volume (**Fig. 11-12**). This was closely followed by more crystal-rich phonolite of somewhat more mafic composition, which is thought to have come from a lower level in the reservoir. The final products were a hybrid group of very crystal-rich ejecta formed by mixing of phonolitic and more mafic components. The disequilibrium features in this last magma are so marked that they must have developed during the turbulent discharge, for they are too great to have persisted for long in a high-temperature magma chamber. The compositional relations are consistent with crystallization of the observed phenocrysts and segregation of the differentiated liquid into the uppermost cupola of a vertically zoned intrusion.

The form of such a body may be indicated by the evidence seen in Puy de Dôme, a late Pleistocene to Recent complex of lavas and viscous domes near the center of a belt of youthful volcanic centers just west of the Rhône Valley in central France (**Fig. 11-13**). Compositions range from basalt to siliceous trachyte, with the more felsic magmas coming last from vents near the center of the complex. One of several hypotheses offered for the temporal and spatial relations of the eruptive sequence postulates a broad, shallow intrusion, the upper part of which melted and assimilated the granite basement rocks. Whatever the origins of the compositional differences, they seem to have resulted in a vertically zoned body not unlike those proposed as sources for similarly graded calc-alkaline magmas. Eruptions tapping the upper levels could have supplied the highly differentiated magmas, whereas magmas drawn from lower parts of the same body supplied the denser, more mafic compositions.

■ Carbonatites and Related Rocks

Of all the unusual rocks of continental alkaline complexes, none are more remarkable than magmatic carbonate rocks and the host of alkali-rich and silica-poor rocks with which they are associated. For many years, most petrologists maintained that carbonatites did not even arise from mantle sources but rather from limestones or dolomites mobilized in the continental crust. Their true magmatic origin was effectively demonstrated when it was shown that carbonatites contain alkalies, P, Ti, Ba, Sr, Nb, Zr, Th, Ce, REE, and other trace elements in concentrations much too great to be derived from common sediments. Their isotopic ratios of oxygen, carbon, and strontium are also distinctly mantle-like. If any question still lingered

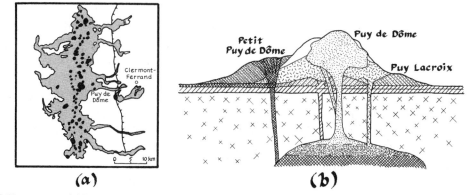

Figure 11-13 (a) The alkaline complex of Puy de Dôme lies on the rim of a graben just east of a line of Pliocene volcanoes in central France. (b) A schematic section through the complex shows a hypothetical relationship between the different extrusive units and a shallow zoned magma chamber. The degree of differentiation of the units is indicated by the darkness of the shading, the lightest being the most felsic. (a) After R. Brousse, 1963, *C. R. Soc. Savantes,* 88; Clermont-Ferrand, 93–114. (b) Modified from original diagram of Y. K. Bentor, 1954, *Bull Serv. Carte Geol. France* 52:373–806.

after these geochemical studies, it disappeared in 1960 when the volcano Oldoinyo L'Engai in Tanzania erupted carbonatite magma with all the morphological aspects of pahoehoe lava. Since that time, it has erupted on several occasions when it could be closely observed and sampled. The lavas were seen to have exceptionally low viscosities, even though the measured temperature ranged between 491°C and 544°C.

Carbonatites, as their name implies, are composed largely of carbonates of Ca, Mg, Na, and Fe. No potassium-rich carbonatite has ever been reported. Typical accessories are apatite, magnetite, and rare minerals, such as monazite and pyrochlore, which in high concentrations are economic sources of the rare-earth elements and several rare metals. True carbonatites range between almost pure calcite at one extreme and sodium carbonate at the other. They normally contain only minor amounts of silica (**Table 11-2**). Those with greater proportions of silicate minerals grade on the one hand into a series of Ca-rich rocks characterized by the mineral melilite and on the other into a Na-rich *ijolite* series with widely ranging proportions of nepheline (**Fig. 11-14**). Even though these series are gradational and divisions between them are somewhat artificial, it is helpful to see how the two main trends vary with changing proportions of their constituent minerals.

Melilite-Bearing Rocks

Pure melilite, it will be recalled, is a solid–solution mineral with two end-members, akermanite ($Ca_2MgSi_2O_7$) and gehlenite ($Ca_2Al_2SiO_7$), but in igneous rocks, another component, $CaNaAlSi_2O_7$, also enters the mineral.

	Some examples of the chemical compositions of carbonatites, alnoite, ijolites, and melilite-bearing rocks. The two carbonatite analyses illustrate extreme				
Table 11-2	sodium-rich and calcium-rich compositions (1 and 2, respectively).				

	1.	2.	3.	4.	5.
SiO_2	0.16	1.07	33.26	37.29	39.92
TiO_2	0.02	0.10	2.15	3.46	2.80
Al_2O_3	–	0.44	5.90	14.41	6.60
Fe_2O_3	0.28	1.88	5.30	4.23	8.90
FeO	–	–	6.54	6.10	5.89
MnO	0.38	0.45	0.15	0.32	0.22
MgO	0.38	0.46	26.41	4.15	12.59
CaO	14.02	52.77	14.47	13.69	15.71
Na_2O	32.22	0.08	1.23	5.61	1.19
K_2O	8.38	0.03	0.82	4.22	0.86
H_2O+	0.56	–	1.91	0.69	2.71
H_2O-	–	3.03	0.09	0.04	1.15
P_2O_5	0.85	1.43	0.76	0.77	0.47
SrO	1.42	0.79	–	1.06	–
BaO	1.39	0.16	0.08	0.22	–
CO_2	31.55	37.19	1.10	3.29	–
Cl	3.40	–	–	0.33	–
F	2.50	–	–	0.32	–
SO_3	4.43	–	0.22	6.07	–

1. Sodium carbonate lava, 1988 eruption of Oldoinyo Lengai Volcano, Tanzania (J. Keller and M. Kraft, 1990, *Bull. Volc.* 52:629–645).
2. Carbonititic dike, Kaiserstuhl alkaline complex, Rhine Graben, Germany (J. Keller, 1981, *Jour. Volc. Geoth. Res.* 9:423–431).
3. Monticellite alnoite, Isle Cadieux (N. L. Bowen, 1922, *Amer. Jour. Sci.* 203:1–34).
4. Ijolite, Magnet Cove, Arkansas (R. L. Erickson and L. V. Blade, 1963, *U. S. Geol. Surv. Prof. Paper 425*).
5. Olivine-melilite-nepheline lava, Kisingiri, Kenya (M. J. Le Bas, 1977, *Carbonatite-Nephelinite Volcanism.* Wiley, 347 p.).

It is convenient to think of these components as equivalents of silica-deficient diopside or anorthite and albite with an extra CaO. Indeed, reaction between basaltic magmas and calcite can lead to melilite replacing plagioclase and pyroxene in just this way. Experimental work has shown that melilite is precipitated together with olivine, usually the Ca-rich variety monticellite, from liquids very rich in Ca and Na but poor in SiO_2.

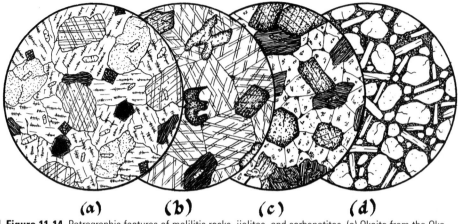

(a)   (b)   (c)   (d)

Figure 11-14 Petrographic features of melilitic rocks, ijolites, and carbonatites. (a) Okaite from the Oka Complex of Quebec contains very abundant melilite with lesser amounts of green biotite, calcite, nepheline, perovskite, apatite, and magnetite. (b) Monticellite carbonatite, also from the Oka Complex, consists mainly of calcite, but monticellite, pyroxene, melanite, and apatite are also important. (c) A mafic variety of ijolite from Oldoinyo L'engai, Tanzania, contains bright green aegirine-augite, nepheline, melanite, green biotite, apatite, and sphene. (d) A carbonatite lava erupted from the same volcano in 1960 consists mainly of sodium carbonate. Rounded clots of phonolitic composition are believed to have exsolved as an immiscible liquid as the magma rose in the conduit of the volcano.

No useful purpose would be served by elaborating on more than the main members of the highly diverse family of melilite-bearing rocks. Typical of many are the *alnoites*, which can be thought of as rocks intermediate between carbonatites and mafic potassium-rich basalts. In addition to melilite, they contain biotite, olivine, pyroxene, calcite, lesser amounts of chromium- and titanium-rich minerals, and a dark-brown, titaniferous, andradite garnet called *melanite*. The alnoites of Ile Cadieux, Quebec, are especially notable, for they contain two forms of olivine, one a common magnesium-iron type and the other the calcium-magnesium variety, monticellite ($CaMgSiO_4$). The latter has been shown experimentally to be the product of subsolidus reaction between normal olivine and melilite to form augite and monticellite.

Melilite and feldspar seem to be mutually exclusive in igneous rocks, for they are never found together. For example, volcanic varieties of melilite-rich rocks, the melilitites, contain no feldspar but only melilite, augite, and various titanium-rich minerals. Olivine, biotite, nepheline and leucite are other possible phases. As these assemblages suggest, the melilitic rocks may be either sodic or potassic, but the latter are less common.

The Ijolite Series

In several continental provinces, such as Magnet Cove, Arkansas, the Monteregian Hills of Quebec, and Kisingiri in western Kenya, carbonatites are closely associated with and gradational into the series of nephelinic rocks referred to collectively as *ijolites* (Fig. 11-14). Again, depending on their proportions of minerals, these range from leucocratic nephelinites, consisting of nepheline and only minor amounts of pyroxene, to melanocratic pyroxenites. Nepheline is the dominant felsic mineral, leucite being much less common, whereas either titaniferous augite or aegirine-augite is the most important mafic component. A wide variety of other mafic minerals is very characteristic. The rare but beautiful *jacupirangites*, for example, are a petrographer's delight. They contain only minor amounts of nepheline but are composed chiefly of bright green aegirine-augite rimmed with purple titaniferous augite and intergrown with differing amounts of magnetite-ilmenite, biotite, melanite, olivine, calcite, apatite, barkevikite, perovskite, pyrite, and zeolites. Some varieties contain melilite and others wollastonite.

While ijolites are characteristically sodic, they grade into a corresponding group of potassic rocks of more gabbroic composition, known as *essexites*. Like ijolites, the latter contain abundant nepheline, but it is accompanied by K-feldspar, and titaniferous augite and amphibole are more characteristic than aegirine. Essexites are found in many of the same igneous complexes as ijolites, but they are more commonly associated with and gradational into plagioclase-bearing rocks of more common gabbroic and syenitic types.

Geological Relationships

Although recorded in 90 or so localities throughout the world, carbonatites, melilitic rocks, and ijolites tend to occur in small bodies, more than half of which are in the eastern and southern parts of Africa. Although characteristic of continental rifting, they are not confined to the continents; carbonatites are reported from the oceanic islands of the Cape Verde Archipelago in the eastern part of the Atlantic Ocean.

The eruptive centers have no visible relationship to crustal structures, nor do they define linear trends with regular progressions of ages, as would be expected if they were produced at hotspots. Most intrusive bodies are ring-complexes emplaced as part of volcanic vents. Their form is typically that of a concentrically zoned cylindrical or conical pipe. The examples from the Monteregian province (**Fig. 11-15**) illustrate the great variety of rock types found in closely associated bodies of a single district. The structural relationships and origins of the radial zonation of these various units have been interpreted in almost as many different ways as there are rock types. In the case of the Monteregian pipes, the contacts with surrounding rocks are

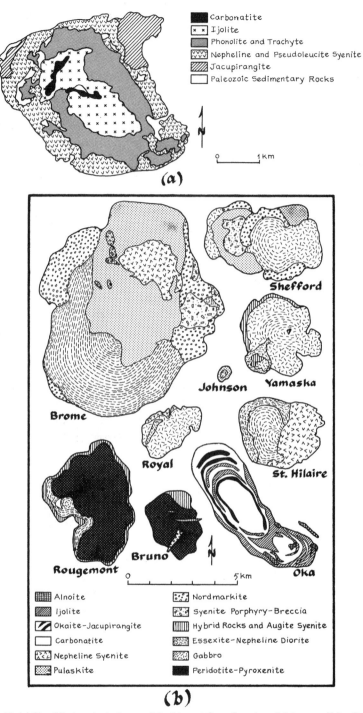

Figure 11-15 (a) Simplified geological map of the Magnet Cove Complex of Arkansas. (After R. L. Erickson and L. V. Blade, 1963, *U. S. Geol. Surv. Prof. Paper* 425.) (b) General features of the main Monteregian intrusions of Quebec. (After compilation by A. R. Philpotts, 1974, *The Alkaline Rocks*, H. Sorensen, ed., pp. 293–310.)

sharp and clearly cross-cutting, and their internal structure has been attributed in some cases to successive intrusions and in others to inward crystallization and differentiation.

Many of these ring-complexes, particularly those with cores of carbonatite, show clear evidence that the outer zones and wall rocks have been strongly altered by volatile-rich fluids. A classic example is the Fen Complex (**Fig. 11-16**) that intrudes Precambrian basement rocks just west of the Oslo Graben in southern Norway. The central carbonate core, originally interpreted as altered limestone, consists of calcite, dolomite, ankerite, and hematite with small bodies of brecciated kimberlite, all of which appear to have been emplaced explosively in a violent volcanic eruption. This core is surrounded by a succession of zones that pass first into mixed carbonates and feldspathoidal rocks of the alnoite variety and then into various types of ijolites and an outer ring of nepheline syenite or phonolite. The granitic gneisses of the walls have been transformed by wholesale metasomatic alteration, resulting, no doubt, from fluids emanating from the intrusion. This process is referred to as *fenitization*, and the syenitic rocks it produced are called *fenites*, after this locality where they were first described. The fenites are essentially syenites that tend to be nepheline-rich close to the intrusion and progressively more silica-saturated outward until they give way to a zone of fractured but otherwise unaltered gneiss.

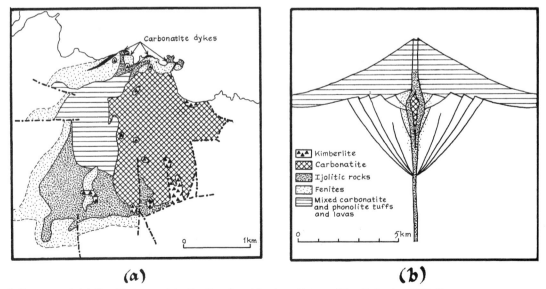

(a)   *(b)*

Figure 11-16 (a) Simplified map of the Fen Complex of Southern Norway. (After E. Saether, 1957, *Kong. Norske Fidensk, Skr.* No. 1.) (b) A schematic diagram showing possible relationships between such a ring-complex and the structure carbonatite volcanoes, such as those of East Africa. (After E. A. Middlemost, 1974, Lithos, 7:255–275.)

The relationship between these ring-complexes and volcanic vents can be seen in the deeply eroded cone of Napak in Uganda, where a nearly circular plug of carbonatite is surrounded by ijolitic rocks. Only parts of the flanks of the volcano are still preserved. They are composed mainly of nephelinitic tuffs and lavas, but the core of the volcano contains a zoned assemblage very much like that of the Fen Complex.

Petrologic Relationships

It is clear that volatile components are an important constituent of alkaline magmas. The effects of carbon dioxide, chlorine, and fluorine are pervasive, not only in the main mineral assemblages but also in alteration zones in and around intrusions. The very low viscosities and explosive character of the magmas are certainly due to their large concentrations of dissolved volatiles. The amounts of these volatiles are too great to be the result of residual concentration by crystal fractionation; they must have been inherent in the magmas at their origin, and they may have been further concentrated in the upper zones of shallow magma chambers.

As the fenites clearly show, volatile transfer must have led to metasomatic alteration of some of the rocks while they were in the crystalline state, but other structural and textural aspects of the rocks have been interpreted as evidence that volatiles can be exchanged in magmas that are still largely if not completely liquid. Perhaps the most controversial of these are the round felsic clots or *ocelli* commonly found in a matrix of mafic minerals. Advocates of gaseous transfer propose a process in which the magma vesiculates on rising and then cools so that the gas pressure in the vesicles declines and a partial vacuum is created that draws a vapor-rich fluid into the vesicle. A second explanation, supported by a growing body of experimental evidence, is that the ocelli were a separate phase exsolved not as a gas but as an immiscible liquid.

The role of liquid immiscibility is particularly well demonstrated in CO_3-rich alkaline liquids. The system Na_2O-CaO-SiO_2-Al_2O_3-CO_2, shown in **Figure 11-17**, has a large two-liquid field at elevated pressures and at most geologically reasonable temperatures. It corresponds remarkably well to the associations found in assemblages of carbonatites and highly undersaturated alkaline rocks. The volcano Oldavai L'Engai mentioned earlier is a vivid example. It erupts lavas and tephra consisting essentially of sodium carbonate but with clots of silicates having the composition of sodic phonolite. Before this immiscible relation was recognized, it was believed that the carbonates were derived from evaporites in the beds of nearby Lake Natron, but several geochemical aspects of the evaporites, particularly their carbon isotopic ratios and fluorine contents, show that the reverse is more likely.

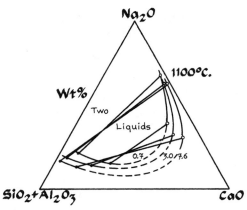

Figure 11-17 The system Na_2O-SiO_2-Al_2O_3-CaO-CO_2 at 1100°C. Pressure of CO_2 (in kbars) is indicated by the numbers on the curves bounding the two-liquid field. Note that the tie lines connecting compositions of coexisting liquids rotate with increasing pressure so that one becomes richer in CaO while the other becomes richer in SiO_2 and Al_2O_3. The first can be thought of as equivalent to calcium carbonate liquids and the latter to alumina-silicate melts. (After I. C. Freestone and D. L. Hamilton, 1980, *Contr. Mineral. Petrol.* 73:105–117.)

The carbonatites erupted from the volcano, being very soluble in rainwater, are quickly dissolved and carried into the groundwater system from whence they are transported via hot springs back to the surface. Over centuries they have accumulated to saturation levels and precipitated in thick evaporite deposits known as *trona*. This process of "meteoric differentiation" leaves insoluble silicates residually enriched in the structure of the volcano and gives the impression that only nephelinitic tephra have been erupted.

■ Lamprophyres

Closely related to the silica-poor and alkali-rich group just discussed is another broad series of mafic and ultramafic alkaline rocks known collectively as lamprophyres. Although more common as minor associates of other rock types than as an independent series of their own, they have no apparent genetic link to the more common rock series. Almost all are hypabyssal porphyritic rocks without plutonic equivalents. They are confined almost exclusively to continental settings, mostly in the deep interior regions of cratons but also in orogenic plutonic belts, where they form dikes, sills, or volcanic pipes. Some have been discharged as coarse heterogeneous tephra or, more rarely, as small flows of dense lava.

Early studies of lamprophyres resulted in a complex classification and nomenclature based on their diverse modal assemblages, but most recent

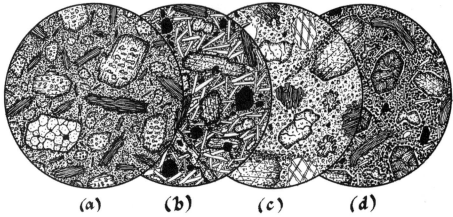

(a)   (b)   (c)   (d)

Figure 11-18 Petrographic features of lamprophyres and kimberlite. (a) A calc-alkaline biotite lamprophyre (minette) from a volcanic neck in the Navajo region of Arizona contains reddish brown biotite with oxidized rims, zoned augite, orthoclase, magnetite, and abundant accessory apatite. A small altered xenolith of granitic basement rock is shown at the left edge. (b) An alkaline keratophyre (camptonite) from the Oregon Coast Range contains phenocrysts of dark-brown barkevikite, Ti-augite, and olivine in a groundmass rich in andesine and apatite. (c) An alnoite from the Oka Complex of Quebec has phenocrysts of biotite, augite, and olivine in a matrix of melilite, carbonates, perovskite, magnetite, and apatite. (d) A kimberlite from the Premier Diamond Pipe of South Africa consists of serpentinized olivine, phlogopite, ilmenite, garnet, and altered rock fragments in a turbid groundmass of calcite, serpentine, and clay.

work has placed greater emphasis on their chemical and petrologic affinities. Four or five main groups are distinguished (**Fig. 11-18**).

Calc-alkaline lamprophyres are typically found in orogenic systems, most often as late-stage dikes cutting felsic plutonic rocks of dioritic to granitic compositions. Their SiO_2 content normally exceeds 48 percent, and they contain more K_2O than Na_2O. In these respects, calc-alkaline lamprophyres are gradational into the potassic subalkaline series discussed earlier. Were it not for their greater contents of water and carbonates, their compositions would differ little from those of mafic shoshonitic lavas. The principal petrographic types are broadly distinguished as *biotite lamprophyres* and *hornblende lamprophyres* according to the nature of the dominant mafic mineral. If the main feldspar is potassic, the former are referred to as *minettes*, if plagioclase *kersantites*. The equivalent hornblende-rich rocks are *vogesites* and *spessartites*, respectively. In most instances, however, the feldspar is fine grained and so thoroughly altered that its identification is at best uncertain. All common types may contain diopsidic augite, serpentinized olivine, apatite, iron oxides, and carbonates. Although most contain both normative and modal olivine, quartz may be common, particularly in the groundmass of biotite-rich types. This is due not so much to a greater SiO_2 content, but to the large amount of K_2O that enters the silica-poor micas, and leaves the remaining assemblage more silica rich than it would be if orthoclase were formed.

Alkaline lamprophyres are common associates of nepheline syenites and carbonatites, chiefly in stable continental interiors and rifts where they may be associated with ultramafic lamprophyres and lamproites. Having less than 48-percent SiO_2 and large amounts of Na_2O, they typically contain 5- to 10-percent normative Ne. Like the calc-alkaline group, they can be divided according to their principal mafic minerals and feldspar, but in all types, the amphibole is the brown sodic variety, barkevikite, or the darker brown to black Ti-rich kaersutite. Biotite, although a common constituent, is less important than Ti-augite. The amphibole-rich *camptonites* normally have labradorite in the groundmass, but the augite-rich *monchiquites* contain little or no feldspar. Instead, their groundmass consists of analcite, nepheline, or pale glass of phonolitic composition. These seem to be textural variations reflecting differing degrees of crystallization of the groundmass rather than inherently different chemical compositions. Other common constituent minerals of the alkaline lamprophyres are olivine, iron-oxides, apatite, zeolites, and carbonates, normally dolomite. A few containing melilite and abundant carbonates are gradational into ultramafic types, such as the alnoites mentioned in a previous section.

One example of a lamprophyre province will serve to illustrate the relationships found in many others. The Navajo-Hopi region of northeastern Arizona was the scene of many small, highly explosive volcanic eruptions during Pliocene time. Today, only lava-capped mesas and towering remnants of volcanic necks (**Fig. 11-19**) remain of what was once a volcanic field covering thousands of square kilometers. The region can be divided into two sub-fields having very distinct lamprophyres, one calc-alkaline and the other alkaline. The rocks of the Navajo region are mainly minettes crowded with xenoliths from all levels of the crust. Those of the adjacent Hopi region are more sodic. They are mainly monchiquites and a gradational series of related rocks, including analcite basanites and olivine-augite basalts. The crustal xenoliths that choke most of the Navajo pipes are much scarcer in the more sodic rocks of the Hopi region.

Early interpretations drew mainly on the abundance of biotite-rich plutonic and metamorphic basement rocks to explain the unusual potassium contents of the Navajo rocks. Breakdown of biotite at shallow levels by their host magma could release a hydrous, potassium-rich fluid that would add many of the essential elements of the calc-alkaline lamprophyres, including the volatile components responsible for their explosive eruptions. Although assimilation of this kind must have been important, especially for the calc-alkaline Navajo rocks, it is not a satisfactory explanation for the

Figure 11-19 An eroded volcanic neck of the Navajo region, Arizona. These towering columns were cores of Pliocene volcanoes that erupted lamprophyric tuffs and lavas over much of northeastern Arizona and adjacent parts of Utah and New Mexico.

alkaline Hopi lamprophyres or for ultrapotassic rocks in general. A deeper origin is needed for the latter group in which there is less evidence for strong contamination. Some of the possibilities can be seen from studies of kimberlites, another group of potassic alkaline rocks that has been studied in greater detail.

■ Kimberlites

Although closely akin to the alkaline lamprophyres, kimberlites are distinctive enough in their associations, geologic occurrence, and petrological importance to merit a special identity of their own. They are volatile-rich, potassic, ultramafic rocks that have risen from deep origins to form dikes, sills, and volcanic pipes, many of them diamondiferous.

Kimberlites have been found in the southern and western parts of Africa, as well as in Quebec, Siberia, India, Brazil, Australia, and the eastern and central United States. None is known from the oceans. Few are of Precambrian age; they seem to have become more abundant with time. In southern Africa, they have reappeared repeatedly in the same provinces at intervals of hundreds of millions of years. Although often cited as examples of continental hotspots, their distribution is not linear except on a local scale, and their scattered recurrence in nearby regions, even after long periods of drift, indicates that the magmas originate in the continental lithosphere, not in the underlying mantle.

Because of their economic importance, diamond-bearing kimberlite pipes have been explored and studied in minute detail. Three main zones are recognized (**Fig. 11-20**). In most places, the shallow "crater zone" has long since been removed by erosion, but the underlying "diatreme zone," which extends to depths of a kilometer or two, has been explored in most of the deeper mines. It is typically a highly brecciated and deeply weathered, steep-sided pipe formed by explosive boring through the shallow crust. It rises from a "root zone" in which the shape tends to be more dike-like and vesiculation less intense. The preservation of sedimentary xenoliths indicates that temperatures of the shallow, brecciated horizons were unusually low, but the dikes and sills at greater depths were emplaced at higher temperatures. They are also more homogeneous and less contaminated.

Brecciated kimberlites are difficult rocks to study because of the deep weathering and hydrothermal alteration that have reduced the minerals to clay and serpentine. Their compositions and textural forms differ widely, even in a single body. The most common petrographic type (Fig. 11-18d) is porphyritic with large amounts of serpentinized olivine, clay, and carbonates, and widely differing proportions of phlogopite, purplish pyrope-rich garnet, ilmenite, chrome diopside, apatite, spinel, perovskite, and monticellite. Some of these minerals are phenocrysts, but many of the largest crystals are xenocrysts far out of equilibrium with their host.

Bulk chemical compositions are similar to those of potassic peridotites. SiO_2 rarely exceeds about 33 percent, but Al_2O_3, TiO_2, P_2O_5, and alkali contents are greater than those of other ultramafic rocks. The ratio K_2O/Na_2O is normally greater than 2.0 and may reach 5.0 or more. Differences in their isotopic ratios and trace element concentrations indicate that most kimberlites are mixtures of magmas derived from at least two distinct sources, one of which is rich in mica and another in carbonates.

In some respects, kimberlites resemble the carbonatites with which they are closely associated in some of the ring complexes described earlier. They contain many of the same minerals, and their characteristic trace-element patterns are similar. Whereas carbonatites are more characteristic of continental rifting, kimberlites are found mainly in tectonically stable regions.

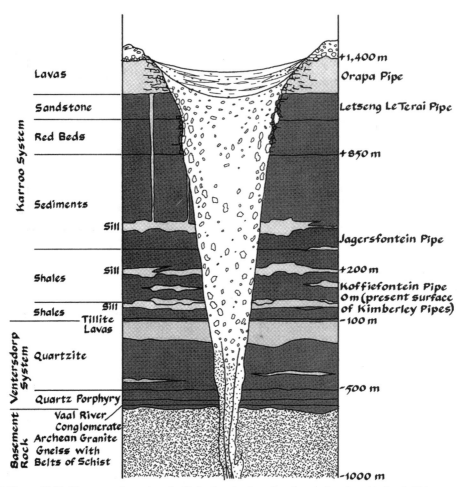

Figure 11-20 Diagrammatic section of a kimberlite pipe based on the relations found in South African diamond mines. The pipes are of Cretaceous age and have been eroded and weathered to differing depths. The narrow feeder consists of multiple unvesiculated dike-like bodies of differing composition. (After K. G. Cox, 1978, *Sci. Amer.* 238:120–132.)

Kimberlitic dikes associated with carbonatite ring complexes are small and lack the variety of xenoliths so characteristic of diamond pipes.

In addition to diamonds, kimberlites contain a wealth of exotic xenoliths and xenocrysts. The compositions of the large megacrysts of olivine, pyroxenes, garnet, and ilmenite tend to be highly variable and must have complex origins. In many ways, the ultramafic xenoliths resemble those found in basaltic magmas, but garnet lherzolites and eclogites are much more common in kimberlites, and the most common depleted type is harzburgite rather than wehrlite. Two textural varieties have been distinguished (Figs. 7-12e and 7-12f), one granular and the other highly deformed. In the latter, the mineral grains are intensely crushed and sheared, and much of the olivine has shadowy "kink bands" resulting from deformation of their crystal lattice. Pyroxene shows less deformation, garnet the least of all.

Eclogites consisting of pyrope garnet and a sodium-rich pyroxene, omphacite, are especially intriguing because some have the bulk chemical composition of common basalts but the high-pressure mineral assemblages of mantle rocks. Their origins are a matter of speculation. They could be products of high-pressure crystallization of basaltic liquids or older rocks brought to mantle depths and recrystallized to a mineral assemblage stable at elevated pressures. Their ages, as determined radiometrically, are generally greater than those of their host, and their isotopic compositions indicate that they could not have been in equilibrium with either the kimberlitic magma or the ultramafic inclusions in the same pipes.

Depths of Origin

The xenoliths found in diamond pipes must come from many levels of the crust and mantle. An estimate of the minimum depths tapped by kimberlites can be gained from the experimentally determined stability relationships of the two forms of carbon, graphite and diamond (**Fig. 11-21**). Note that when the diamond-graphite curve is compared with the estimated geothermal gradient for continents, the two are seen to intersect at pressures of about 35 kbars.

Minute inclusions found in a few diamonds show that coesite was the stable form of SiO_2. The pressure at which quartz inverts to coesite defines a somewhat shallower range of depths than diamond, and if the polymorphic transitions of carbon and silica are projected to the calculated geothermal gradient, they place the minimum depth at around 100 km. The maximum depth to which coesite would be the stable form of SiO_2 is about 250 to 300 km. The absence of the higher-pressure polymorph, stishovite, may place an upper limit on pressure, but it is possible that stishovite was once present and inverted to coesite on relief of pressure.

It will be recalled from Chapter 7 that the compositions of coexisting pyroxenes in ultramafic inclusions can be used to estimate the depths and temperatures at which peridotites last equilibrated in the mantle (Figs. 7-15 and 7-16). The garnet lherzolites in kimberlites define a range of temperatures and pressures (Fig. 11-21) that differ slightly from one pipe to another but indicate temperatures close to, but slightly above, the geothermal gradient estimated independently from heat-flow measurements. Inclusions from the shallowest range of depths have undeformed granular textures, whereas

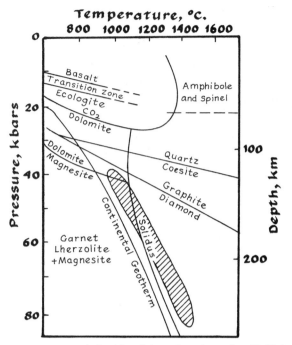

Figure 11-21 Pressure–temperature relationships of the various phases found in kimberlites can be related to the earth's geothermal gradient to define the range of depths from which kimberlites must rise. The diagonally ruled area is the range of temperatures and pressures determined from the compositions of pyroxenes in ultramafic nodules, as shown in Figure 7-20. Also shown is the solidus curve for a garnet lherzolite in the presence of a CO_2-rich fluid. (Solidus curve is taken from D. Canil and C. M. Scarfe, 1990, *Jour. Geoph. Res.* 95:15805–15816.)

almost all those from the greatest depths are highly sheared, as would be expected if they come from the asthenosphere.

The experimentally determined solidus for garnet lherzolite in the presence of a CO_2-rich fluid is shown in the same figure. Note the sharp inflection where the solidus is intersected by a curve marking an abrupt increase of the solubility of CO_2. At this depth, carbonate becomes a stable subsolidus phase. The compositions of the liquids produced by melting in the range of 50 to 70 kbars are found to be very similar to those of natural kimberlites but only within this interval; at shallower depths, liquids are more like melilites, and at greater depths they have more MgO than natural melts. This part of the solidus curve corresponds fairly closely to the range of temperatures and pressures estimated from the pyroxenes in ultramafic xenoliths brought up in kimberlite pipes. Thus, kimberlitic magmas must originate in a "window" between depths of about 150 and 250 km corresponding to the approximate depth of the base of the continental lithosphere.

Melting could be triggered when the solidus temperature is lowered either by upwelling of the mantle or by an influx of a fluid rich in CO_2 and H_2O. The resulting decrease of density can cause the magma to rise diapirically, and because the solidus at this depth has a positive slope that is steeper than the adiabatic gradient, the amount of melt will increase upward. Amphibole, phlogopite, and carbonates become unstable at temperatures only slightly above the solidus and would contribute H_2O and CO_2 to the rising liquid. On reaching a depth of about 80 to 100 km where the solubility of CO_2 decreases abruptly, the magma will suddenly exsolve a volatile phase that greatly accelerates its rise. The highly explosive nature of most alkaline magmas, particularly kimberlites, must be due in large part to this effect, even though most vents are probably initiated by downward propagation of phreatic explosions.

■ Melting Relations at High Pressures

Most recent attempts to understand the origins of kimberlites and related magmas have focused on the role of volatile components at high pressures. It has become increasingly clear that H_2O and CO_2, which are clearly important constituents of these magmas, must have played a key role in their generation and rise to the surface.

As we have seen in earlier chapters, water has a pronounced effect on melting (Fig. 8-17). Note that the solidus temperature where melting begins is greatly reduced by pressures of H_2O up to about 15 kbars, but at greater pressures, it increases again. Because the compressibility of the vapor phase is so much greater than that of the solid, the sign of $\Delta V_{melting}$ is reversed. Increasing pressure of CO_2, on the other hand, has an initial effect of increasing the solidus temperature because CO_2 has such low solubility in the liquid that ΔV is positive. At pressures greater than about 20 kbars, however, this condition begins to disappear; the solubility of CO_2 increases rapidly, probably as a result of a pressure-induced change in the structure of the liquid that causes CO_2 to take the form of $(CO_3)^{2-}$.

The melting relations of lherzolite in the presence of fluids with differing proportions of CO_2 and H_2O are illustrated in **Figure 11-22**. Note that at pressures below the stability range of carbonates, H_2O lowers the solidus much more than CO_2, but at greater pressures, the effect changes abruptly because CO_2 goes into carbonate and the fluid becomes proportionately richer in H_2O. The addition of phlogopite to the system does not change the basic form of the melting relations.

The amount of water contained in the hydrous minerals of peridotites is strictly limited by the bulk composition of the rock. Although amphibole

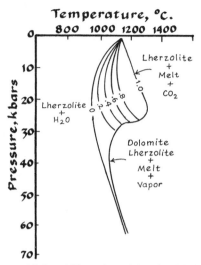

Figure 11-22 Effects of pressure of H_2O and CO_2 on the melting of peridotite. The solidus curve is shown for different proportions of CO_2 and H_2O. The sharp inflection occurs close to the pressures at which the carbonate mineral, dolomite, becomes a stable phase. The intersection of the carbonate stability curve with the solidus migrates according to the proportions of CO_2, as indicated by numbers corresponding to the ratio of CO_2 to $CO_2 + H_2O$. (After P. J. Wyllie, 1979, *Amer. Min.* 64:469–500.)

can contain up to 2- or 3-percent H_2O, the amount of amphibole is restricted by the major elements to about 15 percent of the total mineral assemblage. Hence, H_2O is limited to well under 0.5 percent. Moreover, the stability range of amphibole does not extend to very high temperatures and pressures. It becomes unstable at about the same pressure as spinel, and little, if any, can be present close to the solidus at depths greater than 100 km or so. The stability range of phlogopite is much greater, but the amount of H_2O in phlogopite is limited by the K_2O content of the rock. Because phlogopite requires about 10 percent of this essential component and takes in only 4-percent H_2O, a peridotite with a K_2O content of 0.1 percent can contain only 0.04-percent H_2O if phlogopite is the only hydrous mineral. As we saw in Chapter 8, the nominally anhydrous minerals, particularly pyroxene, can take in small amounts of water, but the amount is much less than that in amphibole or mica.

The limitations on CO_2 are less severe because the major elements needed to form carbonates are present in the main minerals, such as olivine and pyroxene. At pressures greater than about 15 to 25 kilobars, CO_2 combines with olivine and pyroxene to form dolomite, according to the reaction

$$2CO_2 \quad + 2Mg_2SiO_4 + CaMgSi_2O_6 \rightarrow CaMg(CO_3)_2 + 2Mg_2Si_2O_6 \qquad (11\text{-}1)$$

Carbon dioxide   Forsterite   Diopside   Dolomite   Enstatite and at higher pressures of 30 to 40 kilobars, enstatite and dolomite go to magnesite and diopside:

$$Mg_2Si_2O_6 + CaMg(CO_3)_2 \rightarrow CaMgSi_2O_6 + 2MgCO_3 \qquad \text{(11-2)}$$
Enstatite   Dolomite   Diopside   Magnesite

Thus, the mantle is theoretically capable of storing large amounts of CO_2 in the form of carbonate minerals at depths below 100 km or so. It is uncertain how much carbonate is contained in mantle peridotites, but very little dolomite or magnesite is found in mantle xenoliths. Carbonates are certainly subordinate to silicate minerals, and thus, it is doubtful whether the amount of dolomite or magnesite would be limited by the bulk composition. CO_2 in the form of carbonates is probably more abundant at high pressures than H_2O in phlogopite.

The phase relationships of a mantle lherzolite in the presence of small amounts of H_2O and CO_2 are illustrated in **Figure 11-23a**. The change from dolomite to magnesite has little effect on the temperature of the solvus, but a second sharp inflection around 25 kbars marks the depth where amphibole reaches its stability limit, and phlogopite becomes the stable hydrous phase. Thereafter, the temperature of the solidus continues to increase.

The compositions of potassic and carbonatitic liquids on the solidus (**Fig. 11-23b**) define an orderly progression of compositions with increasing pressure. The principal effect is to produce melts that are, first, increasingly silica deficient and alkaline and then more carbonate rich and alkali poor. Note that the pressures at which the liquid compositions cross the series of phase boundaries are less with CO_2 saturation than in the absence of a volatile phase. Although Figure 11-23b was constructed from the phase relationships in a potassic system, the trend for sodic compositions has been shown to be similar, and the general sequence is probably qualitatively valid for most natural magmas.

◼ Ultrapotassic Series

Like their sodic counterparts, highly potassic rocks have a number of unusual features that set them apart from all other rock series. Although an abundance of potassium may be expressed in a variety of minerals, the most characteristic is leucite, $(K,Na)AlSi_2O_6$, or *pseudoleucite*, a faintly birefringent intergrowth of nepheline, K-feldspar, and analcite formed from leucite on cooling. Whereas all rocks with modal nepheline are critically undersaturated in silica, rocks containing the potassic feldspathoid leucite may be alkaline, subalkaline, or peralkaline. This difference is the result of two factors. The first is sim-

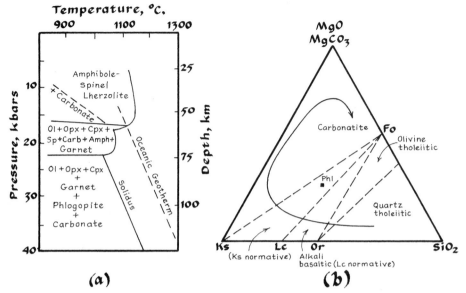

Figure 11-23 (a) Phase relations of lherzolite in the presence of H_2O and CO_2. Molar $CO_2/(CO_2 + H_2O) =$ 0.5. Note that the solidus has two inflexions, one at the stability limit for carbonate and another for amphibole. The kimberlite solidus in Figure 11-22 has only one because H_2O was not a component in the experimental system. The absence of H_2O also causes the temperature to be higher. The oceanic geotherm shown here is for lithosphere that is 30 million years. old. The continental geotherm would be well below the solidus. (After M. Olafsson and D. H. Eggler, 1983, *Earth Planet. Sci. Ltrs.* 64:305–315.) (b) The general trend of potassic liquids with increasing pressure, indicated by an arrow, is essentially the same in the two systems, $KAlSiO_4$-Mg_2SiO_4-SiO_2 and $KAlSiO_4$-MgO-SiO_2-H_2O-CO_2. The pressures at which the liquids cross the field boundaries and become Lc-, Ks-, and magnesite-normative under volatile-free and CO_2-saturated conditions are as follows:

Liquids	Volatile-free	CO_2-saturated
Lc-normative	19 kbars	4 kbars
Ks-normative	34.5 ″	27.5 ″
$MgCO_3$-normative	–	> 29 ″

(Adapted from R. F. Wendlandt and D. H. Eggler, 1980, *Amer. Jour. Sci.* 280:385–420.)

ply the greater silica content of leucite. Unlike nepheline, $(Na,K)AlSiO_4$, which contains only 40- to 44-percent SiO_2, leucite, $(K,Na)AlSi_2O_6$, contains an additional SiO_2 and about 54-percent silica. The second reason is related to the incongruent melting of potassium feldspar. Melting of sanidine or orthoclase at low pressures does not yield a liquid of its own composition but leucite and a liquid somewhat richer in SiO_2 than feldspar (**Fig. 11-24a**). The relationships are much like those of enstatite, which melts incongruently to forsterite and a liquid richer in SiO_2 than pyroxene. Thus, if a liquid that is not undersaturated in SiO_2 precipitates leucite, the remaining liquid can become quartznormative.

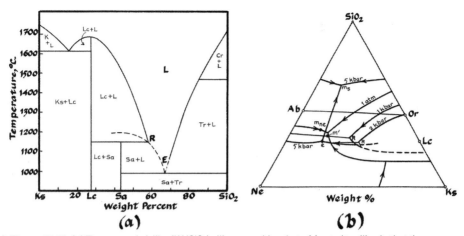

Figure 11-24 (a) The system kalsilite (KAlSiO₄)-silica resembles that of forsterite-silica in that the intermediate compound, orthoclase, melts incongruently to a more silica-poor phase, leucite, and a more silica-rich liquid. (After J. F. Schairer and N. L. Bowen, 1955, *Amer. Jour. Sci.* 253:681–746.) (b) The effect of increasing water pressure on the system nepheline-kalsilite-silica of Figure 11-2 is to reduce the field of leucite, as shown by the boundaries retreating toward the right side of the triangle. At the same time, the boundary between the feldspars and feldspathoids moves farther from silica and the minimum on this boundary eventually becomes an eutectic. (After compilation by J. Gittins, 1979, *The Evolution of the Igneous Rocks: Fiftieth Anniversary Perspectives*, H. S. Yoder, Jr., ed., pp. 351–390.)

As can be seen from the system Ne-Ks-Q (**Fig. 11-24b**), the leucite field contracts with increasing water pressures so that the incongruent melting of the potassium feldspar disappears at pressures slightly in excess of 2 kbars. At the same time, the temperature minimum at m_{ne} becomes a true eutectic, just as the corresponding minimum, m_s, does in the silica-saturated part of the system (Fig. 11-2). The field of leucite, although smaller, persists up to water pressures of about 8.4 kbars and even higher under dry conditions.

Thus, even though it is a feldspathoid, leucite does not necessarily indicate the same degree of silica deficiency as nepheline. Many rocks with modal leucite do not have that mineral in their norm. For this reason, a distinction is often made between leucite-bearing rocks that are critically undersaturated in SiO_2, that is, those with normative Ne, and those that are not alkaline in the strict sense of the term. For example, leucite-bearing basalts that contain olivine are usually distinguished from those that do not. Because the former are more likely to be undersaturated with silica, they are called *leucite basanites*, whereas the latter are referred to as *leucite tephrites*. A similar distinction is made between felsic leucite-rich rocks containing normative hypersthene and those with modal or normative nepheline. The former are given the name *leucite trachytes*, the latter *leucite phonolites*.

The potassic equivalent of nepheline is kalsilite (named from the first five letters of its chemical formula, $KAlSiO_4$). The mineral, which is optically indistinguishable from nepheline, has been reported from only two provinces, the Western Rift of Africa and Italy.

Volcanic Potassic Suites

Some of the first potassic lavas to attract scientific attention were the leucite-rich products of Vesuvius, the volcano responsible for the destruction of Pompeii in 79 A.D. Although it is now known that these lavas are not, as was once thought, unique to the Roman Province of central Italy, leucitic rocks are found in only a few scattered localities outside the Mediterranean. Examples are known from the western United States, Africa, Australia, and Indonesia, but all these occurrences are small and isolated.

If one can speak of such diverse suites as "types" when some occur in only a single, small locality, it is possible to divide potassic rocks into three very broad types—alkaline, peralkaline, and subalkaline—on the basis of their relative contents of alkalies, silica, and alumina. Some of their salient petrographic varieties are illustrated in **Figure 11-25**.

Alkaline potassic rocks are, by definition, critically undersaturated in SiO_2—that is, they have nepheline in their norm. The leucitic lavas of the

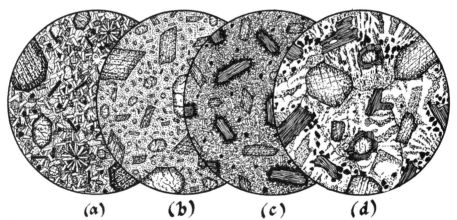

(a)   **(b)**   **(c)**   **(d)**

Figure 11-25 Petrographic features of ultra-potassic rocks. (a) A leucite basanite from Muhavura Volcano, Rwanda, consists of phenocrysts of olivine and zoned Ti-augite in a groundmass rich in small crystals of leucite. (b) Ugandite is a similar rock in that it consists mainly of leucite, augite, and olivine but has a sodium-rich groundmass and is more mafic. (c) Wyomingite from the Leucite Hills, Wyoming, contains phenocrysts of titaniferous phlogopite and slender crystals of greenish augite in a groundmass rich in leucite. (d) Shonkinite from the Shonkin Sag intrusion, Montana, consists mainly of sanidine, augite, biotite, olivine, iron-titanium oxides, and minor nepheline and apatite (diameter of fields 3 mm).

Western Rift of East Africa are perhaps the most important example. They are so rich in potassium and deficient in silica that, in addition to leucite, they contain kalsilite, $KAlSiO_4$, as well as olivine, melilite, biotite, and the usual accessories of these minerals, perovskite and apatite. Typical chemical compositions are illustrated in **Table 11-3**. Even the most mafic varieties, such as

Table 11-3	Chemical compositions of characteristic types of potassic alkaline rocks (lamproites).							
	1.	*2.*	*3.*	*4.*	*5.*	*6.*	*7.*	*8.*
SiO_2	44.08	38.05	41.63	47.67	49.86	53.99	55.02	56.26
TiO_2	1.85	3.84	4.94	1.13	0.76	1.06	2.39	2.54
Al_2O_3	9.93	9.85	7.99	18.14	11.07	17.82	11.10	10.13
Fe_2O_3	3.07	8.41	5.60	0.63	3.64	5.16	3.20	3.56
FeO	8.12	2.80	4.96	6.48	5.52	2.94	1.51	0.83
MnO	0.17	0.21	0.16	–	0.17	0.12	0.06	0.06
MgO	20.34	13.55	16.09	4.19	14.60	4.25	7.12	8.05
CaO	8.77	13.90	10.19	9.01	8.38	6.15	4.80	3.84
Na_2O	1.55	1.31	0.68	2.78	1.96	3.30	1.39	1.29
K_2O	1.74	3.02	7.33	7.47	3.54	4.53	11.58	12.00
P_2O_5	0.37	0.95	0.44	0.50	0.50	0.68	1.83	1.46
CO_2	–	1.47	–	–	–	–	–	–
Cl	–	–	–	–	–	–	0.03	–
F	–	0.27	–	–	–	–	0.50	–
Cr_2O_3	–	0.02	–	–	–	–	0.10	–
SrO	–	0.24	–	–	–	–	0.24	–
BaO	–	0.27	–	–	–	–	1.23	–
Molecular Norms								
Ap	0.75	2.13	0.91	1.05	1.02	1.42	4.11	3.03
Il	2.49	4.71	6.80	1.57	1.03	1.48	2.32	1.37
Tn	–	–	–	–	–	–	0.82	3.22
Pv	–	–	–	–	–	–	0.37	–
Ru	–	–	–	–	–	–	0.45	–
Mt	3.10	–	1.56	0.66	3.72	4.90	–	–
Hm	–	5.93	1.67	–	–	0.33	–	0.16
Or	9.94	0.62	–	13.52	20.44	26.78	58.90	54.92
Ab	4.77	–	–	–	15.00	29.65	–	–
An	14.50	12.22	–	15.09	10.71	20.44	–	–
Di	20.18	33.66	37.72	21.37	21.39	4.51	12.29	3.27

Continued

Table 11-3   Continued

	1.	2.	3.	4.	5.	6.	7.	8.
Hy	–	–	–	–	–	9.49	–	20.45
Ol	39.05	5.76	18.80	6.81	25.36	–	9.73	0.81
Ne	5.21	7.14	0.18	15.11	1.32	–	–	–
Ac	–	–	4.59	–	–	–	8.67	9.20
Lc	–	13.94	8.31	24.80	–	–	–	–
Kp	–	–	19.45	–	–	–	–	–
Ks	–	–	–	–	–	–	2.28	4.65
Ns	–	–	–	–	–	–	0.26	–
Q	–	–	–	–	–	1.00	–	–
Cc	–	3.76	–	–	–	–	–	–
Hl	–	–	–	–	–	–	0.17	–
Fl	–	0.36	–	–	–	–	0.58	–
Cm	–	0.01	–	–	–	–	0.07	–

The following normative minerals may be unfamiliar: Tn—sphene ($CaTiSi_O5$), Ru—rutile (TiO_2), Lc—leucite ($KAlSi2O_6$), Ks—potassium metasilicate (K_2SiO_3), Ns—sodium metasilicate (Na_2SiO_3), Hm—hematite (Fe_2O_3), Cm—chromite (Cr_2FeO_4), Hl—halite (NaCl), Fl—fluorite (CaF_2), Pv—perovskite ($CaTiO_3$) (see Appendix A).

1. Olivine ugandite, Uganda (A. Holmes and F. Harwood, 1937, *Mem. Geol. Surv. Uganda,* 3).
2. Katungite, Uganda (A. D. Edgar and M. Arima, 1981, *N. Jb. Miner. Mh.* 12:539–552).
3. Biotite mafurite, Uganda (A. Holmes, 1942, *Min. Mag.* 26:197–217).
4. Leucite tephrite, 1872 lava of Vesuvius (H. S. Washington, 1906, *Carnegie Inst. Washington, Publ. 57*).
5. Absarokite, Absaroka Mountains, Montana (D. E. Gest and A. R. McBirney, 1979, *Jour. Volc. Geoth. Res.* 6:85–104).
6. Shoshonite, Absaroka Mountains, Montana (D. E. Gest and A. R. McBirney, 1979, *Jour. Volc. Geoth. Res.* 6:85–104).
7. Average wyomingite, Leucite Hills, Montana (Th. G. Sahama, 1974, *The Alkaline Rocks*, H. Sorensen, ed., pp. 96–109).
8. Average orendite, Leucite Hills, Montana (Th. G. Sahama, 1974, *The Alkaline Rocks*, H. Sorensen, ed., pp. 96–109).

ugandite, are exceptionally rich in Ba, Sr, Zr, and REE, as well as TiO_2 and P_2O_5. The name *lamproite* is used as a general term for mafic varieties consisting mainly of phlogopite, Ca-pyroxene or amphibole, olivine, leucite, and sanidine but without primary plagioclase, melilite, nepheline, or melanite.

Peralkaline potassic rocks are exemplified by the lavas of the Leucite Hills, Wyoming. Although highly potassic and rich in leucite, they are not strongly undersaturated in SiO_2. Their deficiency of alumina is reflected in the norm, not only as acmite but also as potassium metasilicate, Ks, a normative mineral with the formula K_2SiO_3. (Ks is calculated from the K and Si remaining after both Al and Fe^{3+} have been exhausted. Although its normative designation is Ks, it should not be confused with kalsilite [$KAlSiO_4$], which, unfortunately, is often abbreviated Ks on phase diagrams. The designation for normative kalsilite is Kp after its polymorphic form, kaliophilite.)

The normative mineral Ks is not a modal component of the rocks. Instead, the lavas are unusually rich in phlogopite. *Wyomingites*, for example, are essentially phlogopite-rich leucite tephrites; similar rocks with abundant sanidine in their groundmass are called *orendites*. The sanidine in the latter is an interesting manifestation of the peralkaline potassic character of the rocks. It is rich in ferric iron, which substitutes for Al in the feldspar lattice, and in that sense it is analogous to acmite, the sodium-ferric iron-pyroxene that characterizes peralkaline sodic rocks.

Subalkaline potassic rocks may have ratios of K_2O to Na_2O ranging from less than 1.0 to 6.0 or more. Those with smaller ratios may be gradational into the more common types of rock series. In Chapter 9, for example, we noted that certain subalkaline series of the continental interiors are somewhat more potassic than the common calc-alkaline series of continental margins (Fig. 10-8) and that they define a "high-K series" consisting of trachybasalts, trachyandesites, and latites, all distinguished petrographically by modal alkali feldspars, usually sanidine or orthoclase. In most respects, these rocks follow the same trends of differentiation as normal calc-alkaline rocks.

Although these mildly potassic calc-alkaline rocks are sometimes said to be *shoshonitic*, true *shoshonites* are richer in alkalies, particularly K_2O, and contain leucite. They are essentially basaltic rocks composed of olivine and augite phenocrysts in a groundmass of orthoclase-rimmed labradorite, olivine, and leucite. With increasing amounts of olivine, they grade into *absarokites*, and with increasing silica into *banakites*. Their compositions are characterized by large concentrations of Ba, Rb, Sr, and most large-ion lithophile elements.

Differentiation of Ultra-Potassic Magmas

Trends of differentiation in these highly potassic series are difficult to define and almost impossible to explain. In many cases, felsic alkaline magmas occur alone or in a bimodal association with mafic rocks and few if any intermediate compositions. Only rarely is the range of compositions extensive, and in most provinces where basic, intermediate, and felsic rocks are found in close association, their genetic relationships are obscure.

The problem is well illustrated by a rare example of a differentiated potassic intrusion, the Shonkin Sag laccolith in the Highwood Mountains of Montana (**Fig. 11-26**). Some 40 to 70 meters thick, the body has chilled upper and lower margins that grade into medium- to coarse-grained mafic rocks consisting mainly of augite, sanidine, olivine, with subordinate biotite, titaniferous magnetite, apatite, zeolites, carbonates, and differing

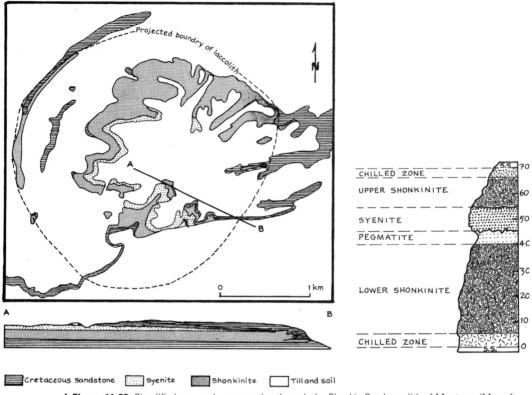

CHILLED ZONE

UPPER SHONKINITE

SYENITE

PEGMATITE

LOWER SHONKINITE

CHILLED ZONE

Cretaceous sandstone   Syenite   Shonkinite   Till and soil

Figure 11-26 Simplified map and cross section through the Shonkin Sag Laccolith of Montana. (Map after C. S. Hurlbut, Jr., and D. Griggs, 1939, *Bull. Geol. Soc. Amer.* 50:1043–1112; section after G. C. Kendrick and C. L. Edmond, 1981, *Geology* 9:615–619.)

amounts of pseudoleucite and minor nepheline. The main mafic mineral, pyroxene, ranges in composition from Ca-rich augite to Na-rich acmite. Sphene and the dark-brown, iron-rich garnet, melanite, are common accessories. These rocks, called *shonkinites* after the Indian name for the Highwood Mountains, encase a core of sanidine-rich syenite. The lower part of the latter is pegmatitic and intermediate in composition between the underlying shonkinite and a more differentiated syenite above. All the rocks are strongly undersaturated with normative Ne reaching values of nearly 30 percent in the syenites, but most of the modal nepheline seems to have been altered to zeolites. The minerals of both the shonkinite and syenite have similar compositions, and the differences between the two rock types are mainly in the proportions, rather than the compositions of dark and light minerals.

The manner in which the body differentiated is difficult to explain in terms of any conventional scheme of crystal fractionation. The chilled margin contains abundant phenocrysts of what is now pseudoleucite and lesser amounts of biotite, and thus, we can assume that both minerals were stable phases in the magma at the time of intrusion. In the more slowly cooled interior, pseudoleucite gives way to sanidine, as would be expected if the early formed crystals reacted with the liquid on cooling. However, when the compositional variations of the various units from the chilled margin to the syenitic core are plotted on a diagram for the system Q-Ne-Kp, they define a nearly linear series (**Fig. 11-27a**). It must be remembered, of course, that in such a plot the mafic components have been ignored so that the bulk compositions are projected from outside the triangle. It is instructive, nevertheless, to compare the observed trends with those that would result from crystal fractionation (Fig. 11-27b).

Note that a potassium-rich liquid, such as X, crystallizing initially in the leucite field, should follow a curved course away from the composition of leucite that becomes increasingly sodic with falling temperature. On reaching that part of the boundary of the leucite field that is a peritectic, that is, the part above the feldspar join, the remaining liquid may react with the leucite crystals to form K-feldspar or, if no reaction takes place, continue to descend toward the granite minimum at m_s. A more sodic liquid, such as Y, reaches the leucite field boundary below the feldspar join, where it is a cotectic. It then precipitates leucite together with K-feldspar, and descends toward the

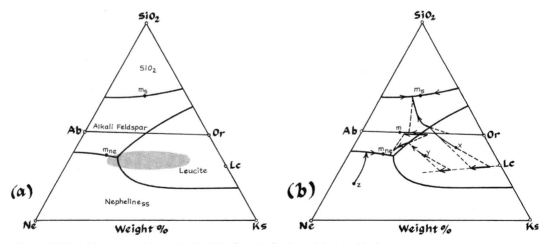

Figure 11-27 (a) The compositions of rocks of the Shonkin Sag Laccolith plotted in the same system as Figure 11-24a. (After W. P. Nash and J. F. G. Wilkinson, 1970, *Contr. Mineral. Petrol.* 25:241–269.) (b) Trends of various liquids crystallizing in the silica-deficient part of the system Ne-Ks-Q. (After S. A. Morse, 1980, *Basalts and Phase Diagrams.*)

other low-temperature end-point at m_{ne}. An even more sodic liquid, such as Z, falls in the field of primary crystallization of nepheline, and its course on cooling would be away from the composition of that mineral. The evolving liquid will eventually reach the nepheline-feldspar boundary and descend along it toward m_{ne}.

The rocks of Shonkin Sag follow none of these trends. Instead, they fall along a line between the composition of leucite and that of the eutectic at m_{ne}. Because of their proximity to the trace of the reaction point R as it advances along the feldspar join with falling pressure (Fig. 11-27b), the rock compositions might be explained as the result of loss of water during degassing of the intrusion were it not for the fact that loss of a water-rich volatile phase causes the point R to migrate in the opposite direction, *increasing* the leucite field and tending to stabilize that mineral. As already noted, sanidine, not leucite, is the stable phase in the late syenitic rocks. Alternatively, if the amount of water had increased, perhaps because the concentration of the volatile components in a diminishing volume of liquid, temperature would also have to increase if the composition of the liquid were to follow the point R toward the right and away from the minimum at m_{ne}.

The fact that the variations correspond to differing proportions of minerals in a matrix of nearly constant phonolitic composition suggests that the differentiated rocks are simply products of gravitational segregation of crystals. Leucite, being very light, might float and accumulate in the upper levels of the reservoir, but leucite is not notably more abundant at the top of the intrusion than it is on the floor.

Another possibility is that volatile transfer was responsible for the differing proportions of K_2O and Na_2O. It was proposed many years ago, for example, that the potassic character of the leucitic lavas of Vesuvius results from selective partitioning of Na into a volatile phase. The volatiles escape into a hydrothermal system that is manifested in sodium-rich hotsprings and fumaroles on the flanks of the volcano. This selective loss of sodium would leave the magma relatively enriched in K.

In the case of the Shonkin Sag, the extensive hydrothermal alteration of the rocks and the wide variation of their alkali ratios lend credence to this interpretation. Evidence of a vapor phase is found in the abundance of zeolites, carbonates, and other products of hydrothermal alteration. Moreover, the proportions of Na and K are consistent with the experimentally determined partitioning of alkalies between feldspars and hydrothermal solutions. The composition of a hydrous vapor phase in equilibrium with alkali feldspars is very sensitive to the composition of the feldspar with which it is in equilibrium (**Fig. 11-28**). It tends to be much richer in sodium than all but the most sodic feldspars. If the feldspars of the rocks plotted in Fig. 11-27a have

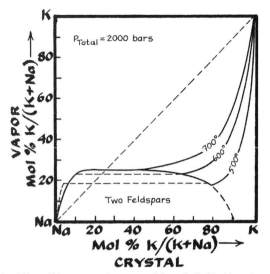

Figure 11-28 The ratio of Na to K in a vapor phase containing alkali chlorides, shown on the vertical axis, is sensitive to that of coexisting alkali feldspar, shown on the horizontal axis. The data shown here are for a total pressure of 2000 bars and a concentration of 2 mole percent total alkali chlorides in the vapor phase. At this pressure, the feldspar solvus limits the range of compositions of the alkali feldspars (Fig. 4-6b). Note that for a broad range of feldspar compositions, the composition of the vapor phase, shown by the curves labeled for temperature, is much richer in sodium than coexisting K-feldspars. Because of these differences, a chloride-rich vapor phase could be an effective agent of fractionation of alkalies. (After P. M. Orville, 1963, *Amer. Jour. Sci.* 261:201–237.)

approximately the same ratios of alkalies as the bulk rocks, a vapor in equilibrium with them would be much richer in sodium, and if separated and deposited in a cooler part of the system, there would be a marked change in the bulk composition of the remaining crystalline assemblage. Thus, a vapor phase may be an effective means of fractionating alkalies, and the observed variations may have less to do with crystal–liquid fractionation than with alkali transfer in a vapor phase exsolved during late stages of cooling.

Another explanation of the linear compositional variations would relate the syenites and shonkinites to two immiscible liquids that separated during the middle and late stages of cooling. This interpretation, which was favored by the early geologists who examined the body, fell into disfavor when N. L. Bowen convinced many American geologists that immiscibility was an unlikely explanation for differentiation of magmas. More recently, however, immiscibility has been shown to be a very real feature of many alkaline liquids, and the fact that the two main rocks of the laccolith contain very similar minerals in different modal proportions is consistent with such a relationship.

■ Mantle Origins of Intra-Plate Alkaline Magmas

Highly alkaline magmas, particularly the potassic varieties, have only recently been accepted as primary melts of the mantle. They were long considered to be the result of assimilation of continental crust or even the products of partial melting of metamorphic rocks rich in biotite or hornblende. Even though very similar rocks were found on oceanic islands, it was argued that these islands must be underlain to be ancient continental crust. We now know that this is impossible.

The volcanic islands of Tristan da Cunha and Gough lie close to the Mid-Atlantic Ridge in the South Atlantic, and both have unusually potassic alkaline rocks. The lavas of Tristan da Cunha, for example, include leucite basanites rich in Rb, Ba, Sr, and Nb; those of Gough are somewhat less potassic but are no less enriched in lithophile elements. The occurrence of carbonatites in the Cape Verde islands was mentioned earlier. Trinidade and Fernando da Noronha, two tiny islands in the western part of the South Atlantic, have nephelinitic lavas and sodalite phonolites that differ in only minor ways from those of East Africa. Xenoliths in the volcanic rocks of Trinidade include coarse-grained rocks similar to mafic varieties of ijolite, such as jacupirangites.

Thus, there is no question that intraplate alkaline magmas with these unusual compositions are derived from the mantle and do not require continental lithosphere to account for their unusual compositions. How then are they produced? As we saw in Chapter 7, the common types of oceanic basalts are readily explained as the products of differing degrees of melting garnet lherzolite. Getting highly potassic magmas, nephelinites, melilites, and the like from such a source is something else. Some other factor must contribute to the compositions of mafic melts that have large amounts of MgO and yet are unusually rich in alkalies and incompatible elements.

High-pressure experiments in which reasonable mantle source rocks, such as garnet- or spinel-lherzolite, are partially melted under a wide range of conditions fail to yield potassic liquids undersaturated in SiO_2. Moreover, when liquids of these compositions are crystallized at high pressures their near-liquidus phases do not include all the essential minerals of normal mantle peridotites—olivine, two pyroxenes, and spinel or garnet. Each composition is found to have different liquidus minerals that crystallize over a wide range of temperature, pressure, and contents of CO_2 and H_2O. In short, there is no simple way in which these magmas could be produced by melting of normal mantle.

The nature of the source rocks and conditions of melting of these magmas can be determined experimentally by comparing the high-pressure liquidus minerals of the lavas to the ultramafic rocks found in them as inclusions. An especially thorough study of this kind was conducted on a suite of highly potassic lavas and associated ultramafic inclusions from the Western Rift of Africa. The lavas are from the ugandite-katungite-mafurite series mentioned in an earlier section (Table 11-3). They have brought to the surface a variety of mafic and ultramafic nodules, but the most common types consist mainly of clinopyroxene, phlogopite, and minor Fe-Ti oxides, sphene, and apatite. They have little or no garnet or olivine and have cross-cutting veins and other textural features suggestive of metasomatic recrystallization.

When the potassic lavas were melted and recrystallized at high pressures in the presence of a water-rich volatile phase, the liquidus phases were found to be Ca-rich pyroxene, phlogopite, and a Fe-Ti oxide. Olivine was an additional mineral in some but not all compositions. These results show that the liquids would have been in equilibrium with a deep source consisting of the same principal minerals found in their associated ultramafic inclusions but not those of normal lherzolites. This conclusion was confirmed by partially melting an ultramafic assemblage with the composition of the average phlogopite-bearing nodule. The liquids obtained in this way have compositions, although not identical, at least very similar to those of the lavas (**Table 11-4**). Moreover, the changes in composition of the liquid with progressive degrees of melting corresponded in a crude fashion to the variations in the series of lavas.

Although experiments of this kind have yet to explore the full range of possible melting conditions, they indicate that under certain conditions the lavas could be derived by partial melting of ultramafic rocks of appropriate compositions. The combination of compositions and melting conditions determined in each set of experiments is unique to an individual lava; each type seems to require different melting conditions and source rocks, few of which correspond to normal mantle compositions. While the phase equilibria seem to explain the association of various types of intraplate alkaline magmas, they leave unanswered the question of how the source rocks were altered mineralogically and enriched in elements that are normally only minor constituents of the mantle.

Mantle Metasomatism

We have already noted that an increase of CO_2 and H_2O can lead to major mineralogical changes, and it is not difficult to imagine these reactions taking place when a volatile-rich fluid or melt invades normal mantle peridotites in advance of a front of melting. A loss of garnet, spinel, olivine, and orthopyroxene and an increase of clinopyroxene and either phlogopite or amphibole are consistent with the effects one would expect from addition of CO_2 and H_2O. Note that both reactions defined by Equations 11-1 and

	Partial melting at thirty kbars of an assemblage of clinopyroxene, phlogopite, and minor Ti-magnetite, sphene, and apatite corresponding to the average composition of ultramafic nodules (No. 1) erupted from volcanoes in the Katwe-Kikorongo and Bunyaruguru district of southwestern Uganda resulted in liquids with the composition shown in column 2. The composition at 25-percent melting is close to that of the average lava in the same district (3). The amount of melt increased from about 20 percent at 1225°C to about 30 percent at 1250°C

Table 11-4

	1.	2.	3.
SiO_2	40.20	39.9	37.1
TiO_2	4.92	5.28	5.55
Al_2O_3	7.27	7.84	6.78
Fe_2O_3	6.69	–	–
FeO	7.12	11.5*	11.9*
MnO	0.09	0.21	0.18
MgO	12.60	10.15	10.1
CaO	12.21	12.35	12.4
Na_2O	1.21	1.95	1.56
K_2O	1.21	4.96	4.06
H_2O+	1.49	2.27	–
CO_2	0.98	0.76	–
P_2O_5	0.70	0.98	1.41
F	0.15	–	–
Cl	<0.02	–	–

* All iron reported as FeO.

Source: F. E. Lloyd, M. Arima, and A. D. Edgar, 1985, *Contrib. Mineral. Petrol.* 91:321–329.

11-2 yield clinopyroxene at the expense of olivine and orthopyroxene. Increased activity of H_2O or K can lead to formation of amphibole or phlogopite at the expense of garnet, spinel, and olivine. Thus, the minerals that are conspicuously absent from the mantle xenoliths are those that would be de-stablized by an increase of CO_2 and H_2O.

In addition to these mineralogical changes, the bulk compositions of ultramafic inclusions in highly alkaline magmas show increased abundances of K, Na, Al, F, S, Cl, Ca, Ti, Fe, Rb, Y, Zr, Nb, Ba, and REE. It is thought that these are introduced as part of the same process. Fluids rich in CO_2 and H_2O at mantle temperatures and pressures are powerful solvents capable of dissolving large amounts of alkalies. For example, at pressures between 10 and 30 kbars and temperatures of 1050°C to 1100°C, such fluids can dissolve as much as 25 weight percent K_2O. It is interesting to note that the solubility of Na is lower than that of K at high pressures because pressure causes the former to enter omphacite, a sodic clinopyroxene that is stable under these conditions. Hence, fluids are likely to have large K/Na ratios at deep levels, whereas the reverse would be true at shallow depths. This may account, at least in part, for the marked differences of enrichment of the two alkalies in magmas.

The physical mechanisms by which these metasomatic effects come about are poorly understood. Whether the active agent is a fluid or a silicate melt may be largely a question of semantics, for under these conditions, the distinction between the too is ill-defined. Having low viscosities and densities, both are highly mobile and should migrate with ease through a crystalline matrix. They would outpace more mafic melts in a rising zone of melting. If the source of the fluid is a crystallizing intrusion or a zone of progressive melting, it may not be able to alter large volumes. The mass required to affect wholesale changes is so great that in these cases the process must operate on a local scale, such as an aureole at the leading edge of an advancing front of melting or along veins where the effect is limited to narrow selvages of the walls. If the source is a convecting asthenosphere giving up mobile components to the lithosphere, the scale of the effect may be vastly greater.

An interesting consequence of the latter possibility is its potentiality for generating felsic magmas. One of the most puzzling aspects of peralkaline and alkaline magmatism in certain provinces is the relatively large proportion of highly differentiated rocks and the scarcity of intermediate compositions. We noted earlier the trachytes and phonolites that make up such a large proportion of the Tertiary sequence of the East African Rift. Similar although perhaps less extreme examples are to be found in many intraplate provinces, both continental and oceanic. The exceptionally large proportions of extreme differentiates are difficult to explain in terms of conventional crystal fractionation of basalt, but they present no obstacle if one postulates melting of a separate source that has been greatly enriched metasomatically in felsic components and incompatible trace elements.

Figure 11-29 illustrates how this may come about. Melts rising along a geotherm will encounter the stability fields of several alkali-rich felsic minerals at depths of 80 to 100 km and pressures near 30 kbars. These phases include kalsilite, jadeite, sanidine, nepheline, and, of course, amphibole. They will crystallize these minerals, and if the process continues, it could contribute enough of these essential components to serve as a source of felsic magma. The sharp inflexions of the solidus at these same depths provide a favorable

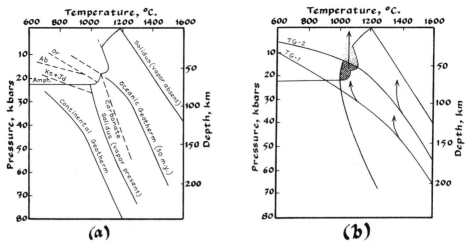

Figure 11-29 (a) Stability fields of felsic minerals in the low-pressure regime of the systems illustrated in Figure 11-23a. The names of the phases are written on the stable sides of their limiting curves (adapted from D. K. Bailey, 1987, Alkaline igneous rocks, *Geol. Soc. Spec. Publ.* No. 30, 1–13). (b) As explained in Chapter 10 (Fig. 10-20), melting along a thermal gradient, such as TG-1, would produce volatile-rich liquids at shallow depths and progressively more advanced melting at deeper levels. The low-density, low-viscosity melts produced at the upper-most levels may rise and encounter the stability fields of amphibole and felsic minerals in the stippled region. They will accumulate and create an enriched zone from which felsic liquids could be generated by a later rise of the thermal gradient (TG-2). Magmas produced by melting of this assemblage in the diagonally ruled region can have very felsic compositions (Fig. 10-23) and could erupt together with more mafic magmas coming from greater depths.

condition for melting, either by a rise of temperature or decrease of pressure. The greater abundance of voluminous felsic alkaline magmas in regions underlain by old continental lithosphere indicates that the process of enrichment proceeds slowly on a regional scale.

Mantle Sources of Enriched Magmas

The magmatic processes we have been considering here have an important bearing on the continuing differentiation of the earth outlined in the opening chapter. If the elements that are characteristically enriched in alkaline magmas are derived almost entirely from mantle sources, as all we have learned about them indicates, they involve a selective transfer of lithophile elements contributing to the long-term evolution of the crust and mantle.

The basic nature of this process remains to be resolved. It may be one that recycles crustal material and maintains a more or less constant balance in the earth as a whole, or it may be part of the continuing segregation of lithophile components from the mantle into the crust.

This problem is similar and no doubt related to the one raised by oceanic hotspots. There, too, mantle-derived magmas contribute incompatible elements to the crust from sources with geochemical characteristics very different from those tapped at midocean ridges. While most petrologists would agree that the sources of these lithophile components lie somewhere beneath the asthenosphere, there is little agreement on their nature or origin. Some advocate recycling between the crust and mantle via subduction of oceanic lithosphere; others assign an important role to an enriched lower mantle that has continued to be a source of lithophile components through much of geologic time. We have already noted (in Chapter 8) that the diverse character of intraplate oceanic basalts must reflect mantle inhomogeneities on a scale ranging from broad regional provinces down to more local domains. On a small scale, we noted that even within a single seamount, lava flows may differ, both isotopically and in their trace-element patterns. On a global scale, the intraplate magmas erupted in the southern hemisphere are consistently more enriched in incompatible elements that those of the northern hemisphere.

If these inhomogeneities stem from parts of the mantle that were depleted by an earlier melting event, their distribution would be consistent with the smaller proportions of continental crust in the southern hemisphere, but they could also reflect crustal material returned to the mantle via subduction. In discussing the calc-alkaline magmas of orogenic systems, we found convincing evidence that at least part of the crustal component of subducted plates returns to the continents in the form of calc-alkaline magmas, but beyond that, nothing more was said of the fate of the remaining oceanic lithosphere that continues to descend into the mantle. Fragments of basaltic crust and remnants of subducted sediments may continue to descend and form a deep reservoir that later serves as a source for hotspot magmas.

Depending on one's preferred model for mantle convection, this subducted material could accumulate at the 670 km discontinuity at the base of the upper mantle, or it could continue to the base of the mantle. In the first case, magma could be generated by heat coming from the lower mantle and from potassium and other heat-producing elements contained in the subducted material itself. At the core–mantle boundary, it could be produced by heat released from the interior of the core or from exothermic reactions between molten nickel–iron and silicate minerals.

It would seem that, with all the petrologic and geochemical tools at our disposal, it would be a simple matter to distinguish between inhomogeneities coming from subducted lithosphere and those from residual primordial material. In theory, the rare gases, such as helium, should be able to do this because, as we noted in Chapter 8, they provide a way of identifying magmas coming from previously untapped sources. The isotopic ratios of helium in intraplate lavas differ from one locality to another and even within a single hotspot chain. Variations of this kind indicate that the mantle has not completely expelled its primordial helium and that degassing has not been uniform, at least at the levels tapped by hotspots. If volcanism is the main mechanism of degassing, we can infer that intraplate magmas are derived, at least in part, from inhomogeneities of the primitive mantle. It is doubtful that helium could be retained by subducted crustal rocks long enough to return to the mantle.

Other geochemical factors, however, indicate that subducted lithosphere does in fact contribute to these magmas. The isotopic ratios of Sr, Nd, and Pb define at least three enriched components found in differing proportions in most intraplate magmas (**Fig. 11-30**). One of these is distinguished by unusually radiogenic Pb ($^{206}Pb/^{204}Pb > 20.5$) derived from sources with elevated ratios of ^{238}U to ^{204}Pb. This group is usually referred to as HIMU after the term μ that is used for this ratio (Appendix D, Fig. D-2). Two other groups, EMI and EMII, are products of enriched mantle. Both have unusual amounts of large-ion lithophile elements and light REE giving the magmas elevated ratios of $^{87}Sr/^{86}Sr$ and low ratios of $^{143}Nd/^{144}Nd$, but this characteristic is more pronounced in EMII ($^{87}Sr/^{86}SR > 0.7065$) than it is in EMI ($^{87}Sr/^{86}Sr = 0.7050$ to 0.7055). With the exception of MORB, the isotopic character of all the major types of basalt can be accounted for as combinations of components from these sources.

The characteristics of these three components have been explained as the results of changes that are thought to take place during dehydration of the subducted lithosphere. It will be recalled from Chapter 9 that the behavior

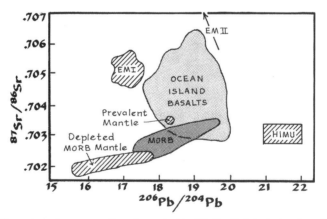

Figure 11-30 Several mantle sources of basalts can be identified by their isotopic ratios of Pb, Sr, and Nd. Two of these (EM-I and EM-II) are more enriched in lithophile elements than the "prevalent mantle"; the high-μ component (HIMU) is distinguished by an unusually large ratio of U to Pb. Note that most ocean-island basalts can be explained as combinations of different proportions of magmas from these three components. Mid-ocean-ridge basalts (MORB) are derived from mantle sources that have been depleted by one or more earlier melting event. (Adapted from A. Zindler and S. Hart, 1986, *Ann Rev. Earth Planet. Sci.* 14:493–571.)

of incompatible trace-elements during this process is governed in large part by their solubilities in a hydrous vapor phase. High-field-strength elements (HFSE), such U, Th, Nb and Ta, are less soluble than large-ion lithophile elements (LILE), such as Cs, Rb, K, Ba, and Sr. Thus, dehydration of the upper part of a subducted slab should result in a loss of the more soluble LILE and residual enrichment of less soluble elements, such as Nb and Ta. Relative to the original midocean ridge basalt, the dehydrated basalt would have small LILE/HFSE ratios, and these ratios should be retained even after the basalt is converted to eclogite. If the elements have very small distribution coefficients, their ratios should not be greatly altered by subsequent partial melting or shallow differentiation.

The parent nuclides of radiogenic Pb, U and Th, have large ionic potentials (U^{4+} 4.5, U^{6+} 8.2, and Th^{4+} 4.3), whereas that of Pb is much smaller, about 1.7. The ratios of U and Th, which are retained, to Pb, which will be preferentially lost, will therefore be increased, and as the radioactive isotopes of U and Th decay, the dehydrated rocks will take on the isotopic character of HIMU. This effect will also be evident in low ratios of LILE, such as Ba and Rb to HFSE, such as Th and Nb.

These same ratios are distinctly larger in basalts with a large component of the EMI-type source because the abundances of HFSE are proportionately lower. Ba, an element that is strongly enriched in pelagic sediments, is especially abundant in basalts of this type. This, together with somewhat more radiogenic Sr ratios, has been taken as evidence that this component comes from the subducted sedimentary layer of oceanic crust. Because the abundances of HFSE are very low in sea water, pelagic sediments absorb relatively less of these insoluble elements. The EMII component is similar and commonly gradational into EMI. It contains much less Ba but more radiogenic Sr and for this reason is thought to reflect the contribution of terrigenous sediments eroded from the continents and mixed with pelagic material in trenches.

It is probably unrealistic to expect clearly identifiable patterns in these relations. Melts from mantle inhomogeneities introduced by subducted crust must be mixtures of all these components and perhaps others as well. Their trace-element ratios may be altered by differing degrees of partial melting, and their isotopic ratios will vary with age. Whatever the merits of these hypotheses, however, they illustrate the fascinating trends of current work and demonstrate the wealth of opportunities for future research.

Selected References

Bonin, B., 1986, *Ring Complex Granites and Anorogenic Magmatism*. North Oxford Academic, 188 p. A concise, well-presented survey of an important group of continental intrusive rocks.

Dawson, J. B., 1980, *Kimberlites and Their Xenoliths.* Springer-Verlag, 252 p. A very thorough treatment of the occurrence, compositions, and modern genetic theories of kimberlites.

Fitton, J. G., and B. G. J. Upton, eds., 1987, Alkaline igneous rocks, *Geol. Soc. Spec. Publ.* No. 30., 568 p. By far the most comprehensive compilation of papers on all aspects of alkaline rocks. The numerous reviews of broad topics are unusually informative and lucid.

Le Bas, M. J., 1977, *Carbonatite: Nephelinite Volcanism,* John Wiley, 347 p. A detailed account of carbonatites and related rocks with emphasis on those of Kenya but with discussions of general petrogenetic relationships.

Pasteris, J. D., 1984, Kimberlites: Complex mantle melts. *Ann. Rev. Earth Planet. Sci.* 12:133–153. A concise summary of recent work on kimberlites with a useful bibliography of relevant literature.

APPENDIX A

Calculation of Normative Minerals

The principles of normative calculations are discussed in Chapter 2. The procedure given here is for the "Molecular" or "Barth-Niggli" norm. The values obtained can be converted to a CIPW norm by multiplying the proportions of minerals by their molecular weights and recalculating to 100 percent. Abbreviations for minerals of the molecular norm are capitalized, whereas those of the CIPW norm are not.

Mineral	Symbol	Chemical Formula	Formula Weight
Acmite	Ac	$Na_2O \cdot Fe2O_3 \cdot 4SiO_2$	462
Albite	Ab	$Na_2O \cdot Al_2O_3 \cdot 6SiO_2$	524
Anorthite	An	$CaO \cdot Al_2O_3 \cdot 2SiO_2$	278
Apatite	Ap	$3.3CaO \cdot P_2O_5$	310
Calcite	Cc	$CaO \cdot CO_2$	100
Chromite	Cm	$FeO \cdot Cr_2O_3$	224
Corundum	C	Al_2O_3	102
Diopside	Di	$CaO \cdot (Mg,Fe)O \cdot 2SiO_2$	217–248*
Fluoite	Fl	CaF_2	78.1
Halite	Hl	$NaCl$	58.4
Hematite	Hm	Fe_2O_3	160
Hypersthene	Hy	$2(Mg,Fe)O \cdot 2SiO_2$	200–264*
Ilmentite	Il	$FeO \cdot TiO_2$	152
Kaliophilite	Kp	$K_2O \cdot Al_2O_3 \cdot 2SiO_2$	316
K-metasilicate	Ks	$K_2O \cdot SiO_2$	154
Leucite	Lc	$K_2O \cdot Al_2O_3 \cdot 4SiO_2$	436
Magnetite	Mt	$FeO \cdot Fe_2O_3$	232
Nepheline	Ne	$Na_2O \cdot Al_2O_3 \cdot 2SiO_2$	284
Na-metasilicate	Ns	$Na_2O \cdot SiO_2$	122
Olivine	Ol	$2(Mg,Fe)O \cdot SiO_2$	141–204
Orthoclase	Or	$K_2O \cdot Al_2O_3 \cdot 3SiO_2$	556
Perovskite	Pf	$CaO \cdot TiO_2$	136
Pyrite	Pr	FeS_2	120
Quartz	Q	SiO_2	60.1
Rutile	Ru	TiO_2	79.9
Sodium carbonate	Nc	$Na_2O \cdot CO_2$	106
Sphene	Tn	$CaO \cdot TiO_2 \cdot SiO_2$	196
Thenardite	Th	$Na_2O \cdot SO_3$	142
Wollastonite	Wo	$CaO \cdot SiO_2$	116
Zircon	Z	$ZrO_3 \cdot SiO_2$	183

* Two values are given for the molecular weights of solid-solution ferromagnesian minerals, the first for the Mg end-member and the second for its Fe counterpart.

The following procedure is a condensed version that does not include all possible components but illustrates the basic principles of the calculation.

A. *Calculating cation percentages*
1. Divide each weight percentage except H_2O by the equivalent weight of the oxide based on a single cation (e.g., $AlO_{1.5}$ rather than Al_2O_3).
2. Multiply each number thus obtained by 1000.
3. Find the total of all the products of step 2. Divide each by this total, and multiply by 100. The proportions obtained in this way should sum to 100.

B. *Calculating provisional norms*
In the following steps, it is necessary to maintain a running balance of the amount of each component left after assigning portions of it to the various normative minerals.
1. Calcite is formed from CO_2 and an equal amount of Ca.
2. Apatite is formed from P and 1.67 times this amount of Ca.
3. Pyrite is formed from S and 0.5 times this amount of Fe^{2+}.
4. Ilmenite is formed from Ti and an equal amount of Fe^{2+}.
5. The alkali feldspars are formed provisionally from K and Na, each being combined with an equal amount of Al and three times as much Si to form Or and Ab, respectively.
6. a. If there is an excess of Al over K + Na, it is assigned to An by combining the remaining Al with an equal amount of Si and half as much Ca.
   b. If there is not enough Ca to combine with Al in this way to make An, the excess Al is assigned to C.
7. a. If in 5 there is an excess of Na over Al, it is combined with an equal amount of Fe^{3+} and twice as much Si to make Ac.
   b. If there is an excess of Fe^{3+} over Na, it is combined with half as much Fe^{2+} to make Mt.
   c. If there is not enough Fe^{2+} to use all of the Fe^{3+}, the remaining Fe^{3+} is assigned to Hm.
   d. Any remaining Fe^{2+} is combined with Mg and Mn and thereafter treated as FM.
8. Di is formed from the Ca left from 1, 2, and 6a by combining it with an equal amount of FM and twice as much Si.
9. Hy is formed by combining the FM remaining from 8 with an equal amount of Si.

C. *Balancing Si*
Having assigned all the cations to provisional normative minerals, we next consider the distribution of Si.
10. a. If an excess of Si remains, it is assigned to Q.
    b. If there is a deficiency of Si, minerals of lesser silica content must be substituted, either wholly or in part. First, Hy is converted to Ol according to the equation:

EXAMPLE   **517**

$$4 \, Hy = 3 \, Ol + 1 \, Q$$

where Q is the amount of the deficiency of Si.

     c. If all Hy has been changed to Ol and Si is still deficient, Ab is converted to Ne according to the equation:

$$5 \, Ab = 3 \, Ne + 2 \, Q$$

where Q is now the deficiency of Si remaining after 10b.

     d. If this still leaves a deficiency of Si, Or is converted to Lc according to the equation

$$5 \, Or = 4 \, Lc + 1 \, Q$$

where Q is the deficiency remaining after 10c.

     e. In rare cases, there is not enough Si to form Lc. Then Kp is formed as follows:

$$4 \, Lc = 3 \, Kp + 1 \, Q$$

■ Example

A. *Calculation of cation proportions (C.P.) and percentages.*

Oxide	Wt %	Mol. Wt.	C.P. × 1000	Cation %
SiO_2	49.10	60.09	817	46.0
TiO_2	3.59	79.90	45	2.5
Al_2O_3	16.21	50.99	318	17.9
Fe_2O_3	2.87	79.85	36	2.0
FeO	6.84	71.85	95	5.3
MnO	0.05	70.85	1	0.1
MgO	5.04	40.31	125	7.0
CaO	8.90	56.08	159	9.0
Na_2O	3.53	30.99	114	6.4
K_2O	2.76	47.10	59	3.3
P_2O_5	0.54	70.98	8	0.5
			1777	100.0

B. *Calculation of Norm*

	Provisional	Final
1. Analysis has no CO_2, so go to step 2.		
2. Ap = 0.5 P + 1.67 × 0.5 Ca =	1.3	1.3
3. Analysis has no S, so go to 4		
4. Il = 2.5 Ti + 2.5 Fe^{2+} =	5.0	5.0
5. Or = 3.3 K + 3.3 Al + 3 × 3.3 Si =	16.5	16.5
Ab = 6.4 Na + 6.4 Al + 3 × 6.4 Si =	32.0 – 4.8	27.2
6. a. An = 4.1 Ca + 8.2 Al + 8.2 Si =	20.5	20.5
7. b. Mt = 2.0 Fe^{3+} + 1.0 Fe^{2+}	3.0	3.0
c. FM = (5.3 + 0.1 + 7.0) – 2.5 – 1.0 = 8.9		
8. Di = 4.1 Ca + 4.1 FM + 8.2 Si =	16.4	16.4
9. Hy = 8.9 FM + 8.9 Si =	17.8 – 17.8	0.0
10. b. Ol = 17.8 – 2.4 =		7.2
c. 4.8 Ab = 2.9 Ne + 1.9 Si		2.9

Running Balance	
Si = 46.0 – 9.9 – 19.2 – 8.2 – 4.1 – 1.8 – 7.1 – 2.4 + 1.9	= 0
Ti = 2.5 – 2.5	= 0
Al = 17.9 – 3.3 – 6.4 – 8.3	= 0
Fe^{3+} = 2.0 – 2.0	= 0
Fe^{2+} = 5.3 – 2.5 –1.0 – 1.8	= 0
Mn = 0.1 – 0.1	= 0
Mg = 7.0 – 7.0	= 0
Ca = 9.0 – 0.8 – 4.1 – 4.1	= 0
Na = 6.4 – 6.4	= 0
K = 3.3 – 3.3	= 0
P = 0.5 – 0.5	= 0

Calculations of Densities and Viscosities of Silicate Melts

■ Density

The calculation given below is based on the method of Bottinga and Weill (1970, *Amer. Jour. Sci.* 269:169–182) but uses more recently determined values of molar volumes. The original analysis, in weight percent, is recalculated to molecular proportions water-free in the manner explained in Appendix A but without renormalizing to 100 percent. $T = (\text{temperature in } °C - 1400) \times 10^{-5}$

$Z1 = SiO_2 \times 27.03$

$Z2 = TiO_2 \times 22.6 \times (1 + T \times 26.7)$

$Z3 = AlO_{1.5} \times 36.63/2 \times (1 + T \times 14.7)$

$Z4 = Fed_{1.5} \times 43.73/2 \times (1 + T \times 12.2)$

$Z5 = (FeO + MnO) \times 13.85 \times (1 + T \times 31.2)$

$Z6 = MgO \times 11.43 \times (1 + T \times 9.4)$

$Z7 = CaO \times 16.32 \times (1 + T \times 38.4)$

$Z8 = NaO_{0.5} \times 14.39 \times (1 + T \times 23.5)$

$Z9 = KO_{0.5} \times 22.965 \times (1 + T \times 24.9)$

Density $= 100/(Z1 + Z2 + Z3 + Z4 + Z5 + Z6 + Z7 + Z8 + Z9)$

■ Viscosity

The calculation below is based on the method of H-R. Shaw (1972, *Amer. Jour. Sci.* 272:870–893). It gives the viscosity, η, of crystal-free silicate melts in poise in log units. Weight percentages of oxide components are first recalculated to 100 percent and then converted to cation proportions. Temperature (T) is in degrees K.

$SI = SiO_2/60.0843$

$TI = TiO_2/79.8988$

$AL = Al_2O_3/50.9806$

$F3 = Fe_2O_3/79.8461$

$F2 = FeO/71.8461$

$MN = MnO/70.9374$

$MG = MgO/40.3044$

$CA = CaO/56.9895$

$NA = Na_2O/30.9895$

$K = K_2O/47.0980$

$H = H_2O/18.0152$

$P = P_2O_5/70.9723$

Normalize to 100

$FM = F3 + F2 + MN + MG$

$NK = NA + K$

$C = SI + TI + AL + FM + CA + (NK + P)/2 + H$

$Ql = AL \times 6.7$

$Q2 = FM \times 3.4$

$Q3 = (CA + TI) \times 4.5$

$Q4 = NK \times 1.4$

$Q5 = H \times 2$

$QT = (Ql + Q2 + Q3 + Q4 + Q5) \times SI/C^2$

$X = [QT/(1 \times SI/C) \times (10000/T - 1.5)] - 6.4$

$\text{Log } \eta = X/2.303, \ \eta = e^X$

The viscosity of partly crystallized magmas can be estimated from the following empirical equation (McBirney and Murase, 1984, *Annu. Rev. Earth Planet. Sci.* 12:337; Murase, McBirney, and Melson, 1985, *Jour. Volc. Geoth. Res.* 24:193).

$$\log \eta_{\text{eff}} = \log \eta_0 + \frac{0.019 D_m}{(1/\phi)^{1/3} - 1}$$

where

η_{eff} = the effective viscosity of the liquid and crystals

η_0 = the viscosity of the liquid alone

D_m = the mean diameter of the crystals in microns

ϕ = the volume fraction of crystals

APPENDIX

C

Error Functions

The two error functions most commonly used in calculations of thermal or chemical diffusion are given in this table. Others are given in standard references (such as J. Crank, 1967, *Mathematics of Diffusion*, Oxford Press).

x	erf x	erfc x	x	erf x	erfc x
0.0	0.0	1.0	0.0	0.0	1.0
0.05	0.056372	0.943628	1.1	0.880205	0.119795
0.1	0.112463	0.887537	1.2	0.910314	0.089686
0.15	0.167996	0.832004	1.3	0.934008	0.065992
0.2	0.222703	0.777297	1.4	0.952285	0.047715
0.25	0.276326	0.723674	1.5	0.966105	0.033895
0.3	0.328627	0.671373	1.6	0.976348	0.023652
0.35	0.379382	0.620618	1.7	0.983790	0.016210
0.4	0.428392	0.571608	1.8	0.989091	0.010909
0.45	0.475482	0.524518	1.9	0.992790	0.007210
0.5	0.520500	0.479500	2.0	0.995322	0.004678
0.55	0.563323	0.436677	2.1	0.997021	0.002979
0.6	0.603856	0.396144	2.2	0.998137	0.001863
0.65	0.642029	0.357971	2.3	0.998857	0.001143
0.7	0.677801	0.322199	2.4	0.999311	0.000689
0.75	0.711156	0.288844	2.5	0.999593	0.000407
0.8	0.742101	0.257899	2.6	0.999764	0.000236
0.85	0.770668	0.229332	2.7	0.999866	0.000134
0.9	0.796908	0.203092	2.8	0.999925	0.000075
0.95	0.820891	0.179109	2.9	0.999959	0.000041
1.0	0.842701	0.157299	3.0	0.999978	0.000022

D Mathematical Functions of Radiogenic Isotopes

As noted in the text, the decay of radioactive elements is not only useful for determining the ages of rocks and minerals, but also yields useful information about the evolution of the crust and mantle and the origins of magmas.

Three isotopic systems are currently of special interest to petrologists: rubidium-strontium, uranium-thorium-lead, and samarium-neodymium. Basic equations governing all three are illustrated by the rubidium-strontium system that was described briefly in Chapters 1 and 9. Strontium has several isotopes, one of which, ^{87}Sr, is produced by decay of ^{87}Rb. Its abundance is measured relative to another isotope, ^{86}Sr, which is stable and is not derived from decay of another element. The rate of decay of any unstable parent nuclide, such as ^{87}Rb, is proportional to the number of atoms remaining at any time, t.

$$\frac{dN}{dt} = -\lambda N \tag{1}$$

where the decay constant, λ, is the statistical proportion of atoms that decay in a given unit of time. The minus sign indicates that the number of atoms decreases with time. Integrating Equation 1 we get

$$-\ln N = \lambda t + C \tag{2}$$

where C is the constant of integration. At t = 0,

$$C = -\ln N_0 \tag{3}$$

where N_0 is the initial number of atoms in a system such as a rock or mineral. Therefore

$$-\ln N = \lambda t - \ln N_0$$

$$\ln \frac{N}{N_0} = -\lambda t$$

$$\frac{N}{N_0} = e^{-\lambda t}$$

$$N = N_0 e^{-\lambda t} \tag{4}$$

This is the standard equation for decay of any radioactive nuclide. The elapsed time at which n = 0.5 N_0 is defined as the half-life of the parent, $t_{1/2}$, and the relationship between t and X in (4) is

$$t_{1/2} = 0.693/\lambda$$

Because the half-life of ^{87}Rb is about 48.8×10^9 years, λ has a value of 1.42×10^{-11} per year.

Because decay is an atom-for-atom process, the initial number of atoms of ^{87}Rb is equal to the present number plus the number of new radiogenic atoms, ^{87}Sr*, produced by decay during time t. Equation 4 can be written

$$^{87}\text{Rb} = (^{87}\text{Rb} + {}^{87}\text{Sr})^{e-\lambda t} \tag{5}$$

rearranging

$$^{87}\text{Sr}^* = {}^{87}\text{Rb}(e^{\lambda t} - 1) \tag{6}$$

The total amount of ^{87}Sr is the sum of the original amount, ^{87}Sr$_0$, and the amount produced by decay in time t:

$$^{87}\text{Sr} = {}^{87}\text{Sr} + {}^{87}\text{Sr}^* \tag{7}$$

Combining Equations 6 and 7 gives

$$^{87}\text{Sr} = {}^{87}\text{Sr}_0 + {}^{87}\text{Rb}(e\lambda t - 1) \tag{8}$$

The required data are obtained by mass spectrometry, a technique capable of measuring only isotopic ratios. Therefore, the quantities are converted to ratios by dividing by ^{86}Sr, which is constant and tends to have the same order of abundance as ^{87}Sr. Thus, we have

$$\frac{^{87}\text{Sr}}{^{86}\text{Sr}} = \left[\frac{^{87}\text{Sr}}{^{86}\text{Sr}}\right]_0 + \frac{^{87}\text{Rb}}{^{86}\text{Sr}}(e^{\lambda t} - 1) \tag{9}$$

Because the small values of λ make exponential terms negligible, the last term, when expanded, reduces to

$$(e^{\lambda t} - 1) = 1 + \lambda t + (\lambda t)^2/2 + \ldots - 1 = \lambda t$$

Thus, Equation 9 becomes

$$\frac{^{87}\text{Sr}}{^{86}\text{Sr}} = \left[\frac{^{87}\text{Sr}}{^{86}\text{Sr}}\right]_0 + \frac{^{87}\text{Rb}}{^{86}\text{Sr}}\lambda t \tag{10}$$

In most analytical procedures, the absolute abundances of Rb and Sr are measured in parts per million by weight, whereas in Equation 1 they are in numbers of atoms. Their weights can be converted to proportions of atoms by dividing each by its atomic weight. (The atomic weights vary with the proportions of isotopes and, for precise calculations, must be calculated for the specific sample.) At a given time, such as the present, the proportion of ^{87}Rb in total Rb is essentially constant at 27.83 percent. Because of radiogenic variability, the proportion of ^{87}Sr differs slightly from one sample to another, but it is typically about 9.87 percent. Therefore

$$\left[\frac{^{87}\text{Rb}}{^{86}\text{Sr}}\right]_{\text{atomic}} = \left[\frac{\text{Rb}}{\text{Sr}}\right]_{\text{wt}} \times \frac{87.62}{85.47} \times \frac{.2783}{.0987} = 2.89 \frac{\text{Rb}}{\text{Sr}} \text{ t} \tag{11}$$

By inserting this factor and multiplying it by the value of λ, Equation 10 can be rewritten as

$$\frac{^{87}\text{Sr}}{^{86}\text{Sr}} = \left[\frac{^{87}\text{Sr}}{^{86}\text{Sr}}\right]_0 + 4.018 \times 10^{-11} \frac{\text{Rb}}{\text{Sr}} \text{ t} \tag{12}$$

which is the equation for a straight line (Fig. D-1) with its origin at the initial ratio $(^{87}\text{Sr}/^{86}\text{Sr})_0$ and a slope proportional to t.

It can be seen that the ratio $^{87}\text{Sr}/^{86}\text{Sr}$ increases with time by an amount that is directly proportional to the ratio of the parent-to-daughter elements, Rb/Sr. A series of rocks or minerals that initially have the same ratio of Sr isotopes but different Rb/Sr ratios will evolve, as shown in **Figure D-I**, to new isotopic ratios that define a straight line known as an *isochron*.

If two or more co-magmatic rocks have compositions that define such a line when their isotopic ratios are plotted against Rb/Sr, they could be interpreted as

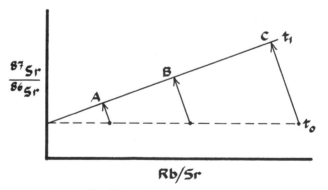

Figure D-1 The rate of increase of $^{87}\text{Sr}/^{86}\text{Sr}$ is directly proportional to the ratio of Rb to Sr (specifically $^{87}\text{Rb}/^{86}\text{Sr}$) so that, with time, a series of rocks or minerals with different Rb and Sr contents will evolve Sr isotopic ratios defining a straight line (as shown in Figure 9-20). Homogenization of the Sr by reheating may reset the isotopic ratios to a uniform value without changing their Rb/Sr ratios so that further decay of Rb would lead to a subsequent isochron that would be a measure of the time elapsed since the secondary event occurred (Fig. 9-23a). A linear array of points on such a diagram may not necessarily be a true isochron; it could also result from mixing. An intermediate value, such as B, could be produced from end-members A and C.

having evolved to their present isotopic ratios over the period of time defined by the isochron. If the rocks are old, one can calculate the isotopic ratio they had at the time they were erupted by substituting their known age in the decay equation and solving for the initial ratio. If this correction results in a flat line, that is, a zero age, then the magmas would be interpreted as having evolved in an isotopically uniform reservoir in which they differentiated shortly before erupting. If the line still has a slope, however, it could be interpreted in two ways. It could indicate that the magmas evolved independently and, after differentiating, remained isolated from one another for a corresponding period of time before they were erupted. But a similar linear array of compositions could also result from simple mixing of two end-members, one with low ratios of the isotopes and Rb/Sr and the other with high ratios. Such a relation would be a "pseudo-isochron" and would reflect the amount of mixing rather than elapsed time.

The other isotopic systems, U/Th-Pb and Sm-Nd, evolve in similar ways. The rare-earth element samarium-147 (^{147}Sm) decays to another rare earth (^{143}Nd) with a half-life of 106 billion years ($\lambda = 6.34 \times 10^{-12}$), and the abundance of ^{143}Nd is compared to that of the essentially stable isotope, ^{144}Nd, just as ^{87}Sr is measured relative to ^{86}Sr. Pb, however, has three radiogenic isotopes, ^{206}Pb derived from ^{238}U, ^{207}Pb from ^{235}U, and ^{208}Pb from ^{232}Th. Each of these nuclides has its own decay curve with time. As explained in Chapter 10, the abundances of the radiogenic isotopes of Pb are normally compared to that of the stable, nonradiogenic isotope ^{204}Pb and, when the ratios are plotted against each other, they define curves that reflect the differing half-lives and changing abundances of the two parent isotopes of U (**Fig. D-2**). ^{207}Pb has increased only slightly in the last four billion years, because much of its parent, ^{235}U, decayed early in the Earth's history.

Because the parent/daughter ratios of the mantle are not uniform, the growth curves of the Pb isotopes differ from one region to another, and these differences cause the igneous rocks of one province to differ more or less systematically from those of another. At any given time, however, the isotopic ratios of Pb from sources with different parent/daughter ratios would define an isochron, just as they do for Sr. As Figure D-2 shows, the present isochron (often referred to as the zero-age isochron for the Earth, or the "Geochron") should govern the lead ratios of modern, mantle-derived rocks. Unfortunately, they do not always do so. Other events in the earth's history appear to have redistributed U and Pb at different times and by different amounts so that the isotopic character of a given suite of rocks tends to be governed by the history of the crust and mantle in a particular region.

As explained in the opening chapter, the Nd system resembles that of Sr but is in a sense complementary. With time, the ratio of ^{143}Nd to ^{144}Nd increases as ^{147}Sm decays to ^{143}Nd with a decay constant, λ, of 6.54×10^{-12} yr^{-1}. The rate of increase of the Nd ratio is a direct function of the relative amounts of Sm and Nd, but in this case the parent–daughter ratio, Sm/Nd, is smaller in crustal rocks than in the mantle (even though the absolute abundances of both of these rare-earth elements are greater in the crust). In a similar way, the ratio of Sm to Nd is smaller in partial melts than in their source rocks, whereas the opposite is true of Rb/Sr.

The isotopic ratios of Nd and Sr are often compared by relating them to standard values, which are normally taken as the estimated ratios for the bulk earth. For ^{143}Nd/^{144}Nd this is 0.51264, and for ^{87}Sr/^{86}Sr, it is taken as 0.7047. Ratios are

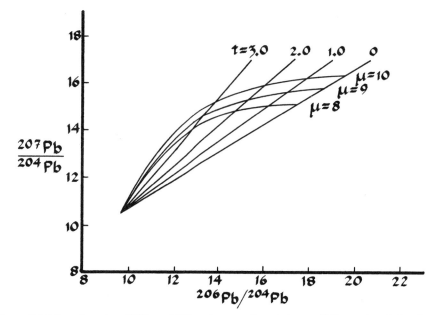

Figure D-2 When the ratios of ^{207}Pb and ^{206}Pb to the nonradiogenic isotope ^{204}Pb are plotted, they define a curve with a slope that varies according to [the changing parent/daughter ratios The ratio of ^{238}U to ^{204}Pb, which determines the rate of increase of ^{206}Pb/^{204}Pb, is referred to as μ. The ratio of ^{235}U to ^{204}Pb, which governs the rate of increase of ^{206}Pb/^{204}Pb, has a value of μ/137.8. Thus the rates of increase of both radiogenic isotopes can be related to μ, and, as illustrated by the curves for different values of this factor, the evolution of the isotopes with time would have a different curve according to μ. Points on two or more curves at a given time define a straight line or isochron. The specific zero-age isochron for the present day is called the "Geochron."

reported in terms of ε_{Nd} and ε_{Sr}, the fractional deviation from the reference value in units of 0.01 percent. Thus, for neodymium

$$\varepsilon_{Nd} = \left[\frac{\left(^{143}Nd/^{144}Nd \right)_{sample}}{0.51264} - 1 \right] \times 10^4$$

and for strontium

$$\varepsilon_{Sr} = \left[\frac{\left(^{87}Sr/^{86}Sr \right)_{sample}}{0.7047} - 1 \right] \times 10^4$$

Atomic and Molecular Weights and Radii

Atomic and molecular weights and radii of the major and minor elements most commonly used in petrologic calculations. Oxide weights, where appropriate, are given for the single-cation formula.

Element	Charge	Atomic Radius	Atomic Wt.	Molecular Wt. of Oxide
Si	+4	0.42	28.09	60.09
Ti	+4	0.68	47.90	79.90
Al	+3	0.51	26.98	37.49
Fe	+3	0.64	55.85	51.92
Fe	+2	0.74	55.85	71.85
Mn	+2	0.80	54.94	70.94
Mg	+2	0.66	24.31	30.31
Ca	+2	0.99	40.08	56.08
Na	+1	0.97	22.99	30.99
K	+1	1.33	39.10	47.10
P	+5	0.35	30.97	70.98
H	+1	1.54	1.01	9.01
C	+4	0.16	12.01	44.01
Ba	+2	1.34	137.33	153.33
Cr	+3	0.63	52.00	76.00
Li	+1	0.78	6.94	14.94
Nb	+5	0.62	92.91	132.91
Ni	+2	0.69	58.70	74.70
Rb	+1	1.48	85.47	93.47
Sc	+3	0.83	44.96	68.96
Sr	+2	1.12	87.62	103.62
Th	+4	1.10	232.04	264.04
Zr	+4	0.79	91.22	123.22
La	+3	1.22	138.91	154.91
Ce	+4	1.63	140.12	172.12
Sm	+3	1.13	150.40	170.40
Lu	+3	0.99	174.97	194.97
F	-1	1.33	19.00	-
Cl	-1	1.81	35.45	-
S	-2	1.84	32.06	-
S	+4	0.37	32.06	64.06

F

Distribution Coefficients and Normalizing Factors

(a) Typical partition coefficients of trace elements between crystals and liquids. The values shown here are only approximate and are intended only to illustrate general trace-element behavior. For more precise values, see the database in www.earthref.org.

				Basaltic Liquids				
Element	Oliv	Opx	Cpx	Amph	Garnet	Spinel	Plag	Mica
Ni	4–10*	8.3	2.5	6.0	0.5	5.0	0.05	7.6
Cr	0.2	2.0	11.5	5.2	2.0	10.0	0.06	7.0
Co	3.9	2.4	1.0	6.5	3.2	2.0	0.05	1.1
V	0.04	0.24	0.36	9.0	1.5	26.7	0.04	79.5
Zn	0.7	–	0.4	0.4	0.6	11.6	0.01	2.8
Sc	0.2	1.2	2.7	3.5	3.4	2.0	0.03	3.0
Y	0.003	0.03	0.05	1.1	2.8	0.004	0.03	0.02
Sr	0.01	0.003	0.11	0.6	<0.1	<0.1	2.1	0.1
Ba	0.02	0.05	0.02	0.4	<0.1	<0.1	0.38	<0.1
Rb	0.02	0.006	0.03	0.4	<0.1	<0.1	0.09	2.1
Zr	0.01	0.03	0.14	1.2	0.8	0.97	0.01	2.5
Th	0.001	0.006	0.044	0.11	0.001	0.029	0.006	0.12
Nb	0.01	0.002	0.012	0.2	0.01	1.002	0.01	0.09
La	0.00	0.001	0.013	0.17	0.01	0.003	0.112	0.03
Eu	0.002	0.013	0.05	1.1	0.3	0.002	0.043	1.08
Lu	0.08	0.027	0.096	0.6	5.5	0.023	0.023	–

* The value for Ni in olivine varies with oxygen fugacity from about 4 near the wustite-magnetite buffer to 10 approaching the fayalite-magnetite-quartz buffer.

			Rhyolitic Liquids			
Element	Opx	Cpx	Amph	K-Feld	Biotite	Plag
K	<0.1	<0.1	0.08	–	–	0.1
Rb	<0.1	<0.1	0.14	0.3	4.1	0.04
Sr	<0.1	0.5	0.02	3.9	0.17	4.4
Ba	0.09	0.1	0.10	4.6	9.7	0.3
Zr	0.00	0.00	0.7	0.09	0.5	0.03
Nb	0.00	0.00	3.4	1.00	9.1	0.032
Th	0.19	0.14	0.16	0.01	1.3	0.02
La	0.00	0.001	0.2	0.05	0.3	0.26
Eu^{2+}	0.041	12.27	3.2	2.6	0.8	6.49
Lu	0.00	0.00	1.8	0.02	0.6	0.47

(b) Distribution coefficients for trace elements in coexisting mafic and felsic immiscible liquids. Ranges of values are given for the weight ratio of the element in the mafic liquid to the amount in a co-existing felsic liquid (after E. B. Watson, 1976, *Contrib. Mm. Petrol.* 56:119–134).

Ba	1.37–1.63	Lu	4.18–7.13	Sm	4.19–5.15
Ca	1.79–3.14	Mg	2.04–2.23	Ta	4.18–4.42
Cr	3.22–4.29	Mn	2.86–3.09	Ti	2.87–3.32
Cs	0.30–0.31	P	8.34–13.33	Zr	2.28–2.54
La	3.70–4.31	Sr	1.33–1.84		

(c) Distribution coefficients for trace elements in immiscible sulfide liquids coexisting with liquids of basaltic composition. Values are weight ratios of the concentration in the sulfide to the concentration in the basaltic liquid (after V. Rajamani and A. Naldrett, 1978, *Econ. Geol.* 478:82–93).

	Olivine	Basalt		Andesite
	Basalt	1255°C	1305°C	1255°C
Ni	231	274	257	460
Cu	333	245	180	243
Co	n.d.	80	61	n.d.

(d) Abundances of rare-earth elements in the Leedey chondritic meteorite. Dividing the abundances in other rocks by these amounts yields normalized values that plot as smoother curves that are more readily compared and interpreted (after Masuda, et al., 1973, *Geoch. Cosmoch. Acta.* 37:234–248, and G. G. Goles, personal communication. Atomic radii are for six-coordination and are taken from Whittaker and Muntus, 1970, *Geoch. Cosmoch. Acta.* 34:945–956).

Element	Symbol	Atomic No.	Ionic Radius	Abundance
Lanthanum	La	57	1.13	0.330
Cerium	Ce	58	1.09	0.880
Praseodymium	Pr	59	1.08	0.112
Neodymium	Nd	60	1.06	0.600
Samarium	Sm	62	1.04	0.181
Europium	Eu	63	1.03	0.0069
Gadolinium	Gd	64	1.02	0.249
Terbium	Tb	65	1.00	0.047
Dysprosium	Dy	66	0.99	0.325
Holmium	Ho	67	0.98	0.070
Erbium	Er	68	0.97	0.200
Thulium	Tm	69	0.96	0.030
Ytterbium	Yb	70	0.95	0.200
Lutetium	Lu	71	0.94	0.034

(e) Standard values for normalizing minor and trace elements in spidergrams (after A. Pearce, 1983, in *Continental Basalts and Mantle Xenoliths,* Hawksworth and Norry, eds., Shiva Publ.; and Pearce, et al., 1987, *Phil. Irons. Roy. Soc. London* A300:299–317. Values for La, Lu, and Ni from G. G. Goles, personal communication). Values are in parts per million except for K_2O, P_2O_5, and TiO_2, which are in weight percent of the oxide. A more extensive table of average abundances in rocks commonly used as reference standards is given in (f).

Element	Abundance	Element	Abundance	Element	Abundance
Sr	120	K_2O	0.15	Rb	2
Ba	20	Th	0.2	Ta	0.18
Nb	3.5	La	3.8	Ce	10
P_2O_5	0.12	Zr	90	Hf	2.4
Sm	3.3	TiO_2	1.5	Y	30
Yb	3.4	Lu	0.52	Sc	40
Cr	250	Ni	150		

(f) Element concentrations (in ppm) for chondrites, normal- and enriched-type mid-ocean-ridge basalts (N- and E-MORB), and ocean island basalt (OIB) (after S. S. Sun and W. F. McDonough, 1989, *Magmatism in the Ocean Basins,* A. D. Saunders and M. J. Norry, eds., *Geol. Soc. Amer. Spec. Publ. No.* 42:313–345).

Element	C1 Chondrite	Primitive mantle	N-type MORB	E-type MORB	OIB
Cs	0.188	0.032	0.0070	0.063	0.387
Tl	0.140	0.005	0.0014	0.013	0.077
Rb	2.32	0.635	0.56	5.04	31.0
Ba	2.41	6.989	6.30	57	350
W	0.095	0.020	0.010	0.092	0.560
Th	0.029	0.085	0.120	0.60	4.00
U	0.008	0.021	0.047	0.18	1.02
Nb	0.246	0.713	2.33	8.30	48.0
Ta	0.014	0.041	0.132	0.47	2.70
K	545.0	250	600	2100	12.000
La	0.237	0.687	2.50	6.30	37.0
Ce	0.612	1.775	7.50	15.0	80.0
Pb	2.47	0.185	0.30	0.60	3.20
Pr	0.095	0.276	1.32	2.05	9.70
Mo	0.92	0.063	0.31	0.47	2.40
Sr	7.26	21.1	90	155	660
P	1220.0	95	510	620	2700
Nd	0.467	1.354	7.30	9.00	38.5
F	60.7	26	210	250	1150
Sm	0.153	0.444	2.63	2.60	10.0
Zr	3.87	11.2	74	73	280
Hf	0.1066	0.309	2.05	2.03	7.80
Eu	0.058	0.168	1.02	0.91	3.00
Sn	1.72	0.170	1.1	0.91	3.00
Sb	0.16	0.005	0.01	0.01	0.03
Ti	445	1300	7600	6000	17.200
Gd	0.2055	0.596	3.680	2.970	7.620
Tb	0.0374	0.108	0.670	0.530	1.050
Dy	0.2540	0.737	4.550	3.550	5.600
Li	1.57	1.60	4.3	3.5	5.6
Y	1.57	4.55	28	22	29
Ho	0.0566	0.164	1.01	0.790	1.06
Er	0.1655	0.480	2.97	2.31	2.62
Tm	0.0255	0.074	0.456	0.356	0.350
Yb	0.170	0.493	3.05	2.37	2.16
Lu	0.0254	0.074	0.455	0.354	0.300

Glossary of Rock Names

The brief definitions given here are for general use and quick reference. More precise definitions can be found in the text and in standard references, such as

American Geological Institute, 1972, *Glossary of Geology*, Washington, D. C., 858 p.

Irvine, T. N., and W. R. A. Baragar, 1971, A guide to the chemical classification of common volcanic rocks. *Can. Jour. Earth Sci.* 8:523–548.

Holmes, A., 1920, *The Nomenclature of Petrology,* Alien & Unwin, 284 p. Reprinted in 1972 by Hafner Publishing Co.

Streckeisen, A., 1978, Classification and nomenclature of volcanic rocks, lamprophyres, carbonatites and melilitic rocks. *Neues Jahrb. Miner. Abhandl.* 134:1–14.

Absarokite: a potassium-rich, olivine basalt characterized by plagioclase rimmed with orthoclase; leucite may also be present. With decreasing olivine grades into shoshonite.

Achondrite: a stony meteorite that lacks chondules. Most are coarser grained than chondrites, and they contain little if any metallic nickel-iron. Many are aggregates of fragments, but of all the types of meteorites, they are most like terrestrial rocks.

Adakite: a magnesium-rich variety of andesite that is thought to be a product of partial melting of eclogite.

Adamellite: see quartz monzonite.

Alaskite: leuco-granite, normally with a granular texture and less than a few percent mafic minerals.

Alkali dolerite: dolerite with normative nepheline and other characteristics of alkaline basalts; often differentiates to small pods or veins of syenite.

Alnoite: a silica-poor alkaline lamprophyre, normally with melilite; phlogopite, olivine, carbonates, and perovskite are common constituents.

Andesite: a calc-alkaline volcanic rock with 53- to 63-percent SiO_2 and a color index of less than 30; andesite with 53- to 56-percent SiO_2 is usually referred to as basaltic andesite.

Ankaramite: basalt, normally of alkaline character, with abundant large phenocrysts of augite and olivine.

Anorthosite: a plutonic rock consisting almost solely of plagioclase, normally labradorite. One type of anorthosite is found in large, layered intrusions, another in unstratified plutons, or "massifs."

Aplite: a fine-grained, felsic rock often found in shallow dikes and sills. Textures referred to as "aplitic" are granular or "sugary" with few of the crystals showing well-developed crystal faces.

Banakite: a potassic volcanic rock similar to shoshonite but with more than about 57-percent SiO_2.

Basanite: basalt with a modal feldspathoid, normally nepheline. Leucite basanite must also contain olivine; otherwise, it is a tephrite.

Benmoreite: a mildly alkaline rock intermediate between mugearite and trachyte, normally with subequal amounts of alkali feldspar and andesine or oligoclase.

Boninite: an unusually Mg-rich andesite with little or no feldspar.

Calc-alkaline: a type of subalkaline magma series characterized in the middle stages of differentiation by andesites. Chemical compositions are hypersthene normative, have little iron enrichment in the middle stages of differentiation, and generally have more than 16.5-percent Al_2O_3.

Camptonite: an amphibole-rich alkaline lamprophyre.

Carbonatite: a magma or igneous rock consisting mainly of carbonates. The most common minerals are calcite, magnesite, and sodium carbonate.

Chondrite: a common type of stony meteorite characterized by chondrules in a fine-grained matrix of silicate minerals and metallic nickel-iron. Their compositions differ somewhat, but on average, they are probably close to the original, undifferentiated compositions of the terrestrial planets.

Chondrule: a spherical grain or aggregate, often with radially oriented crystals of olivine and pyroxene, found in many stony meteorites. Chondrules are thought to have formed from silicate droplets during accretion of the parent bodies of meteorites.

Chromitite: a rock composed mainly of chromite.

Comendite: a peralkaline rhyolite containing a sodic pyroxene or amphibole. Similar to pantellerite, but more felsic and richer in silica, although not necessarily alkalies.

Dacite: a felsic, subalkaline rock with 63- to 68- or 69-percent SiO_2, a color index less than 20, and at least 10-percent normative or modal quartz.

Diabase: see dolerite.

Diorite: an intermediate plutonic rock (color index usually between 40 and 30) in which neither quartz nor alkali feldspar account for more than 10 percent of the volume.

Dolerite: a medium-grained, hypabyssal rock of basaltic composition; commonly found in dikes and sills. In America, synonymous with diabase.

Dunite: a coarse-grained, ultramafic rock composed primarily of olivine.

Eclogite: a high-pressure assemblage of sodic pyroxene (omphacite) and pyrope-rich garnet. Ca-poor pyroxene, and olivine may also be present but are not major constituents. The bulk chemical composition is similar to that of basalt.

Essexite: an alkaline gabbro having orthoclase as an essential constituent.

Eucrite: gabbro consisting of plagioclase (bytownite or anorthite), augite, and olivine.

Ferro-basalt: a basalt, normally tholeiitic, that has evolved to a high degree of iron enrichment with more than 12-percent total iron oxides.

Ferro-gabbro: the plutonic equivalent of ferro-basalt.

Gabbro: a coarse-grained rock of basaltic composition composed of calcic plagioclase, pyroxene, and possibly olivine, opaque oxides, and in some cases hornblende.

Granite: a felsic, plutonic rock consisting of about equal parts quartz, potassium feldspar, and sodic plagioclase. The name is also used as a general term for all leucocratic, silica-rich, plutonic rocks.

Granodiorite: a felsic, plutonic rock similar to granite but with less potassium feldspar (less than a third of the total feldspar) and a higher color index (usually between 20 and 30).

Granophyre: a medium- to fine-grained felsic rock of rhyolitic composition; commonly found as differentiates of hypabyssal intrusions. Textures referred to as "granophyric" are characterized by intimate intergrowths of quartz and alkali feldspar.

Harzburgite: a coarse-grained, ultramafic rock composed primarily of olivine and enstatite or bronzite; "saxonite" is considered synonymous with harzburgite, but some petrologists reserve the former for harzburgites containing opaque oxide minerals.

Hawaiite: an andesine-basalt with normative nepheline.

Hyperite: gabbro with more orthopyroxene than clinopyroxene. Equivalent to augite-bearing norite.

Icelandite: an intermediate member of the tholeiitic series similar to andesite but with more iron and less alumina (usually < 16.5-percent Al_2O_3); associated with oceanic and other types of anorogenic subalkaline series.

Ignimbrite: a fragmental volcanic rock laid down by a pyroclastic flow; most are siliceous, and many are welded; however, these are not essential elements of the definition.

Ijolite: an alkaline, plutonic rock consisting chiefly of nepheline and aegirine-augite, normally without feldspar; most contain a variety of other minerals, such as melanite, apatite, green biotite, sphene, and calcite.

Jacupirangite: a mafic variety of ijolite.

Kakortokite: a nepheline syenite containing the red zirconium mineral, eudialyte, and a sodic pyroxene, usually acmite.

Katungite: a melilite-bearing, ultrapotassic, volcanic rock, usually with olivine and minor amounts of leucite but no augite.

Kersantite: a calc-alkaline lamprophyre with biotite and plagioclase.

Komatiite: an ultramafic lava consisting almost entirely of olivine and pyroxene and characterized by acicular crystals that grew rapidly from the low-viscosity, high-temperature melt to give the rock a distinctive "spinifex" texture.

Lahar: a type of fragmental volcanic material deposited by debris flows; characterized by unsorted clasts with a wide range of sizes and compositions.

Lamproite: a general term for mafic volcanic or hypabyssal ultrapotassic rocks.

Lamprophyre: a diverse group of mafic hypabyssal or volcanic rocks, normally rich in alkalies, hydrous phases, P_2O_5, CO_2, and certain minor or trace elements, such as Ti, Ba, and Sr. Some varieties lack feldspar.

Larvikite: an augite syenite characterized by coarse plagioclase with brilliant schiller structure.

Latite: a subalkaline, volcanic rock of intermediate silica content containing potassium feldspar; a potassic variety of andesite.

Leucitite: a rock consisting mainly of leucite and pyroxene and usually lacking feldspar.

Leucogranite: a granite of very low color index (< 10).

Lherzolite: a peridotite consisting of olivine, diopsidic augite, enstatite, and possibly spinel or garnet.

Lujavrite: a textural variety of kakortokite distinguished by slender plagioclase and acicular pyroxene.

Mafurite: an ultrapotassic rock consisting chiefly of kalsilite, olivine, and augite, with increasing silica grades into leucitite.

Maskelynite: isotropic plagioclase formed by intense shock. Found in meteorites and terrestrial rocks shocked by meteorite impact or a nuclear explosion.

Melilitite: a rock containing at least 50-percent melilite.

Metaluminous: a magma or igneous rock in which molecular $Na_2O + K_2O + CaO > Al_2O_3 > Na_2O + K_2O$.

Migmatite: a composite rock composed of igneous-looking crystals and metamorphic minerals of similar mineralogical composition; may be formed by ultrametamorphism or proximity to a magma.

Minette: a calc-alkaline lamprophyre with biotite and potassium feldspar.

Monchiquite: an alkaline lamprophyre rich in augite but with little or no feldspar.

Monzonite: a plutonic rock of intermediate color index and subequal alkali feldspar and plagioclase; has less than 10-percent quartz.

Mugearite: an oligoclase basalt, usually with normative nepheline.

Naujaite: a sodalite-rich nepheline syenite with microcline, albite, acmite, and a sodic amphibole.

Nephelinite: a nepheline-rich rock lacking feldspar but normally containing a sodic or titaniferous pyroxene.

Nordmarkite: a quartz-bearing alkali syenite, usually with biotite and aegirine.

Norite: a type of gabbro containing hypersthene and labradorite; augite and iron-oxide minerals are also possible phases.

Okaite: a very silica-poor alkaline rock consisting mainly of melilite, hauyne, Ti-biotite, and a wide variety of accessory minerals.

Ophiolite: a collective name applied to basalts, gabbros, ultramafic rocks, and pelagic sediments that were originally part of the oceanic lithosphere.

Orendite: a leucite-bearing volcanic rock rich in diopsidic augite and phlogopite; unlike wyomingite, its groundmass contains sanidine.

Palagonite: hydratated basaltic glass, usually with a golden orange color, formed by subaqueous or subglacial eruptions.

Pantellerite: a peralkaline rhyolite with sodic pyroxene or amphibole; slightly less felsic than comendite.

Pegmatite: any very coarse-grained igneous rock.

Peralkaline: a magma or igneous rock in which molecular $Na_2O + K_2O > Al_2O_3$; can be identified by acmite (Ac) in the norm.

Peraluminous: a magma or igneous rock in which molecular $Al_2O_3 > Na_2O + K_2O + CaO$; can be identified by corundum (C) in the norm.

Peridotite: an ultramafic, plutonic rock consisting of at least 90-percent combined olivine and pyroxene.

Plagiogranite: a felsic, plutonic rock consisting mainly of quartz and sodic plagioclase but lacking K-feldspar; synonym, trondhjemite.

Phonolite: a felsic, alkaline rock composed of alkali feldspar (typically anorthoclase or sanidine) and nepheline; the volcanic or hypabysssal equivalent of nepheline syenite.

Pulaskite: a leucocratic syenite with perthitic or antiperthitic feldspar.

Quartz monzonite: a monzonite with 10- to 20-percent quartz; also called adamellite.

Rhyodacite: a volcanic rock intermediate between dacite and rhyolite.

Rhyolite: a felsic, volcanic rock with more than 68-percent SiO_2; broadly equivalent to granite.

Saxonite: see harzburgite.

Seriate: a texture in which the crystals have a continuous, gradational range of sizes.

Shonkinite: a potassium-rich, mafic syenite consisting mainly of sanidine, augite, biotite, olivine, apatite, and zeolitized nepheline.

Shoshonite: a potassium-rich basalt characterized by plagioclase rimmed with orthoclase; leucite may also be present.

Siderite: a type of meteorite composed mainly of native nickel-iron, commonly with minor amounts of the iron sulfide, troilite. Also a mineral with the composition $FeCO_3$.

Siderolite: a stony-iron meteorite composed of crystals of olivine or pyroxene in a matrix of metallic nickel-iron.

Sideromelane: clear, basaltic glass.

Sovite: a leucocratic, coarse-grained carbonatite.

Spessartite: a calc-alkaline lamprophyre containing hornblende and plagioclase.

Syenite: a felsic, plutonic rock with subequal amounts of sodic plagioclase and K-feldspar, but less than 10-percent quartz.

Tachylite: opaque basaltic glass.

Tephra: a general term for all types of pyroclastic deposits.

Tephrite: a leucite basalt lacking olivine.

Teschenite: analcite dolerite.

Tholeiite: a type of subalkaline magma characterized by normative hypersthene and strong iron enrichment in the middle stages of differentiation. The same name is applied to the basaltic parent of such a series.

Tonalite: a plutonic rock in which more than 90 percent of the feldspar is plagioclase, and quartz accounts for at least 15 percent of the volume; gradational into trondhjemite with decreasing color index.

Trachyandesite: an intermediate volcanic rock with abundant plagioclase, usually andesine, and at least 10-percent alkali feldspar; intermediate between andesite and latite. The name is not favored in modern usage.

Trachybasalt: a variety of weakly alkaline basalt containing abundant plagioclase, usually labradorite, and at least 10-percent alkali feldspar; has little or no feldspathoid.

Trachyte: a felsic, volcanic rock with no nepheline and less than 10-percent quartz or tridymite; the volcanic or hypabyssal equivalent of syenite.

Tristanite: an intermediate member of the potassic alkaline series.

Troctolite: a gabbroic rock rich in calcic plagioclase and containing olivine but little or no pyroxene.

Trondhjemite: a felsic, plutonic rock with sodic plagioclase and quartz but little or no K-feldspar; synonym, plagiogranite.

Ugandite: a leucite-bearing, silica-deficient, potassic volcanic rock rich in mafic minerals, mainly olivine, Ti-augite, and Ti-magnetite.

Uralite: a pale green amphibole, commonly the product of alteration of pyroxene.

Vogesite: a calc-alkaline lamprophyre containing hornblende and potassium feldspar.

Websterite: an ultramafic plutonic rock composed primarily of Ca-rich and Ca-poor pyroxene and little or no olivine; plagioclase, spinel, or garnet may be present in accessory amounts.

Wyomingite: a potassic volcanic rock with phenocrysts of phlogopite and possibly diopsidic augite in a leucite-rich groundmass.

Xenocryst: a crystal of foreign origin included in a magma or igneous rock.

Xenolith: a rock of foreign origin included in a magma or igneous rock.

Index